MEASURE AND INTEGRATION

MEASURE AND INTEGRATION

Sterling K. Berberian

The University of Texas at Austin

Chelsea Publishing Company

Bronx, New York

The present work is a reprint of a work first published
by The Macmillan Company, New York, in 1965.
It is published at New York, 1970, and is printed
on alkaline (acid-free) paper.

Preliminary edition, *A First Course in Measure and Integration,*

Library of Congress catalog card number 74-128871

International Standard Book Number 0-8284-0241-8

Printed in the United States of America

To the memory of my grandmother Nazli

Preface

The urge to write this book grew out of my experiences in teaching two courses, entitled *Measure Theory* and *Integration over Locally Compact Spaces*, at the University of Iowa. Accordingly, the book has two main divisions. *Part I* deals with the general theory of measure and integration over abstract measure spaces (Chapters 1 through 7). In *Part II* this theory is put to work in locally compact topological spaces and topological groups (Chapters 8 and 9).

The material in Part I is conceived as the content of a one-semester terminal course in measure and integration for the general mathematics student, at the early graduate level. The material in Part II is more specialized, inclining towards functional analysis; it is suitable for a second course in measure and integration at the advanced graduate level, but will have to be augmented in order to fill out a semester.

Prerequisites for reading Part I are a familiarity with ε, δ arguments, and with the language of naïve set theory. For Part II the reader must also have a background in general topology, as well as a nodding acquaintance with general algebraic structures.

The stuff of Part I is standard repertory: measures and outer measures, measurable functions, integrable functions, convergence theorems for sequences of integrable functions, Fubini's theorem on iterated integration, and the Radon-Nikodym theorem. The experienced reader will find no surprises here, excepting possibly some of the refinements of Fubini's theorem discussed in Section 42.

In Part II the general theory is specialized to the context of locally compact spaces (Chapter 8), and then applied to the study of invariant integration over locally compact groups (Chapter 9). The climax of Chapter 8 is the Riesz-Markoff representation theorem, while Chapter 9 culminates in the construction of the group algebra of a locally compact group.

Throughout the book, I have indulged my fondness for order theoretic arguments by casting the Monotone Convergence Theorem in the leading role. This theorem is included in the basic chapter on integration, Chapter 4. Other classical convergence theorems are taken up in Chapter 5, but they are not used in any of the later chapters. This arrangement has the defect of making Chapter 5 appear to be the most omissible chapter of the book, whereas these convergence theorems are really indispensable in applications; nevertheless, it is hoped that the isolation of these convergence theorems will help to understand their true role in those situations where they must be used. A good example is the role of the "$\mathscr{L}^1$ completeness theorem" in the group algebra of a locally compact group; it is needed only as the last touch prior to invoking the general theory of Banach algebras.

It is surely not an exaggeration to say that the key concept in the Lebesgue theory of integration is that of *null set*. Much of the convenience of the theory comes from the ease of manipulating null sets; some of its limitations are due to the difficulty of ascertaining whether certain sets are null sets. Thus, null sets are at once the blessing and the curse of the theory, and the tone of the whole exposition is set by how one treats them. If one is careful about null sets, the subject is easy and the problems are discernible; if one is careless about null sets, the subject is mysterious and the problems are invisible. A feature of the present exposition is an extraordinary tidiness with regard to null sets; it is hoped that the resulting slight increase in tedium is outweighed by a gain in clarity.

Probably the most novel feature of the book is that the terms *measurable* and *integrable* are applied only to functions with finite real values. We thus exclude (1) functions with possibly infinite values, (2) functions with complex values, and (3) functions with vectorial values. My own view, based on experimentation in class, is that infinite values are an unnecessary complication in the arithmetic of point functions; moreover, infinite values contribute essentially nothing to understanding integration, since an integrable function, by almost anyone's definition, will be finite valued almost everywhere. While the extension of the theory from real-valued functions to complex-valued functions is an elementary exercise, no deeper than the resolution of a complex number into its real and imaginary parts, the theory of measure and integration for vector-valued functions seems to me a different subject altogether, far removed from the Monotone Convergence Theorem, and therefore quite beyond the scope of this book.

Another feature of the exposition is that none of the theorems require the underlying space itself to be a measurable set; that is, we deal exclusively with σ-rings rather than σ-algebras. For example, the assumptions of finiteness and σ-finiteness will be invoked whenever they are appropriate, but total finiteness and total σ-finiteness will not be needed. Measurability of the underlying space is dispensed with by requiring each measurable function to vanish outside some measurable set, and by systematically employing *locally measurable* sets, that is, sets whose intersection with every measurable set is measurable.

The text is meant to be self-contained, and independent of the exercises. Most of the exercises, which are in the form of statements to be proved, have simple solutions based on the text. The internal reference system works as follows: "Exercise 62.3" refers to the third exercise at the end of Section 62, whereas "62.3" refers to the third theorem in Section 62. Certain sections are starred; the material they contain is more specialized, and is not needed for the unstarred sections. Starred exercises contain questions which I asked myself and am willing to confess I could not answer; some of these are probably trivial, but others may be interesting questions for research. I am indebted to Roy A. Johnson for many valuable insights on product measure, and for destarring numerous star-bearing exercises in an earlier draft; in particular, the most interesting exercises on product measure are due to him, and will appear in his thesis.

Following the preface there is a chart showing the logical interdependence of Chapters 4 through 9. Considerable flexibility is possible in arranging a course, since Chapters 5 through 8 can be taken up in any order. For example, a one-semester course emphasizing locally compact groups could be based on Chapters 1–4, 6, 8, and 9. I consider the central core of the book to be Chapters 1–4, 6, and 8. A course taking up only the unstarred sections of Chapters 1–4 and 6 would be, in my view, a respectable course.

Finally, I am under no illusions as to originality, for the subject of measure theory is an old one which has been worked over by many experts. My contribution can only be in selection, arrangement, and emphasis. I am deeply indebted to Paul R. Halmos, from whose textbook I first studied measure theory; I hope that these pages may reflect their debt to his book without seeming to be almost everywhere equal to it.

I am very grateful to my colleague Allen T. Craig, who encouraged me to offer courses in measure theory, and stocked them with many able and stimulating graduate students. How much this book owes its existence to him, only the two of us can know.

S. BERBERIAN

Iowa City

Logical interdependence of Chapters 4–9

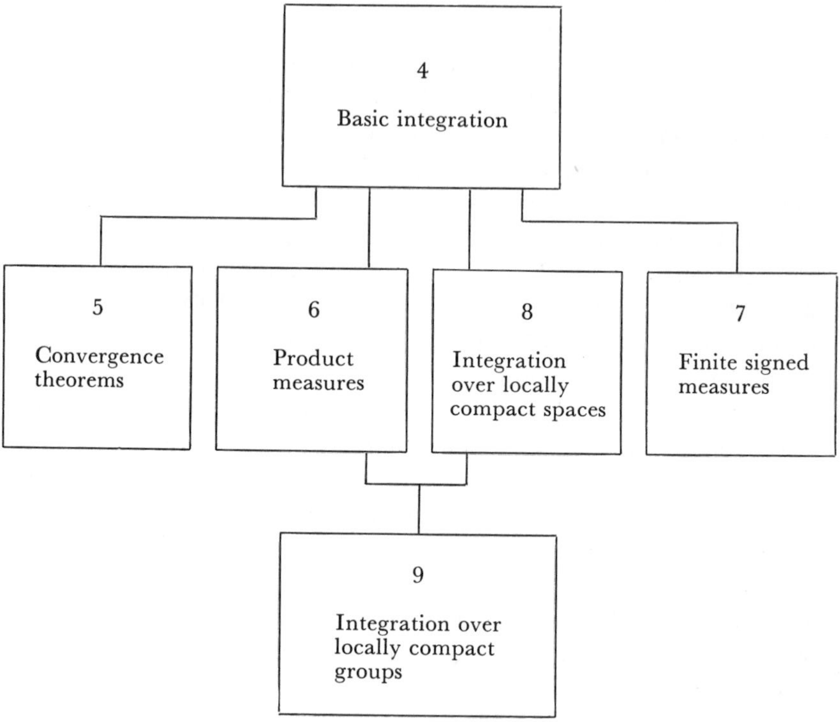

Index of Symbols

Contents

MEASURE AND INTEGRATION

C H A P T E R 1

Measures

The fundamental result of the chapter is that a measure defined on a ring of sets can be extended to the generated σ-ring, the extension being unique when the original measure is σ-finite.

0. Set Theoretic Notations and Terminology

This section should be skipped at the first sign of boredom. We write $x \in A$ to indicate that x is an element of the set A, while $x \notin A$ means that x is *not* an element of the set A. The set whose only elements are $x_1, \ldots, x_n$ is denoted

$$\{x_1, \ldots, x_n\};$$

we allow the possibility that $x_i = x_j$ even if $i \neq j$. The set whose only element is x is denoted $\{x\}$, and is called a *singleton*. The *empty set* is denoted $\varnothing$.

We write either $A \subset B$ or $B \supset A$ to indicate that A is a *subset* of B. Two sets A and B are said to be *equal* in case both $A \subset B$ and $B \subset A$; one then writes $A = B$. If for each element x of a set X there is given a proposition $P(x)$, then

$$\{x \in X: P(x)\}$$

stands for the subset of X consisting of all those elements x for which the corresponding proposition $P(x)$ is true; when there is no doubt as to the identity of the underlying set X, we may write briefly

$$\{x: P(x)\}.$$

For example, if R is the set of all real numbers, $a \in R$, $b \in R$, and $a \leq b$, the various types of finite intervals in R are denoted thus:

$$\begin{aligned} (a, b) &= \{x: a < x < b), \\ [a, b] &= \{x: a \leq x \leq b\}, \\ [a, b) &= \{x: a \leq x < b\}, \\ (a, b] &= \{x: a < x \leq b\}. \end{aligned}$$

A set whose elements are sets will be called a *class*. If X is a set, the class of all subsets of X is called the *power set* of X and is denoted $\mathscr{P}(X)$. If A and B are subsets of X, we write

$$\begin{aligned} A \cup B &= \{x : x \in A \quad \text{or} \quad x \in B\}, \\ A \cap B &= \{x : x \in A \text{ and } x \in B\}, \\ A - B &= \{x : x \in A \text{ and } x \notin B\}, \\ A \triangle B &= (A - B) \cup (B - A); \end{aligned}$$

the terms *union*, *intersection*, *difference*, and *symmetric difference*, respectively, are used to indicate these combinations of A and B. The *complement* of A (in X) is denoted $\complement A$, thus,

$$\complement A = \{x : x \in X \text{ and } x \notin A\};$$

that is, $\complement A = X - A$. The symbol A' will also occasionally be used to denote the complement of A.

We write $f\colon X \to Y$ to indicate that f is a *mapping* of the set X into the set Y. The terms *function* and *transformation* are synonymous with "mapping." The set X is called the *domain* of f; we always assume that the domain of a mapping is **nonempty**. If A is a subset of X, and B is a subset of Y, we write

$$\begin{aligned} f(A) &= \{y : y = f(x) \text{ for some } x \in A\}, \\ f^{-1}(B) &= \{x : f(x) \in B\}, \end{aligned}$$

for the *direct image* of A under f, and the *inverse image* of B under f, respectively. The *range* of f is $f(X)$. If $f(X) = Y$, the mapping f is said to be *surjective* (or *onto*). If $f(x_1) = f(x_2)$ implies $x_1 = x_2$, the mapping f is said to be *injective* (or *one-one*). If f is both surjective and injective, it is called *bijective.*

If $f\colon I \to X$, we may write $x_i = f(i)$ and speak of a *family* $(x_i)_{i \in I}$ of elements of X, *indexed* by the set I (whose elements are then called "indices"). If, moreover, the indexing set I is finite, we speak of a *finite family* of elements of X. The concept "empty family" is excluded by the above conventions. When a nonempty class $\mathscr{A}$ of sets is referred to as a *family* of sets, we assume the "identical indexing of $\mathscr{A}$ by itself," that is, $f\colon \mathscr{A} \to \mathscr{A}, f(A) = A$ for all $A \in \mathscr{A}$.

If $(A_i)_{i \in I}$ is a family of subsets of a set X, we write

$$\bigcup_{i \in I} A_i = \{x : x \in A_i \text{ for some } i \in I\},$$

$$\bigcap_{i \in I} A_i = \{x : x \in A_i \text{ for all } i \in I\},$$

for the *union* and *intersection*, respectively, of the family. If, moreover, the set I of indices is finite, we speak of $\bigcup_{i \in I} A_i$ as a *finite union* of subsets of X, and of $\bigcap_{i \in I} A_i$ as a *finite intersection* of subsets of X. In particular, when $I = \{1, \ldots, n\}$, the notations $A_1 \cup \cdots \cup A_n$ and $A_1 \cap \cdots \cap A_n$ are also used, or sometimes $\bigcup_1^n A_i$ and $\bigcap_1^n A_i$. The notations $\bigcup_1^\infty A_i$ and $\bigcap_1^\infty A_i$ are used when I is the set of all positive integers; we then speak of a *countable union* and a *countable intersection* of subsets of X. The "union of an empty family of subsets of X" and "intersection of an empty family of subsets of X" are excluded by our conventions (in the theory of sets they are usually defined to be $\varnothing$ and X, respectively).

1. Rings and σ-Rings

If X is a set, a nonempty class $\mathscr{R}$ of subsets of X is called a **ring** if it is closed under the formation of set theoretic differences and finite unions; in other words, $A - B$ and $A \cup B$ belong to $\mathscr{R}$ whenever A and B do. If, moreover, X belongs to $\mathscr{R}$, then $\mathscr{R}$ is called an **algebra** of subsets of X. If A and B belong to a ring $\mathscr{R}$, then so do the empty set

$$\varnothing = A - A,$$

the symmetric difference

$$A \triangle B = (A - B) \cup (B - A),$$

and the intersection

$$A \cap B = (A \cup B) - (A \triangle B).$$

If $\mathscr{R}$ is an algebra, and A belongs to $\mathscr{R}$, then so does its complement

$$\complement A = X - A;$$

conversely, any ring with this property is an algebra, in view of the relation $X = A \cup \complement A$.

The reason for this terminology is as follows. The class $\mathscr{P}(X)$ of *all* subsets of X (that is, the power set of X) is a ring in the usual algebraic sense, provided sums and products are defined by the formulas

$$A + B = A \triangle B$$

and

$$AB = A \cap B.$$

For example, the distributive law

$$A \cap (B \triangle C) = (A \cap B) \triangle (A \cap C)$$

is readily verified. This ring is evidently associative and commutative; since $AA = A$ for every A, it is called a *Boolean* ring. Since $XA = AX = A$, X is a unity element for the ring $\mathscr{P}(X)$. Now, a nonempty subclass $\mathscr{R}$ of $\mathscr{P}(X)$ is a ring in the earlier sense if and only if it is a subring of $\mathscr{P}(X)$ in the algebraic sense; $\mathscr{R}$ is then an algebra of sets if and only if it contains the unity element X.

If $(\mathscr{R}_i)$ is any family of rings of subsets of X, the intersection of the $\mathscr{R}_i$ is clearly a ring. Suppose $\mathscr{E}$ is any class of subsets of X. There exists at least one ring containing $\mathscr{E}$, namely, $\mathscr{P}(X)$. The intersection of the family of all rings containing $\mathscr{E}$ is a ring containing $\mathscr{E}$; it is denoted $\mathscr{R}(\mathscr{E})$, called the ring *generated* by $\mathscr{E}$, and is uniquely determined by the following properties:

Theorem 1. *If $\mathscr{E}$ is any class of subsets of X, then $\mathscr{R}(\mathscr{E})$ is a ring, and $\mathscr{E} \subset \mathscr{R}(\mathscr{E})$; if $\mathscr{R}$ is any ring such that $\mathscr{E} \subset \mathscr{R}$, necessarily $\mathscr{R}(\mathscr{E}) \subset \mathscr{R}$. Briefly: $\mathscr{R}(\mathscr{E})$ is the smallest ring containing $\mathscr{E}$.*

Suppose $\mathscr{E}$ is any class of subsets of X. Consider the class $\mathscr{R}$ of all sets $A \subset X$ such that A is contained in the union of a finite number of sets in $\mathscr{E}$. Evidently $\mathscr{R}$ is a ring, and $\mathscr{R}$ contains $\mathscr{E}$, hence $\mathscr{R}$ contains $\mathscr{R}(\mathscr{E})$. Summarizing,

Theorem 2. *Every set in $\mathscr{R}(\mathscr{E})$ is contained in a finite union of sets in $\mathscr{E}$.*

A nonempty class $\mathscr{S}$ of subsets of X is called a **σ-ring** if it is closed under the formation of set theoretic differences and countable unions; in other words, if A, B, and A_n $(n = 1, 2, 3, \ldots)$ are sets in $\mathscr{S}$, then so are $A - B$ and $\bigcup_1^\infty A_n$. If moreover X belongs to $\mathscr{S}$, then $\mathscr{S}$ is called a **σ-algebra.** Since

$$A \cup B = A \cup B \cup \varnothing \cup \varnothing \cup \ldots,$$

it is clear that every σ-ring is a ring; conversely, any ring which is closed under countable unions is a σ-ring.

A σ-ring is closed under countable intersections, in view of the identity

$$\bigcap_1^\infty A_n = A_1 - \bigcup_1^\infty (A_1 - A_n).$$

If $(\mathscr{S}_i)$ is any family of σ-rings of subsets of X, the intersection of the $\mathscr{S}_i$ is also a σ-ring. Of course $\mathscr{P}(X)$ is a σ-ring. It follows at once that if $\mathscr{E}$ is any class of subsets of X, the intersection of all σ-rings containing $\mathscr{E}$ is a σ-ring containing $\mathscr{E}$; it is denoted $\mathfrak{S}(\mathscr{E})$, called the σ-ring **generated** by $\mathscr{E}$, and is uniquely determined by the following properties:

Theorem 3. *If $\mathscr{E}$ is any class of subsets of X, then $\mathfrak{S}(\mathscr{E})$ is a σ-ring, and $\mathscr{E} \subset \mathfrak{S}(\mathscr{E})$; if $\mathscr{S}$ is any σ-ring such that $\mathscr{E} \subset \mathscr{S}$, necessarily $\mathfrak{S}(\mathscr{E}) \subset \mathscr{S}$. Briefly, $\mathfrak{S}(\mathscr{E})$ is the smallest σ-ring containing $\mathscr{E}$.*

Analogous to Theorem 2, one sees easily:

Theorem 4. *Each set in $\mathfrak{S}(\mathscr{E})$ is contained in a countable union of sets in $\mathscr{E}$.*

The following example is of central importance from Chapter 2 onward. Let X be the set R of all real numbers, and $\mathscr{E}$ the class of all semiclosed intervals $[a, b)$; the σ-ring generated by $\mathscr{E}$ is called the class of **Borel sets** in R. There is no delicate reason here for using intervals $[a, b)$; intervals $(a, b]$, or (a, b), or $[a, b]$ would work as well—the generated σ-ring is the same in each case (this is essentially shown in the proof of 12.2).

The next results will occasionally be useful:

Theorem 5. *If $f: X \to Y$, and $\mathscr{T}$ is a σ-ring of subsets of Y, then the class of all sets of the form $f^{-1}(B)$, where B is in $\mathscr{T}$, is a σ-ring of subsets of X.*

Proof. $f^{-1}(B_1) - f^{-1}(B_2) = f^{-1}(B_1 - B_2)$, and

$$\bigcup_1^\infty f^{-1}(B_n) = f^{-1}\left(\bigcup_1^\infty B_n\right). \quad \blacksquare$$

Corollary. *If $\mathscr{S}$ is a σ-ring of subsets of X, and $E \subset X$, then the class of all sets $E \cap A$, where A is in $\mathscr{S}$, is a σ-ring.*

Proof. If $f: E \to X$ is the identity injection $f(x) = x$, then $f^{-1}(A) = E \cap A$. ∎

Theorem 6. *If $f: X \to Y$, and $\mathscr{S}$ is a σ-ring of subsets of X, then the class of all sets $B \subset Y$, such that $f^{-1}(B)$ is in $\mathscr{S}$, is a σ-ring of subsets of Y.*

Proof. $f^{-1}(B_1 - B_2) = f^{-1}(B_1) - f^{-1}(B_2)$, and

$$f^{-1}\left(\bigcup_1^\infty B_n\right) = \bigcup_1^\infty f^{-1}(B_n). \quad \blacksquare$$

Corollary. *If $\mathscr{S}$ is a σ-ring of subsets of X, and $E \subset X$, then the class of all sets $A \subset X$ such that $E \cap A$ is in $\mathscr{S}$, is a σ-ring.*

Proof. The class of all subsets of E which belong to $\mathscr{S}$ is a σ-ring of subsets of E (it is the intersection of the σ-rings $\mathscr{P}(E)$ and $\mathscr{S}$); our assertion follows at once on applying Theorem 6 to the identity injection $f: E \to X$. ∎

EXERCISES

1. The class of all finite subsets of an infinite set X is a ring, but not a σ-ring.

2. The class of all countable subsets of an uncountable set X is a σ-ring but not a σ-algebra.

3. If $\mathscr{R}$ is a ring of subsets of X, then the algebra generated by $\mathscr{R}$ is the class of all sets E such that either E or $X - E$ belongs to $\mathscr{R}$.

4. A ring which is closed under countable disjoint unions is a σ-ring.

5. A ring which is closed under countable intersections is not necessarily a σ-ring.

6. If $\mathscr{R}$ is the ring generated by the class of all semiclosed intervals

$$[a, b) = \{x: a \le x < b\},$$

then every set in $\mathscr{R}$ can be expressed as a disjoint union of finitely many such intervals.

7. If $\mathscr{R}$ is a ring of subsets of X, write $\mathscr{R}_\lambda$ for the class of all subsets B of X such that $A \cap B \in \mathscr{R}$ for all A in $\mathscr{R}$.

(*i*) $\mathscr{R}_\lambda$ is an algebra of sets, and $\mathscr{R} \subset \mathscr{R}_\lambda$.
(*ii*) If $\mathscr{R}$ is an algebra, then $\mathscr{R} = \mathscr{R}_\lambda$.
(*iii*) If $\mathscr{R}$ is a σ-ring, then $\mathscr{R}_\lambda$ is a σ-algebra.
(*iv*) One has $\mathfrak{S}(\mathscr{R}_\lambda) \subset (\mathfrak{S}(\mathscr{R}))_\lambda$.
(*v*) One has $\mathfrak{S}(\mathscr{R}_\lambda) = (\mathfrak{S}(\mathscr{R}))_\lambda = \mathfrak{S}(\mathscr{R})$ if and only if X is the union of a sequence of sets in $\mathscr{R}$.

2. The Lemma on Monotone Classes

If A_n is an increasing sequence of sets, that is, if $A_n \subset A_{n+1}$ for all n, we shall write $A_n\uparrow$; if moreover the union of the A_n is A, we write $A_n \uparrow A$. Similarly, the symbol $A_n \downarrow A$ means that A_n is a decreasing sequence of sets whose intersection is A. A **monotone class** is a class $\mathscr{M}$ of sets which is closed under countable increasing unions, and countable decreasing intersections; that is, if A_n is a sequence of sets in $\mathscr{M}$, then $A_n \uparrow A$ implies A is in $\mathscr{M}$, and $A_n \downarrow A$ implies A is in $\mathscr{M}$. Every σ-ring contains countable unions and countable intersections, and hence is surely a monotone class. On the other hand, every monotone ring $\mathscr{R}$ is a σ-ring; for, if A_n is a sequence of sets in $\mathscr{R}$ with union A, then $B_n = A_1 \cup \cdots \cup A_n$ defines a sequence in $\mathscr{R}$ such that $B_n \uparrow A$.

Since $\mathscr{P}(X)$ is a monotone class, and since the intersection of any family of monotone classes is also a monotone class, it is clear that given any class $\mathscr{E}$ of subsets of X, there exists a smallest monotone class containing $\mathscr{E}$; this is called the monotone class *generated* by $\mathscr{E}$, denoted $\mathfrak{M}(\mathscr{E})$, and is uniquely determined by the following properties:

Theorem 1. *If $\mathscr{E}$ is any class of subsets of X, then $\mathfrak{M}(\mathscr{E})$ is a monotone class, and $\mathscr{E} \subset \mathfrak{M}(\mathscr{E})$; if $\mathscr{M}$ is any monotone class such that $\mathscr{E} \subset \mathscr{M}$, necessarily $\mathfrak{M}(\mathscr{E}) \subset \mathscr{M}$.*

Since a ring is a σ-ring if and only if it is a monotone class, it is not surprising that the monotone class generated by a ring $\mathscr{R}$ coincides with the σ-ring generated by $\mathscr{R}$, yet the proof is surprisingly delicate:

Theorem 2. *If $\mathscr{R}$ is a ring of subsets of a set X, then $\mathfrak{M}(\mathscr{R}) = \mathfrak{S}(\mathscr{R})$.*

Proof. Since $\mathfrak{S}(\mathscr{R})$ is a monotone class containing $\mathscr{R}$, we have $\mathfrak{S}(\mathscr{R}) \supset \mathfrak{M}(\mathscr{R})$.

Now, $\mathfrak{M}(\mathscr{R}) \supset \mathscr{R}$; to show that $\mathfrak{M}(\mathscr{R}) \supset \mathfrak{S}(\mathscr{R})$, it will suffice to show that $\mathfrak{M}(\mathscr{R})$ is a σ-ring. Let us write $\mathscr{M} = \mathfrak{M}(\mathscr{R})$ for brevity. If E and F are subsets of X, let us say that E "collaborates" with F in case the sets $E - F$, $F - E$, and $E \cup F$ all belong to $\mathscr{M}$. Evidently this relation is symmetric: E collaborates with F if and only if F collaborates with E. For each $F \subset X$, let us write $\mathscr{K}(F)$ for the class (possibly empty) of all collaborators of F. Clearly $E \in \mathscr{K}(F)$ if and only if $F \in \mathscr{K}(E)$.

For each $F \subset X$, $\mathscr{K}(F)$ is a monotone class. For, let E_n be a sequence in $\mathscr{K}(F)$, and suppose $E_n \uparrow E$; since $E_n - F$, $F - E_n$, and $E_n \cup F$ belong to the monotone class $\mathscr{M}$, the relations

$$E_n - F \uparrow E - F, \quad F - E_n \downarrow F - E, \quad \text{and} \quad E_n \cup F \uparrow E \cup F$$

show that $E - F$, $F - E$, and $E \cup F$ belong to $\mathscr{M}$, that is, E collaborates with F. Similarly if $E_n \downarrow E$.

If $F \in \mathscr{R}$, then $\mathscr{M} \subset \mathscr{K}(F)$; for, clearly $\mathscr{K}(F)$ contains $\mathscr{R}$, is a monotone class, and hence contains $\mathfrak{M}(\mathscr{R}) = \mathscr{M}$.

If $E \in \mathscr{M}$, then $\mathscr{M} \subset \mathscr{K}(E)$. For, if $F \in \mathscr{R}$, then $E \in \mathscr{K}(F)$ by the preceding result, hence $F \in \mathscr{K}(E)$. Thus, the monotone class $\mathscr{K}(E)$ contains $\mathscr{R}$, hence it contains $\mathscr{M}$. We have shown for every pair $E, F \in \mathscr{M}$ that the sets $E - F$, $F - E$, and $E \cup F$ belong to $\mathscr{M}$, thus $\mathscr{M}$ is a ring; being monotone, it is a σ-ring. ▌

The form in which we shall use this result, often and profitably, is as follows:

Corollary. *If $\mathscr{R}$ is a ring, $\mathscr{M}$ is a monotone class, and $\mathscr{R} \subset \mathscr{M}$, then $\mathfrak{S}(\mathscr{R}) \subset \mathscr{M}$.*

The proof is instant: $\mathscr{M}$ is a monotone class containing $\mathscr{R}$, hence it contains $\mathfrak{M}(\mathscr{R}) = \mathfrak{S}(\mathscr{R})$. This result will be referred to as the "Lemma on Monotone Classes," briefly, (LMC).

A typical application of the (LMC) runs as follows. We are given a ring

$\mathscr{R}$ and wish to prove that every set in $\mathfrak{S}(\mathscr{R})$ possesses a certain property. It is sufficient to show that every set in $\mathscr{R}$ possesses the property, and that the class $\mathscr{M}$ of all sets which possess the property is a monotone class.

EXERCISE

1. Suppose $\mathscr{R}$ is a ring, and ν is a real-valued set function on $\mathfrak{S}(\mathscr{R})$ such that $\nu(E_n) \to \nu(E)$ whenever $E_n \uparrow E$ or $E_n \downarrow E$. If $\nu \geq 0$ on $\mathscr{R}$, then $\nu \geq 0$ on $\mathfrak{S}(\mathscr{R})$.

3. Set Functions, Measures

The set of all real numbers will be denoted R. The set of *extended real numbers* is the result of adjoining to R the symbols ∞ and $-\infty$; it will be denoted R_e, thus

$$R_e = R \cup \{\infty, -\infty\}.$$

The usual ordering of R is extended to R_e by specifying that

$$-\infty < c < \infty$$

for every real number c. Every nonempty set of extended real numbers will then have a least upper bound and a greatest lower bound. Briefly, R_e is a *complete lattice*. The abbreviations LUB and GLB will be used. If α_n is an increasing sequence of extended real numbers, that is, if $\alpha_n \leq \alpha_{n+1}$ for all n, we shall write $\alpha_n\uparrow$; if moreover $\alpha = \text{LUB } \alpha_n$, we write $\alpha_n \uparrow \alpha$. Dually, $\alpha_n\downarrow$ means that $\alpha_n \geq \alpha_{n+1}$ for all n, and $\alpha_n \downarrow \alpha$ means that moreover $\alpha = \text{GLB } \alpha_n$.

The multiplication in R is extended to R_e as follows: one defines

$$0 \cdot \infty = \infty \cdot 0 = 0(-\infty) = (-\infty)0 = 0,$$

while all other products $\alpha\beta$ are defined in the expected way (sample: $(-2)\infty = -\infty$). The symbols $\infty + (-\infty)$ and $(-\infty) + \infty$ are *not defined*, while all other sums $\alpha + \beta$ of extended real numbers are defined in the expected way (sample: $(-2) + \infty = \infty$).

A **set function** is a function ν whose domain of definition is a nonempty class $\mathscr{E}$ of sets; for our purposes, the values of ν will always be extended real numbers, that is,

$$\nu\colon \mathscr{E} \to R_e.$$

A set function is said to be **finite** if its values are real numbers.

Suppose ν is a set function on $\mathscr{E}$. If

$$\nu(E) \geq 0$$

for all E in $\mathscr{E}$, ν is said to be **positive**; briefly, $\nu \geq 0$. The set function ν is said to be **additive** in case

$$\nu(E \cup F) = \nu(E) + \nu(F)$$

whenever E and F are disjoint sets in $\mathscr{E}$ whose union is also in $\mathscr{E}$. If

$$\nu(E) \leq \nu(F)$$

whenever E and F are sets in $\mathscr{E}$ such that $E \subset F$, ν is said to be **monotone.** If $\varnothing$ belongs to $\mathscr{E}$, ν is monotone, and $\nu(\varnothing) = 0$, evidently ν is positive. On the other hand, an additive positive set function ν defined on a *ring* is necessarily monotone, since $E \subset F$ implies

$$F = (F - E) \cup E,$$

and hence

$$\nu(F) = \nu(F - E) + \nu(E) \geq \nu(E).$$

A **measure** is a set function μ whose domain of definition is a ring $\mathscr{R}$, such that:

(1) μ is additive.
(2) μ is positive.
(3) $\mu(\varnothing) = 0$.
(4) If E_n is an increasing sequence in $\mathscr{R}$ whose union E is also in $\mathscr{R}$, then $\mu(E) = \text{LUB } \mu(E_n)$.

Suppose μ is a measure defined on the ring $\mathscr{R}$. Since μ is additive and positive, μ is monotone; it follows that if E_n is an increasing sequence of sets in $\mathscr{R}$, we have $\mu(E_n)\uparrow$. Thus, condition (4) may be expressed as follows: if

$$E_n \uparrow E,$$

where the E_n and E belong to $\mathscr{R}$, then

$$\mu(E_n) \uparrow \mu(E).$$

We express this condition by saying that μ is **continuous from below.** We emphasize that E_n is assumed to be a *sequence* (indexed by the natural numbers). Some of the most important first properties of a measure are listed in the following:

Theorem 1. *Let μ be a measure on a ring $\mathscr{R}$. Then:*

(5) *μ is* **monotone**: *that is, $\mu(E) \leq \mu(F)$ whenever E and F are sets in $\mathscr{R}$ such that $E \subset F$.*

(6) *μ is (conditionally)* **subtractive**: *that is,*

$$\mu(F - E) = \mu(F) - \mu(E)$$

provided E and F are sets in $\mathscr{R}$ such that $F \supset E$ and $\mu(E)$ is finite.

(7) μ *is* **finitely additive**: *that is, if* $E_1, \ldots, E_n$ *are mutually disjoint sets in* $\mathscr{R}$, *then*

$$\mu\left(\bigcup_1^n E_k\right) = \sum_1^n \mu(E_k).$$

(8) μ *is* **countably additive**: *that is, if* E_k *is a sequence of mutually disjoint sets in* $\mathscr{R}$, *such that* $\bigcup_1^\infty E_k$ *is in* $\mathscr{R}$, *then*

$$\mu\left(\bigcup_1^\infty E_k\right) = \sum_1^\infty \mu(E_k),$$

in the sense that the LUB of the (increasing) sequence of partial sums $\sum_1^n \mu(E_k)$ *is equal to* $\mu(\bigcup_1^\infty E_k)$.

Proof. Monotonicity was shown earlier. Subtractivity follows from $\mu(F) = \mu(F - E) + \mu(E)$ and the fact that the real number $\mu(E)$ may be transposed. Finite additivity follows from additivity by induction.

Suppose now that E_k is a disjoint sequence in $\mathscr{R}$ whose union E is also in $\mathscr{R}$. Setting

$$F_n = \bigcup_1^n E_k,$$

we have $F_n \uparrow E$, and hence $\mu(F_n) \uparrow \mu(E)$ by (4). But

$$\mu(F_n) = \sum_1^n \mu(E_k)$$

by (7). ▮

Occasionally, the following criterion is useful in verifying that a given set function is a measure:

Theorem 2. *If* $\mathscr{R}$ *is a ring, and* μ *is an extended real valued set function on* $\mathscr{R}$ *which is positive, countably additive, and satisfies the condition* $\mu(\varnothing) = 0$, *then* μ *is a measure.*

Proof. We must verify additivity, and continuity from below. If E and F are disjoint sets in $\mathscr{R}$, then

$$\mu(E \cup F) = \mu(E) + \mu(F)$$

results from the relation

$$E \cup F = E \cup F \cup \varnothing \cup \varnothing \cup \cdots$$

and the assumed properties of μ.

Suppose now that $E_n \uparrow E$, where E and the E_n are sets in $\mathscr{R}$. Define $F_1 = E_1, F_2 = E_2 - E_1, \ldots, F_n = E_n - E_{n-1}, \ldots$. The F_n are mutually disjoint, and their union is E, hence

$$\mu(E) = \sum_1^\infty \mu(F_k) = \mathrm{LUB}_n \sum_1^n \mu(F_k) = \mathrm{LUB}_n\, \mu\left(\bigcup_1^n F_k\right) = \mathrm{LUB}_n\, \mu(E_n). \quad ▮$$

Many of the fundamental theorems in the sequel require further assumptions on the measure in question. We will have occasion to use just two such conditions, finiteness and σ-finiteness. A measure μ on a ring $\mathscr{R}$ is said to be **finite** in case $\mu(E) < \infty$ for every E in $\mathscr{R}$. A measure μ on a ring $\mathscr{R}$ is said to be **σ-finite** in case: given any E in $\mathscr{R}$, there exists a sequence E_n in $\mathscr{R}$ such that $E \subset \bigcup_1^\infty E_n$ and $\mu(E_n) < \infty$ for all n; since

$$E = \bigcup_1^\infty E \cap E_n \qquad \text{and} \qquad \mu(E \cap E_n) \leq \mu(E_n) < \infty,$$

we see that σ-finiteness is equivalent to the requirement that every set in $\mathscr{R}$ be expressible as a countable union of sets in $\mathscr{R}$ with finite measure.

The fundamental results of this chapter will be as follows. If μ is a measure on a ring $\mathscr{R}$, it is always possible to extend μ, by the technique of "outer measure," to a measure on the σ-ring $\mathfrak{S}(\mathscr{R})$ generated by $\mathscr{R}$; if, moreover, μ is σ-finite, we may be sure that such an extension is unique (see Sect. 6).

Incidentally, a finite measure is called *totally* finite if its domain of definition is an algebra of sets (that is, contains X). Similarly, a σ-finite measure is called *totally* σ-finite if its domain of definition is an algebra of sets. We shall have no need to use these concepts.

EXERCISES

1. Let $\mathscr{R}$ be any ring of subsets of a set X. If $E \in \mathscr{R}$, define $\mu(E)$ to be the number of points of E when E is finite, and $+\infty$ when E is infinite. Then μ is a measure on $\mathscr{R}$; it is called *discrete measure* on $\mathscr{R}$.

2. If μ is a measure on a ring $\mathscr{R}$, and E, F are sets in $\mathscr{R}$ such that $\mu(E \triangle F) = 0$, then

$$\mu(E) = \mu(F) = \mu(E \cup F) = \mu(E \cap F).$$

4. Some Properties of Measures

In the theorems of this section, we consider a fixed measure μ on a ring $\mathscr{R}$.

Theorem 1. *μ is* **countably subadditive**; *that is, if E_n is any sequence of sets in $\mathscr{R}$ whose union is also in $\mathscr{R}$, then*

$$\mu\left(\bigcup_1^\infty E_n\right) \leq \sum_1^\infty \mu(E_n).$$

Proof. The idea of the proof is to "disjointify" the sequence E_n, so that the countable additivity of μ may be invoked. First define

$$F_n = \bigcup_1^n E_k;$$

then

$$F_n \uparrow E = \bigcup_1^\infty E_n.$$

Define $G_1 = F_1, G_2 = F_2 - F_1, \ldots, G_n = F_n - F_{n-1}, \ldots$. Evidently $G_n \subset E_n$, hence $\mu(G_n) \leq \mu(E_n)$. Since the G_n are mutually disjoint, and their union is E, we have

$$\mu(E) = \sum_1^\infty \mu(G_n) \leq \sum_1^\infty \mu(E_n). \quad \blacksquare$$

Corollary. *If E_n is a sequence of sets in $\mathscr{R}$, and E is a set in $\mathscr{R}$ such that $E \subset \bigcup_1^\infty E_n$, then*

$$\mu(E) \leq \sum_1^\infty \mu(E_n).$$

Proof. Since $E = \bigcup_1^\infty E \cap E_n$, and $\mu(E \cap E_n) \leq \mu(E_n)$, our assertion is immediate from the theorem. $\blacksquare$

Theorem 2. *μ is* **conditionally continuous from above**; *that is, if E_n is a decreasing sequence of sets in $\mathscr{R}$, whose intersection E is in $\mathscr{R}$ (thus $E_n \downarrow E$), and if $\mu(E_1) < \infty$, then*

$$\mu(E_n) \downarrow \mu(E).$$

Proof. Clearly $\mu(E) \leq \mu(E_n) \leq \mu(E_1) < \infty$ for all n. Since

$$E_1 - E_n \uparrow E_1 - E,$$

and μ is continuous from below, we have

$$\mu(E_1 - E_n) \uparrow \mu(E_1 - E).$$

Quoting 3.1,

$$\mu(E_1) - \mu(E_n) \uparrow \mu(E_1) - \mu(E);$$

since all of these numbers are finite, evidently $\mu(E_n) \downarrow \mu(E)$. $\blacksquare$

The next result is a useful tool in reducing discussions about σ-finite measures to the finite case. If F is a fixed set in $\mathscr{R}$, we may define a set function μ_F on $\mathscr{R}$ by the formula $\mu_F(E) = \mu(F \cap E)$. The set function μ_F is called the **contraction** of μ by F.

Theorem 3. *If μ_F is the contraction of μ by a fixed set F in $\mathscr{R}$, then μ_F is a measure on $\mathscr{R}$. If, moreover, $\mu(F) < \infty$, then μ_F is a finite measure.*

Proof. Suppose $E_n \uparrow E$, where E and the E_n are in $\mathscr{R}$. Then $F \cap E_n \uparrow F \cap E$, hence

$$\mu(F \cap E_n) \uparrow \mu(F \cap E);$$

in other words $\mu_F(E_n) \uparrow \mu_F(E)$. Similarly, the additivity of μ_F results at once from the additivity of μ. If, moreover, $\mu(F) < \infty$, then

$$\mu_F(E) = \mu(F \cap E) \leq \mu(F) < \infty$$

for all E in $\mathscr{R}$, hence μ_F is a finite measure. ▌

EXERCISE

1. A nonnegative extended real valued set function μ defined on a ring $\mathscr{R}$ is a measure if and only if it is (*i*) additive, (*ii*) countably subadditive, and (*iii*) $\mu(\varnothing) = 0$.

5. Outer Measures

To motivate the discussion of outer measures, let us consider a problem of extension. Suppose we are given a measure μ on a ring $\mathscr{R}$. We wish to define "measure" for a more extensive class of sets; at the very least we wish to extend μ to the σ-ring $\mathfrak{S}(\mathscr{R})$ generated by $\mathscr{R}$. Suppose, then, that E is a set in $\mathfrak{S}(\mathscr{R})$. According to 1.4, we have $E \subset \bigcup_1^\infty E_n$ for a suitable sequence E_n of sets in $\mathscr{R}$. A crude "measure" of E is $\sum_1^\infty \mu(E_n)$; a more satisfying "measure" is

$$\text{GLB} \sum_1^\infty \mu(E_n),$$

as we vary all possible countable coverings of E by sets in $\mathscr{R}$. It turns out to be convenient to apply this technique to the class of *all* sets A which can be countably covered by sets in $\mathscr{R}$. Let us denote this class by $\mathscr{H}$. Evidently $\mathscr{H}$ is closed under countable unions. Moreover, the class $\mathscr{H}$ has the property that if A is a set in $\mathscr{H}$ and $B \subset A$, then B is also in $\mathscr{H}$; such a class is called **hereditary**. In particular it is clear that $\mathscr{H}$ is a ring, and hence is a σ-ring.

We are thus led to the concept of a **hereditary σ-ring**. Since for any set X, the class $\mathscr{P}(X)$ is a hereditary σ-ring, and since the intersection of any family of hereditary σ-rings is clearly a hereditary σ-ring, it follows that given any class $\mathscr{E}$ of subsets of X, there is a smallest hereditary σ-ring containing $\mathscr{E}$. This is denoted $\mathscr{H}(\mathscr{E})$, and is called the hereditary σ-ring *generated* by $\mathscr{E}$. It is easy to see that $\mathscr{H}(\mathscr{E})$ is simply the class of all sets A which can be countably covered by sets in $\mathscr{E}$.

Let us return to our discussion of a measure μ on a ring $\mathscr{R}$:

Theorem 1. *Let μ be a measure on a ring $\mathscr{R}$, and define an extended real valued set function μ^* on $\mathscr{H}(\mathscr{R})$ by the formula*

$$\mu^*(A) = \text{GLB}\left\{\sum_1^\infty \mu(E_n)\colon A \subset \bigcup_1^\infty E_n,\ E_n \in \mathscr{R}\ (n = 1, 2, \ldots)\right\}.$$

Then:

(1) *μ^* is positive.*
(2) $\mu^*(\varnothing) = 0$.
(3) *μ^* is monotone.*
(4) *μ^* is countably subadditive.*
(5) *μ^* extends μ.*

Proof. (3) Suppose B is in $\mathscr{H} = \mathscr{H}(\mathscr{R})$, and $A \subset B$. If E_n is any sequence of sets in $\mathscr{R}$ such that $B \subset \bigcup_1^\infty E_n$, then also $A \subset \bigcup_1^\infty E_n$, hence

$$\mu^*(A) \le \sum_1^\infty \mu(E_n)$$

by the definition of μ^*; taking GLB over all such coverings of B, we have $\mu^*(A) \le \mu^*(B)$.

(4) Suppose A_n is a sequence of sets in $\mathscr{H}$, and $A = \bigcup_1^\infty A_n$. The problem is to show that

$$\mu^*(A) \le \sum_1^\infty \mu^*(A_n).$$

This is trivial if $\mu^*(A_n) = \infty$ for some n; let us assume $\mu^*(A_n) < \infty$ for all n. Given any $\varepsilon > 0$, it will suffice to show that

$$\mu^*(A) \le \sum_1^\infty \mu^*(A_n) + \varepsilon.$$

For each n, $\mu^*(A_n) < \mu^*(A_n) + \varepsilon/2^n$, hence there exists a sequence $E_{n1}, E_{n2}, E_{n3}, \ldots$ of sets in $\mathscr{R}$ such that

$$A_n \subset \bigcup_k E_{nk} \quad \text{and} \quad \sum_k \mu(E_{nk}) \le \mu^*(A_n) + \varepsilon/2^n.$$

Then $A \subset \bigcup_{n,k} E_{nk}$, hence

$$\mu^*(A) \le \sum_{n,k} \mu(E_{nk}) = \sum_n \sum_k \mu(E_{nk})$$
$$\le \sum_n [\mu^*(A_n) + \varepsilon/2^n] = \sum_n \mu^*(A_n) + \varepsilon.$$

(5) Assuming E is in $\mathscr{R}$, the problem is to show that $\mu^*(E) = \mu(E)$. The relation $E \subset E \in \mathscr{R}$ already shows that $\mu^*(E) \leq \mu(E)$. On the other hand, suppose E_n is any sequence of sets in $\mathscr{R}$ such that $E \subset \bigcup_1^\infty E_n$. By the corollary of 4.1,

$$\mu(E) \leq \sum_1^\infty \mu(E_n);$$

taking GLB over all such countable coverings, $\mu(E) \leq \mu^*(E)$. Thus $\mu^*(E) = \mu(E)$, and in particular $\mu^*(\varnothing) = 0$. ▮

Consider again the problem of extending a measure μ defined on a ring $\mathscr{R}$. According to Theorem 1, we may consider the set function μ^* on the (usually larger) class $\mathscr{H}(\mathscr{R})$. However, μ^* is generally not a measure, since it may fail to be additive. So to speak, we have extended μ too far, and the problem now is to back off μ^* to a subclass $\mathscr{M}$ of $\mathscr{H}(\mathscr{R})$ on which μ^* is again a measure. The miracle is that a subclass $\mathscr{M}$ exists, such that $\mathscr{M}$ is a σ-ring, the restriction of μ^* to $\mathscr{M}$ is a measure, and $\mathscr{M}$ includes $\mathscr{R}$. It turns out that in defining $\mathscr{M}$, proving that it is a σ-ring, and showing that the restriction of μ^* to $\mathscr{M}$ is a measure, we make use only of properties (1) through (4).

Accordingly, it is convenient, and useful in applications, to isolate properties (1) through (4): an **outer measure** is a set function ν whose domain of definition is a hereditary σ-ring, and which is positive, monotone, countably subadditive, and satisfies the condition $\nu(\varnothing) = 0$.

The rest of the section is devoted to the study of a fixed "abstract" outer measure ν (as distinguished from a "concrete" one induced by a measure) defined on a hereditary σ-ring $\mathscr{H}$. A set E in $\mathscr{H}$ is said to be **ν-measurable** in case

$$\nu(A) = \nu(A \cap E) + \nu(A \cap E')$$

for all A in $\mathscr{H}$, where E' denotes the complement of E. So to speak, a set E in $\mathscr{H}$ is ν-measurable if it splits every other set in $\mathscr{H}$ additively. Let us denote by $\mathscr{M}$ the class of all ν-measurable sets. It will be shown that $\mathscr{M}$ is a σ-ring, and the restriction of ν to $\mathscr{M}$ is a measure.

If E is any set in $\mathscr{H}$, then the subadditivity of ν implies

$$\nu(A) \leq \nu(A \cap E) + \nu(A \cap E')$$

for every A in $\mathscr{H}$. Thus in testing for the ν-measurability of E, it is sufficient to show that

$$(*) \qquad \nu(A) \geq \nu(A \cap E) + \nu(A \cap E')$$

for every A in $\mathscr{H}$.

Lemma. *If ν is an outer measure, the class $\mathscr{M}$ of ν-measurable sets is a ring.*

Proof. If $E = \varnothing$, condition $(*)$ above reduces to $\nu(A) \geq \nu(\varnothing) + \nu(A)$, and this results from $\nu(\varnothing) = 0$. Thus $\varnothing$ is ν-measurable.

Suppose E and F are sets in $\mathscr{M}$, and A is in $\mathscr{H}$. The problem is to show that

$$\nu(A) = \nu[A \cap (E - F)] + \nu[A \cap (E - F)'],$$
$$\nu(A) = \nu[A \cap (E \cup F)] + \nu[A \cap (E \cup F)'];$$

in other words, since $(E - F)' = (E \cap F')' = E' \cup F$ and $(E \cup F)' = E' \cap F'$, it must be shown that

$$(i) \qquad \nu(A) = \nu(A \cap E \cap F') + \nu[A \cap (E' \cup F)],$$
$$(ii) \qquad \nu(A) = \nu[A \cap (E \cup F)] + \nu(A \cap E' \cap F').$$

Since F splits $A \cap (E' \cup F)$ additively,

$$\begin{aligned}\nu[A \cap (E' \cup F)] &= \nu[A \cap (E' \cup F) \cap F] + \nu[A \cap (E' \cup F) \cap F'] \\ &= \nu(A \cap F) + \nu(A \cap E' \cap F');\end{aligned}$$

since E splits $A \cap F$ additively, this may be written

$$\begin{aligned}\nu[A \cap (E' \cup F)] &= \nu(A \cap F \cap E) + \nu(A \cap F \cap E') + \nu(A \cap E' \cap F') \\ &= \nu(A \cap F \cap E) + \nu(A \cap E'),\end{aligned}$$

the last equality resulting from the fact that F splits $A \cap E'$ additively. Then,

$$\begin{aligned}&\nu(A \cap E \cap F') + \nu[A \cap (E' \cup F)] \\ &\qquad = \nu(A \cap E \cap F') + \nu(A \cap F \cap E) + \nu(A \cap E') \\ &\qquad = \nu(A \cap E) + \nu(A \cap E') \\ &\qquad = \nu(A),\end{aligned}$$

where we have used successively the fact that F splits $A \cap E$ additively, and E splits A additively. This establishes (i).

Since E splits $A \cap (E \cup F)$ additively,

$$\begin{aligned}\nu[A \cap (E \cup F)] &= \nu[A \cap (E \cup F) \cap E] + \nu[A \cap (E \cup F) \cap E'] \\ &= \nu(A \cap E) + \nu(A \cap F \cap E');\end{aligned}$$

hence

$$\begin{aligned}&\nu[A \cap (E \cup F)] + \nu(A \cap E' \cap F') \\ &\qquad = \nu(A \cap E) + \nu(A \cap F \cap E') + \nu(A \cap E' \cap F') \\ &\qquad = \nu(A \cap E) + \nu(A \cap E') \\ &\qquad = \nu(A),\end{aligned}$$

where we have used the fact that F splits $A \cap E'$, and E splits A. This establishes (ii). ∎

Theorem 2. *If ν is an outer measure on a hereditary σ-ring $\mathscr{H}$, and $\mathscr{M}$ is the class of all ν-measurable sets, then:*

(1) *$\mathscr{M}$ is a σ-ring.*

(2) *If E_n is a sequence of mutually disjoint sets in $\mathscr{M}$, whose union is E, then*

$$\nu(A \cap E) = \sum_1^\infty \nu(A \cap E_n)$$

for every A in $\mathscr{H}$.

(3) *The restriction of ν to $\mathscr{M}$ is a measure.*

Proof. For each A in $\mathscr{H}$, let ν_A be the set function on $\mathscr{M}$ defined by the formula

$$\nu_A(E) = \nu(A \cap E);$$

in view of 3.2, the assertion in (2) is that ν_A is a measure on $\mathscr{M}$.

Since $\mathscr{M}$ is already known to be a ring by the lemma, it will suffice, in proving (1), to show that $\mathscr{M}$ is closed under the formation of countable disjoint unions (see for example the proof of 4.1 for the technique of replacing a countable union by a countable disjoint union). Suppose, then, that E_n is a sequence of disjoint sets in $\mathscr{M}$, with union E, and suppose that A is in $\mathscr{H}$; let us show simultaneously that E is in $\mathscr{M}$ and that (2) holds.

Since $E_1 \cap E_2 = \varnothing$, that is, $E_2 \cap E_1' = E_2$, and since E_1 splits $A \cap (E_1 \cup E_2)$ additively, we have

$$\begin{aligned}\nu[A \cap (E_1 \cup E_2)] &= \nu[A \cap (E_1 \cup E_2) \cap E_1] + \nu[A \cap (E_1 \cup E_2) \cap E_1'] \\ &= \nu(A \cap E_1) + \nu(A \cap E_2);\end{aligned}$$

thus

$$\nu_A(E_1 \cup E_2) = \nu_A(E_1) + \nu_A(E_2).$$

It follows at once that ν_A is finitely additive on the ring $\mathscr{M}$. Defining $F_n = \bigcup_1^n E_k$, we have

$$\nu_A(F_n) = \sum_1^n \nu_A(E_k).$$

Since $F_n \uparrow E$, one has $F_n' \downarrow E'$, and in particular $A \cap F_n' \supset A \cap E'$ for all n. Since F_n is ν-measurable, and ν is monotone,

$$\begin{aligned}\nu(A) = \nu(A \cap F_n) + \nu(A \cap F_n') &\geq \nu(A \cap F_n) + \nu(A \cap E') \\ &= \nu_A(F_n) + \nu(A \cap E') \\ &= \sum_1^n \nu_A(E_k) + \nu(A \cap E');\end{aligned}$$

since n is arbitrary, this yields

$$(i) \qquad \nu(A) \geq \sum_1^\infty \nu_A(E_n) + \nu(A \cap E').$$

But

$$\sum_1^\infty \nu_A(E_n) = \sum_1^\infty \nu(A \cap E_n) \geq \nu\left(\bigcup_1^\infty A \cap E_n\right) = \nu(A \cap E)$$

by countable subadditivity; substituting in (i),

$$(*) \qquad \nu(A) \geq \nu(A \cap E) + \nu(A \cap E'),$$

and this establishes the ν-measurability of E. It follows that

$$\nu(A) \geq \sum_1^\infty \nu_A(E_n) + \nu(A \cap E') \geq \nu(A \cap E) + \nu(A \cap E') = \nu(A),$$

thus

$$\nu(A) = \sum_1^\infty \nu_A(E_n) + \nu(A \cap E').$$

Replacing A by $A \cap E$, this yields

$$\nu(A \cap E) = \sum_1^\infty \nu_A(E_n) + 0.$$

This establishes (2). In particular, setting $A = E$, we have $\nu(E) = \sum_1^\infty \nu(E_n)$, thus ν is countably additive on $\mathscr{M}$, and it follows from 3.2 that the restriction of ν to $\mathscr{M}$ is a measure.

Incidentally,

$$\nu(B \cap E) = \sum_1^\infty \nu(B \cap E_n)$$

for *every* subset B of the underlying set X; simply apply (2) to the set $A = B \cap E$. Thus, the set function ν_B on $\mathscr{M}$, defined by the formula $\nu_B(E) = \nu(B \cap E)$, is also a measure. ∎

If E is a set in $\mathscr{H}$ such that $\nu(E) = 0$, then E is ν-measurable; the condition

$$(*) \qquad \nu(A) \geq \nu(A \cap E) + \nu(A \cap E')$$

is fulfilled for every A in $\mathscr{H}$ because

$$\nu(A \cap E) \leq \nu(E) = 0 \qquad \text{and} \qquad \nu(A) \geq \nu(A \cap E').$$

It follows from monotonicity that every subset of E is also ν-measurable. Thus, the measure obtained by restricting ν to $\mathscr{M}$ satisfies the following condition: if E is in the domain of definition of the measure, and the measure of E is 0, then every subset of E is also in the domain of definition of the measure. Such a measure is said to be *complete*. The hypothesis of completeness will never be needed in the sequel.

There remains the question of how extensive is the class $\mathscr{M}$—conceivably it is $\{\varnothing\}$. The answer is quite satisfactory for the outer measure μ^* induced by a measure μ, as we shall see in the next section.

6. Extension of Measures

We take up again, and solve, the problem of extending a measure μ on a ring $\mathscr{R}$ to a measure on the generated σ-ring $\mathfrak{S}(\mathscr{R})$:

Theorem 1. *If μ is a measure on a ring $\mathscr{R}$, and $\mathscr{M}$ is the class of all μ^*-measurable sets, then*

$$\mathfrak{S}(\mathscr{R}) \subset \mathscr{M},$$

and the restriction of μ^* to $\mathfrak{S}(\mathscr{R})$ is a measure $\bar{\mu}$ extending μ.

Proof. In view of 5.1 and 5.2, it will suffice to prove that $\mathscr{R} \subset \mathscr{M}$; for then $\mathfrak{S}(\mathscr{R}) \subset \mathscr{M}$ by 1.3, and μ^* extends μ by 5.1. Thus, given E in $\mathscr{R}$, A in $\mathscr{H}(\mathscr{R})$, and $\varepsilon > 0$, it will suffice to show that

$$\mu^*(A) + \varepsilon \geq \mu^*(A \cap E) + \mu^*(A \cap E').$$

This is trivial if $\mu^*(A) = \infty$; let us assume $\mu^*(A) < \infty$. Let E_n be a sequence of sets in $\mathscr{R}$ such that

$$A \subset \bigcup_1^\infty E_n \quad \text{and} \quad \sum_1^\infty \mu(E_n) \leq \mu^*(A) + \varepsilon.$$

Then

$$A \cap E \subset \bigcup_1^\infty E_n \cap E$$

and

$$A \cap E' \subset \bigcup_1^\infty E_n \cap E'$$

hence

$$\mu^*(A \cap E) \leq \sum_1^\infty \mu(E_n \cap E)$$

and

$$\mu^*(A \cap E') \leq \sum_1^\infty \mu(E_n \cap E').$$

Since μ is additive on $\mathscr{R}$, we have

$$\mu(E_n) = \mu(E_n \cap E) + \mu(E_n \cap E'),$$

hence

$$\begin{aligned}\mu^*(A) + \varepsilon \geq \sum_1^\infty \mu(E_n) &= \sum_1^\infty \mu(E_n \cap E) + \sum_1^\infty \mu(E_n \cap E') \\ &\geq \mu^*(A \cap E) + \mu^*(A \cap E'). \quad \blacksquare\end{aligned}$$

Indeed, the restriction of μ^* to $\mathscr{M}$ is a complete measure extending μ, but for our purposes the extension of μ to $\mathfrak{S}(\mathscr{R})$ will be far enough. It turns out that when μ is σ-finite, *any* extension of μ to a measure on $\mathfrak{S}(\mathscr{R})$ must coincide with μ^*; this is shown in Theorem 2 below, and is an immediate consequence of the following:

Lemma. *Let $\mathscr{R}$ be a ring, and suppose that μ_1 and μ_2 are measures on the σ-ring $\mathfrak{S}(\mathscr{R})$ generated by $\mathscr{R}$, such that*

$$\mu_1(E) = \mu_2(E)$$

for all E in $\mathscr{R}$. Assume, moreover, that the restriction of μ_i to $\mathscr{R}$ is σ-finite. Then $\mu_1 = \mu_2$ on $\mathfrak{S}(\mathscr{R})$.

Proof. Let us consider first the case that μ_1 and μ_2 are finite measures on $\mathfrak{S}(\mathscr{R})$. Let $\mathscr{A}$ be the class of all sets E in $\mathfrak{S}(\mathscr{R})$ such that $\mu_1(E) = \mu_2(E)$. We are assuming that

$$\mathscr{R} \subset \mathscr{A} \subset \mathfrak{S}(\mathscr{R}).$$

The assertion is that $\mathfrak{S}(\mathscr{R}) \subset \mathscr{A}$; it will suffice, by the (LMC), to show that $\mathscr{A}$ is a monotone class. If $E_n \uparrow E$, where E_n is a sequence of sets in $\mathscr{A}$, then

$$\mu_i(E_n) \uparrow \mu_i(E)$$

by the definition of a measure; since

$$\mu_1(E_n) = \mu_2(E_n)$$

for all n, we conclude that $\mu_1(E) = \mu_2(E)$, thus E is in $\mathscr{A}$. On the other hand, if $E_n \downarrow E$, where E_n is a sequence of sets in $\mathscr{A}$, then

$$\mu_i(E_n) \downarrow \mu_i(E)$$

by conditional continuity from above (4.2), hence again $\mu_1(E) = \mu_2(E)$.

Consider now the general case. Given E in $\mathfrak{S}(\mathscr{R})$, let us show that $\mu_1(E) = \mu_2(E)$. By 1.4, there is a sequence F_n of sets in $\mathscr{R}$ such that

$$E \subset \bigcup_1^\infty F_n,$$

and by the assumed σ-finiteness of the restriction of μ_i to $\mathscr{R}$, we may suppose further that $\mu_i(F_n) < \infty$ for all n. Replacing F_n by $\bigcup_1^n F_k$, we can assume that $F_n\uparrow$, and hence

$$F_n \cap E \uparrow E;$$

it follows that

$$\mu_i(F_n \cap E) \uparrow \mu_i(E),$$

thus it will suffice to show that $\mu_1(F_n \cap E) = \mu_2(F_n \cap E)$ for all n. By 4.3, the contraction $(\mu_i)_{F_n}$ of μ_i by F_n is a finite measure on $\mathfrak{S}(\mathscr{R})$. Since it is clear that

$$(\mu_1)_{F_n} = (\mu_2)_{F_n} \quad \text{on} \quad \mathscr{R},$$

we have $(\mu_1)_{F_n} = (\mu_2)_{F_n}$ on $\mathfrak{S}(\mathscr{R})$ by the first part of the proof. In particular,

$$(\mu_1)_{F_n}(E) = (\mu_2)_{F_n}(E),$$

thus $\mu_1(F_n \cap E) = \mu_2(F_n \cap E)$ for all n. ∎

The above lemma will be referred to as the "Unique Extension Theorem," briefly (UET).

Theorem 2. *If μ is a σ-finite measure on a ring $\mathscr{R}$, then there exists a unique measure $\bar{\mu}$ on $\mathfrak{S}(\mathscr{R})$ which extends μ. Moreover, $\bar{\mu}$ is σ-finite, and $\bar{\mu}(E) = \mu^*(E)$ for all E in $\mathfrak{S}(\mathscr{R})$.*

Proof. According to Theorem 1, the formula $\bar{\mu}(E) = \mu^*(E)$ defines a measure $\bar{\mu}$ on $\mathfrak{S}(\mathscr{R})$ which extends μ. Every other extension of μ to a measure on $\mathfrak{S}(\mathscr{R})$ must coincide with $\bar{\mu}$ by the (UET).

Now, every set in $\mathfrak{S}(\mathscr{R})$ can be covered by a countable union of sets in $\mathscr{R}$ (1.4); moreover, every set in $\mathscr{R}$ is a countable union of sets in $\mathscr{R}$ on which μ, and therefore $\bar{\mu}$, is finite. Thus, every set in $\mathfrak{S}(\mathscr{R})$ can be covered by a countable union of sets in $\mathscr{R}$ on which $\bar{\mu}$ is finite, hence $\bar{\mu}$ is surely σ-finite. ∎

EXERCISES

1. The (UET) is easily generalized as follows. Suppose $\mathscr{R}$ is a ring, and μ_1 and μ_2 are measures on $\mathfrak{S}(\mathscr{R})$ such that (i) $\mu_1(E) \leq \mu_2(E)$ for all E in $\mathscr{R}$, and (ii) the restriction of μ_i to $\mathscr{R}$ is σ-finite. Then $\mu_1 \leq \mu_2$ on $\mathfrak{S}(\mathscr{R})$.

2. If μ is a finite measure on a ring $\mathscr{R}$, and $\bar{\mu}$ is its unique extension to $\mathfrak{S}(\mathscr{R})$, then $\bar{\mu}$ is finite if and only if μ is bounded.

3. If μ is a measure on a ring $\mathscr{R}$, and $A \subset E$, where $E \in \mathfrak{S}(\mathscr{R})$ and $\mu^*(E) < \infty$, then A is μ^*-measurable if and only if

$$\mu^*(E) = \mu^*(A) + \mu^*(E - A).$$

4. If μ is a measure on a ring $\mathscr{R}$, and $\bar{\mu}$ is the restriction of μ^* to $\mathfrak{S}(\mathscr{R})$, then $\mathscr{R}$ and $\mathfrak{S}(\mathscr{R})$ generate the same hereditary σ-ring, and the set functions μ^* and $(\bar{\mu})^*$ are identical.

5. Let $\mathscr{R}$ be a ring, and suppose μ is a measure on $\mathfrak{S}(\mathscr{R})$ whose restriction to $\mathscr{R}$ is σ-finite. If $E \in \mathfrak{S}(\mathscr{R})$ has finite measure, then given any $\varepsilon > 0$ there is a set F in $\mathscr{R}$ such that $\mu(E \triangle F) \leq \varepsilon$.

6. Suppose ν is a measure on a σ-ring $\mathscr{S}$, $\mathscr{R}$ is a subring of $\mathscr{S}$ such that $\mathfrak{S}(\mathscr{R}) = \mathscr{S}$, and μ is the restriction of ν to $\mathscr{R}$. If μ is σ-finite, then $\nu(E) = \mu^*(E)$ for all E in $\mathscr{S}$.

7. If μ is a measure on a σ-ring $\mathscr{S}$, then

$$\mu^*(A) = \text{GLB}\,\{\mu(E) : A \subset E,\ E \in \mathscr{S}\}$$

for every A in $\mathscr{H}(\mathscr{S})$.

8. Suppose μ is a σ-finite measure on a σ-ring $\mathscr{S}$, and $\mathscr{M}$ is the class of all μ^*-measurable sets. If ν is any measure on $\mathscr{M}$ such that $\nu = \mu$ on $\mathscr{S}$, then necessarily $\nu = \mu^*$ on $\mathscr{M}$. (*Hint:* 8.2.)

9. The uniqueness part of Theorem 2 extends as follows. Suppose μ is a σ-finite measure on a ring $\mathscr{R}$, and $\mathscr{M}$ is the class of all μ^*-measurable sets. If ν is any measure on $\mathscr{M}$ such that $\nu = \mu$ on $\mathscr{R}$, then necessarily $\nu = \mu^*$ on $\mathscr{M}$.

10. If μ is a measure on a ring $\mathscr{R}$ of subsets of X, and B is *any* subset of X, then the correspondence

$$E \to \mu^*(B \cap E)$$

defines a measure on $\mathscr{R}$.

*7. Lebesgue Measure

The classical example of a measure is defined for certain subsets of the real line, and generalizes the idea of interval length. This is Lebesgue measure, whose construction we shall now sketch.

All of the proofs in this section are omitted, for the following reasons: (1) some or all of the details can easily be supplied by the reader, (2) there are many excellent and complete accounts in print, (3) the construction of Lebesgue measure is also implicit in each of Chapters 8 and 9 (Riesz-Markoff theorem, Haar's theorem), and (4) Lebesgue measure plays no role in the general theory to be developed in the rest of the text (its role in analysis, needless to say, is enormous).

Lebesgue measure is initially defined for intervals $[a, b)$, where a and b are real numbers such that $a \le b$, and assigns to such an interval the measure $b - a$; it is then extended to the ring $\mathscr{R}$ generated by such intervals, by means of the following two elementary lemmas:

Lemma 1. *Every set E in $\mathscr{R}$ can be written in the form*

$$E = \bigcup_{1}^{r} [a_i, b_i),$$

where $a_1 \le b_1 \le a_2 \le b_2 \le \cdots \le a_r \le b_r$ (and hence the intervals $[a_i, b_i)$ are mutually disjoint).

Lemma 2. *If E is a set in $\mathscr{R}$, and*

$$E = \bigcup_1^r [a_i, b_i) = \bigcup_1^s [c_j, d_j)$$

are two representations of E having the property indicated in Lemma 1, then

$$\sum_1^r (b_i - a_i) = \sum_1^s (d_j - c_j).$$

In view of these lemmas, one may unambiguously define the measure of E in $\mathscr{R}$ by the formula

$$m(E) = \sum_1^r (b_i - a_i),$$

where E is decomposed as in Lemma 1. (Incidentally, arguments which are similar to those needed for Lemma 1 are given in 34.1 and 58.1; in proving Lemma 2, it is convenient to consider first the case that E is itself an interval, and then treat the general case by "meshing" the two partitions of E.)

It is immediate from the definition that m is finitely additive on $\mathscr{R}$; since it is nonnegative, it is also monotone and subtractive. To show that m is a (finite) measure on $\mathscr{R}$, it is enough, by 3.2, to show that m is countably additive. Again we omit the proof, which can be based on the following version of the Heine-Borel theorem: if (a_k, b_k) is a sequence of open intervals which cover the closed interval $[a, b]$, that is, if

$$[a, b] \subset \bigcup_1^\infty (a_k, b_k),$$

then

$$[a, b] \subset \bigcup_1^n (a_k, b_k)$$

for a suitable integer n.

The general theory of extension (6.1, 6.2) can now be applied to the measure m on the ring $\mathscr{R}$. Of course the hereditary σ-ring generated by $\mathscr{R}$ is simply the class of all subsets of the real line R. The class $\mathscr{M}$ of all m^*-measurable sets is called the class of **Lebesgue measurable** sets, and the restriction of m^* to $\mathscr{M}$ is called **Lebesgue measure.** (Incidentally, it can be shown, using the Axiom of Choice, that not every set of real numbers is Lebesgue measurable.)

Recall that the σ-ring $\mathscr{B} = \mathfrak{S}(\mathscr{R})$ generated by $\mathscr{R}$ is called the class of *Borel sets* in R. It follows from 6.1 that every Borel set is Lebesgue measurable. Thus

$$\mathscr{R} \subset \mathscr{B} \subset \mathscr{M},$$

and the restriction of m^* to $\mathscr{B}$ is a measure which extends m; this measure, which we shall also denote by m, will be referred to as the **Borel restriction** of Lebesgue measure. The set theoretic relation between Lebesgue measure and its Borel restriction will be explained in Sect. 9. Evidently both of these measures are σ-finite. In view of Lemma 1 and the (UET), we may summarize as follows:

Theorem 1. *There exists a unique measure m defined on the class of all Borel sets of real numbers, such that*

$$m([a, b)) = b - a$$

for every interval $[a, b)$.

EXERCISE

1. Let m be the Borel restriction of Lebesgue measure. If c and d are fixed real numbers, $c \neq 0$, and T is the transformation on real numbers defined by $Tx = cx + d$, then

$$m(T(E)) = |c|\, m(E)$$

for every Borel set E.

*8. Measurable Covers

Let μ be a measure defined on a σ-ring $\mathscr{S}$, and let μ^* be the outer measure induced by μ on the hereditary σ-ring $\mathscr{H}$ generated by $\mathscr{S}$ (5.1). We are concerned in this section with the relationship between μ and μ^*.

Evidently $\mathscr{H}$ is the class of all subsets of the sets in $\mathscr{S}$, and it is easy to see from the countable subadditivity of μ that

$$\mu^*(A) = \text{GLB}\,\{\mu(E) : A \subset E,\ E \in \mathscr{S}\}$$

for each A in $\mathscr{H}$.

If $A \in \mathscr{H}$ and $E \in \mathscr{S}$, we shall say that E is a **measurable cover** of A in case: (*i*) $A \subset E$, and (*ii*) if $F \in \mathscr{S}$ and $F \subset E - A$, then $\mu(F) = 0$. A suggestive notation for this relation is

$$A \overset{m}{\subset} E,$$

and a convenient verbalization is "A is covered measurably by E."

Let us say that a set $A \in \mathscr{H}$ *has σ-finite outer measure* in case there exists a sequence A_n in $\mathscr{H}$ such that $\mu^*(A_n) < \infty$ for all n and

$$A \subset \bigcup_1^\infty A_n.$$

The main result of the section is that every set A with σ-finite outer measure possesses a measurable cover.

Lemma 1. *If $A \subset E \subset F$, where $E, F \in \mathcal{S}$, and $A \overset{m}{\subset} F$, then also $A \overset{m}{\subset} E$.*

Proof. $E - A \subset F - A$. ▌

Lemma 2. *If $A_n \overset{m}{\subset} E_n$ $(n = 1, 2, 3, \ldots)$, then also*

$$\bigcup_1^\infty A_n \overset{m}{\subset} \bigcup_1^\infty E_n.$$

Proof. Let $A = \bigcup_1^\infty A_n$ and $E = \bigcup_1^\infty E_n$. Assuming $F \in \mathcal{S}$ and $F \subset E - A$, the problem is to show that $\mu(F) = 0$. Defining $F_n = F \cap E_n$, it follows from the relation $F \subset E$ that $F = \bigcup_1^\infty F_n$. Since

$$\mu(F) \leq \sum_1^\infty \mu(F_n),$$

it will suffice to show that $\mu(F_n) = 0$ for each n. But this follows from the relation

$$F_n \subset E_n - A \subset E_n - A_n,$$

and the fact that $A_n \overset{m}{\subset} E_n$. ▌

Lemma 3. *If $A \overset{m}{\subset} E_1$ and $A \overset{m}{\subset} E_2$, then $\mu(E_1 \triangle E_2) = 0$, and so*

$$\mu(E_1) = \mu(E_2) = \mu(E_1 \cup E_2) = \mu(E_1 \cap E_2).$$

Proof. Since $A \subset E_1 \cap E_2 \subset E_1$, and therefore $E_1 - E_1 \cap E_2 \subset E_1 - A$, it follows from the relation $A \overset{m}{\subset} E_1$ that $\mu(E_1 - E_1 \cap E_2) = 0$, that is, $\mu(E_1 - E_2) = 0$. Similarly $\mu(E_2 - E_1) = 0$, and so $\mu(E_1 \triangle E_2) = 0$. It then follows from the relation

$$E_1 \cup E_2 = (E_1 \triangle E_2) \cup (E_1 \cap E_2)$$

that $\mu(E_1 \cup E_2) = \mu(E_1 \cap E_2)$. Finally, it follows from $E_1 \cap E_2 \subset E_i \subset E_1 \cup E_2$ that $\mu(E_1 \cap E_2) = \mu(E_i) = \mu(E_1 \cup E_2)$. ▌

Lemma 4. *If $\mu^*(A) = \mu(E) < \infty$, where $A \subset E$, $A \in \mathcal{H}$, and $E \in \mathcal{S}$, then $A \overset{m}{\subset} E$.*

Proof. Suppose $F \in \mathcal{S}$, $F \subset E - A$. Then $A \subset E - F$, and so

$$\mu(E) = \mu^*(A) \leq \mu^*(E - F) = \mu(E - F) \leq \mu(E);$$

then

$$\mu(E) = \mu(E - F) = \mu(E) - \mu(F)$$

by finiteness, thus $\mu(F) = 0$. ▌

Lemma 5. *If $\mu^*(A) < \infty$, there exists a set E in $\mathcal{S}$ such that $A \overset{m}{\subset} E$, and $\mu^*(A) = \mu(E)$.*

Proof. Choose a sequence E_n in $\mathscr{S}$ such that $A \subset E_n$, $\mu(E_n) \downarrow \mu^*(A)$, and $\mu(E_n) < \infty$ for all n. Let $E = \bigcap_1^\infty E_n$; replacing E_n by $E_1 \cap \cdots \cap E_n$, we may assume that $E_n \downarrow E$. Since $\mu(E_n) \downarrow \mu(E)$ by finiteness, we have $\mu^*(A) = \mu(E)$, and so $A \mathrel{\underline{\overset{m}{\subset}}} E$ by Lemma 4. ∎

Lemma 6. *If $A \mathrel{\underline{\overset{m}{\subset}}} E$, then $\mu^*(A) = \mu(E)$.*

Proof. Since $\mu^*(A) \leq \mu^*(E) = \mu(E)$, the assertion is trivial when $\mu^*(A) = \infty$. Suppose $\mu^*(A) < \infty$. By Lemma 5 there exists an F in $\mathscr{S}$ such that $A \mathrel{\underline{\overset{m}{\subset}}} F$ and $\mu^*(A) = \mu(F)$. Then $\mu(E) = \mu(F)$ by Lemma 3. ∎

Theorem 1. *Every A in $\mathscr{H}$ with σ-finite outer measure has a measurable cover, that is, there exists a set E in $\mathscr{S}$ such that $A \mathrel{\underline{\overset{m}{\subset}}} E$. Necessarily $\mu^*(A) = \mu(E)$.*

Proof. By assumption there is a sequence A_n in $\mathscr{H}$ such that $\mu^*(A_n) < \infty$ for all n and $A \subset \bigcup_1^\infty A_n$. Replacing A_n by $A \cap A_n$, we may assume (by the monotonicity of μ^*) that $A = \bigcup_1^\infty A_n$. By Lemma 5 there exists, for each n, a set E_n in $\mathscr{S}$ such that $A_n \mathrel{\underline{\overset{m}{\subset}}} E_n$. Defining $E = \bigcup_1^\infty E_n$, we have $A \mathrel{\underline{\overset{m}{\subset}}} E$ by Lemma 2, and $\mu^*(A) = \mu(E)$ by Lemma 6. ∎

If in particular μ is σ-finite, it is clear that every A in $\mathscr{H}$ has σ-finite outer measure, and so:

Corollary. *If μ is σ-finite, then every A in $\mathscr{H}$ possesses a measurable cover.*

In the rest of the section, we examine the connection between measurable covers and the concept of μ^*-measurability.

Lemma 7. *If $A \mathrel{\underline{\overset{m}{\subset}}} E$, where A is μ^*-measurable and $\mu^*(A) < \infty$, then $\mu^*(E - A) = 0$.*

Proof. Since μ^* is additive on the class of all μ^*-measurable sets (5.2), and since $\mu^*(A) = \mu(E)$ by Lemma 6, we have

$$\mu^*(E - A) = \mu^*(E) - \mu^*(A) = \mu(E) - \mu^*(A) = 0$$

by 3.1. ∎

Theorem 2. *If A is a μ^*-measurable set of σ-finite outer measure, there exists a set E in $\mathscr{S}$ such that $A \mathrel{\underline{\overset{m}{\subset}}} E$ and $\mu^*(E - A) = 0$.*

Proof. By assumption there is a sequence B_n in $\mathscr{H}$ such that $A \subset \bigcup_1^\infty B_n$ and $\mu^*(B_n) < \infty$ for all n. For each n, choose any set F_n in $\mathscr{S}$ such that $B_n \subset F_n$ and $\mu(F_n) < \infty$. Define

$$A_n = A \cap F_n.$$

Evidently $A = \bigcup_1^\infty A_n$, the A_n are μ^*-measurable, and $\mu^*(A_n) < \infty$. By Lemma 5, there is a set E_n in $\mathcal{S}$ such that $A_n \, \textcircled{m} \, E_n$, and one has $\mu^*(E_n - A_n) = 0$ by Lemma 7. Defining $E = \bigcup_1^\infty E_n$, we have $A \, \textcircled{m} \, E$ by Lemma 2, and $\mu^*(E - A) = 0$ results from the relation

$$E - A \subset \bigcup_1^\infty E_n - A_n$$

and the countable subadditivity of μ^*. ∎

In particular:

Corollary 1. *If μ is σ-finite, then every μ^*-measurable set A possesses a measurable cover E such that $\mu^*(E - A) = 0$.*

Corollary 2. *If μ is σ-finite, and A is any μ^*-measurable set, there exist sets E and F in $\mathcal{S}$ such that $F \subset A \subset E$ and $\mu(E - F) = 0$.*

Proof. By Corollary 1, we may choose E in $\mathcal{S}$ so that $A \subset E$ and $\mu^*(E - A) = 0$. Similarly we may choose G in $\mathcal{S}$ so that $E - A \subset G$ and $\mu^*(G - (E - A)) = 0$. Defining $F = E - G$, we have $F \subset A$, and

$$\begin{aligned} A - F = A - (E - G) &= A \cap G \subset (A \cap G) \cup (G - E) \\ &= G - (E - A); \end{aligned}$$

thus

$$\mu^*(A - F) \leq \mu^*(G - (E - A)) = 0.$$

Finally, $\mu(E - F) = 0$ results from the relation

$$E - F = (E - A) \cup (A - F). \quad ∎$$

The next result is a useful criterion for the μ^*-measurability of sets of finite outer measure:

Theorem 3. *If $\mu^*(A) < \infty$, the following conditions on A are equivalent:*

(a) *A is μ^*-measurable.*

(b) *For every set E in $\mathcal{S}$ such that $A \subset E$, one has*

$$\mu(E) = \mu^*(A) + \mu^*(E - A).$$

(c) *There exists a set E in $\mathcal{S}$ such that $A \subset E$, $\mu(E) < \infty$, and*

$$\mu(E) = \mu^*(A) + \mu^*(E - A).$$

Proof. (a) *implies* (b): This is immediate from the additivity of μ^* on the class of μ^*-measurable sets.

(b) *implies* (c): It suffices to observe that there exists a set E in $\mathcal{S}$ such that $A \subset E$ and $\mu(E) < \infty$, and this is immediate from the definition of $\mu^*(A)$.

(c) *implies* (a): Suppose $E \in \mathscr{S}$ satisfies the condition in (c). By Lemma 5, we may choose F in $\mathscr{S}$ so that $A \overset{m}{\subset} F$ and $\mu^*(A) = \mu(F)$; replacing F by $E \cap F$, we may assume (Lemma 1) that $A \subset F \subset E$. Similarly, there exists a set G in $\mathscr{S}$, $E - A \subset G \subset E$, such that $E - A \overset{m}{\subset} G$ and $\mu^*(E - A) = \mu(G)$. By the assumption (c), $\mu(F) + \mu(G) = \mu(E)$.

We assert that $\mu(F \cap G) = 0$. Indeed, it is clear that $F \cup G = E$, and so

$$\mu(E) = \mu(F \cup G) = \mu(F) + \mu(G) - \mu(F \cap G) = \mu(E) - \mu(F \cap G),$$

and it follows from the finiteness of $\mu(E)$ that $\mu(F \cap G) = 0$.

Now, $A = (E - G) \cup (A \cap G)$, where $E - G \in \mathscr{S}$ and $A \cap G \subset F \cap G$. Since $\mu(F \cap G) = 0$, it follows that $\mu^*(A \cap G) = 0$, and so $A \cap G$ is μ^*-measurable (see Sect. 5); thus A is the union of two μ^*-measurable sets, and is therefore μ^*-measurable (5.2 and 6.1). ∎

EXERCISES

1. If $\mu^*(E - A) = 0$, where $E \in \mathscr{S}$ and $A \subset E$, then $A \overset{m}{\subset} E$.

2. If B is a μ^*-measurable set such that $\mu^*(B) < \infty$, and A is a subset of B such that

$$\mu^*(B) = \mu^*(A) + \mu^*(B - A),$$

then A is also μ^*-measurable.

3. If μ is a measure defined on a ring $\mathscr{R}$, and $\bar{\mu}$ is the restriction of μ^* to $\mathfrak{S}(\mathscr{R})$, then $\mathscr{R}$ and $\mathfrak{S}(\mathscr{R})$ generate the same hereditary σ-ring, and the set functions μ^* and $(\bar{\mu})^*$ are identical. If, moreover, μ is σ-finite, then every A in $\mathscr{H}(\mathscr{R})$ has σ-finite outer measure.

4. The concept of μ^*-measurability can be based on criterion (c) of Theorem 3.

5. If A_n is an increasing sequence of sets in $\mathscr{H}$ of σ-finite outer measure, there exists an increasing sequence E_n in $\mathscr{S}$ such that $A_n \overset{m}{\subset} E_n$ for all n.

6. If $A \overset{m}{\subset} E$ and $\mu^*(E - A) = 0$, then A is μ^*-measurable.

7. If $A_n \uparrow A$, where A_n is a sequence of sets in $\mathscr{H}$, then $\mu^*(A_n) \uparrow \mu^*(A)$.

*9. Completion of a Measure

A measure μ defined on a σ-ring $\mathscr{S}$ is said to be **complete** in case the relations $E \in \mathscr{S}$, $\mu(E) = 0$, and $N \subset E$ imply that $N \in \mathscr{S}$. For example, if ν is an outer measure defined on a hereditary σ-ring $\mathscr{H}$, then the restriction of ν to the class of all ν-measurable sets is a complete measure, as noted in Sect. 5.

If μ is a complete measure on a σ-ring $\mathscr{S}$, then $\mathscr{S}$ contains every set of the form $E \triangle N$, where $E \in \mathscr{S}$, and N is a subset of a set in $\mathscr{S}$ of measure zero; this is immediate from the fact that such a set N necessarily belongs to $\mathcal{S}$. When μ is an *arbitrary* measure on a σ-ring $\mathcal{S}$, the consideration of such sets $E \triangle N$ leads to an extension of μ which is complete:

Theorem 1. *If μ is a measure on a σ-ring $\mathcal{S}$, and $\hat{\mathcal{S}}$ is the class of all sets of the form $E \triangle N$, where $E \in \mathcal{S}$ and N is a subset of a set in $\mathcal{S}$ of measure zero, then (i) $\hat{\mathcal{S}}$ is a σ-ring, and (ii) the set function $\hat{\mu}$ on $\hat{\mathcal{S}}$, defined by the formula*

$$\hat{\mu}(E \triangle N) = \mu(E),$$

is a complete measure.

Proof. From the relation

$$(E_1 \triangle N_1) \triangle (E_2 \triangle N_2) = (E_1 \triangle E_2) \triangle (N_1 \triangle N_2),$$

we see that $\hat{\mathcal{S}}$ contains symmetric differences. Since

$$\begin{aligned} (E_1 \triangle N_1) \cap (E_2 \triangle N_2) & \\ &= (E_1 \cap E_2) \triangle (E_1 \cap N_2) \triangle (N_1 \cap E_2) \triangle (N_1 \cap N_2) \\ &= (E_1 \cap E_2) \triangle N, \end{aligned}$$

where $N \subset N_1 \cup N_2$, it is clear that $\hat{\mathcal{S}}$ contains finite intersections. It then follows from the identity

$$A \cup B = (A \triangle B) \triangle (A \cap B)$$

that $\hat{\mathcal{S}}$ contains finite unions. Moreover, the identity

$$A - B = A \cap (A \triangle B)$$

shows that $\hat{\mathcal{S}}$ contains differences. Summarizing, $\hat{\mathcal{S}}$ is a ring.

In showing that $\hat{\mathcal{S}}$ is a σ-ring, it is convenient to observe that $\hat{\mathcal{S}}$ is precisely the class of all sets $E \cup N$, where $E \in \mathcal{S}$ and N is a subset of a set in $\mathcal{S}$ of measure zero. Indeed, if $E \in \mathcal{S}$, $F \in \mathcal{S}$, $\mu(F) = 0$, and $N \subset F$, the equivalence of these two descriptions of $\hat{\mathcal{S}}$ is immediate from the formulas

$$\begin{aligned} E \triangle N &= (E - F) \cup [(E \cap F - N) \cup (N - E)], \\ E \cup N &= (E \cup F) \triangle [(F - E) - N]. \end{aligned}$$

The fact that $\hat{\mathcal{S}}$ is closed under countable unions now follows at once from the relation

$$\bigcup_1^\infty (E_k \cup N_k) = \left(\bigcup_1^\infty E_k\right) \cup \left(\bigcup_1^\infty N_k\right).$$

To show that the set function $\hat{\mu}$ is well defined, suppose $E_1 \triangle N_1 = E_2 \triangle N_2$. Then $E_1 \triangle E_2 = N_1 \triangle N_2$, and since $N_1 \triangle N_2$ is contained in a set in $\mathscr{S}$ of measure zero, we have $\mu(E_1 \triangle E_2) = 0$. Then $\mu(E_1) = \mu(E_2)$, as shown in the proof of Lemma 3 in Sect. 8.

To see that $\hat{\mu}$ is countably additive, suppose A_k is a sequence of mutually disjoint sets in $\hat{\mathscr{S}}$, and $A = \bigcup_1^\infty A_k$. For each k we may write $A_k = E_k \cup N_k$, with $E_k \in \mathscr{S}$ and N_k contained in a set in $\mathscr{S}$ of measure zero; since

$$A = \left(\bigcup_1^\infty E_k\right) \cup \left(\bigcup_1^\infty N_k\right),$$

and since the E_k are necessarily mutually disjoint, the desired relation

$$\hat{\mu}(A) = \sum_1^\infty \hat{\mu}(A_k)$$

follows at once from the countable additivity of μ.

Summarizing, $\hat{\mu}$ is a measure (3.2) on the σ-ring $\hat{\mathscr{S}}$. Finally, if $A \in \hat{\mathscr{S}}$ and $\hat{\mu}(A) = 0$, it is easy to see that A, and therefore each of its subsets, is contained in a set in $\mathscr{S}$ of measure zero; thus every subset of A belongs to $\hat{\mathscr{S}}$, and so $\hat{\mu}$ is a complete measure. ▌

With notation as in Theorem 1, the measure $\hat{\mu}$ is called the **completion** of μ. The sets in $\hat{\mathscr{S}}$ will be called **$\hat{\mu}$-measurable.** We shall see in the next theorem that when μ is σ-finite, the class of all μ^*-measurable sets is precisely $\hat{\mathscr{S}}$, and the restriction of μ^* to $\hat{\mathscr{S}}$ is $\hat{\mu}$.

Lemma. *If μ is a measure on a σ-ring $\mathscr{S}$, then every set A in $\hat{\mathscr{S}}$ is μ^*-measurable, and $\mu^*(A) = \hat{\mu}(A)$.*

Proof. If $A \in \hat{\mathscr{S}}$, it is clear from the proof of Theorem 1 that we may write $A = E \cup N$, where $E \in \mathscr{S}$, N is a subset of a set in $\mathscr{S}$ with measure zero, and, moreover, $E \cap N = \varnothing$. As observed in Sect. 5, N is μ^*-measurable; it then follows from 6.1 that $E \cup N = A$ is also μ^*-measurable, and that

$$\mu^*(A) = \mu^*(E) + \mu^*(N) = \mu(E) + 0 = \hat{\mu}(A). \quad ▌$$

Theorem 2. *If μ is a σ-finite measure defined on a σ-ring $\mathscr{S}$, then the class $\hat{\mathscr{S}}$ of all $\hat{\mu}$-measurable sets coincides with the class $\mathscr{M}$ of all μ^*-measurable sets, and the restriction of μ^* to $\hat{\mathscr{S}}$ is $\hat{\mu}$.*

Proof. In view of the lemma, it will suffice to show that each μ^*-measurable set A belongs to $\hat{\mathscr{S}}$. By Corollary 1 of 8.2, there exists a set E in $\mathscr{S}$ such that $A \subset E$ and $\mu^*(E - A) = 0$. By 8.1 there is a set F in $\mathscr{S}$ such that $E - A \subset F$ and $\mu(F) = 0$. Thus $E - A \in \hat{\mathscr{S}}$, and it is clear from the formula $A = E - (E - A)$ that A also belongs to $\hat{\mathscr{S}}$. ▌

Corollary. *If μ_0 is a σ-finite measure defined on a ring $\mathscr{R}$, if $\mathscr{S}$ is the σ-ring generated by $\mathscr{R}$, and if μ is the restriction of μ_0^* to $\mathscr{S}$ (see* 6.1), *then the completion of μ is identical with the restriction of μ_0^* to the class of all μ_0^*-measurable sets.*

Proof. It is clear that $\mathscr{R}$ and $\mathscr{S}$ generate the same hereditary σ-ring; moreover, since μ extends μ_0 (6.1), it is easy to see that the set functions μ_0^* and μ^* are identical. Since μ is also σ-finite (6.2), our assertion is immediate from Theorem 2. ∎

In particular, *Lebesgue measure is the completion of its Borel restriction* (see Sect. 7).

10. The LUB of an Increasingly Directed Family of Measures

The material in this section could have immediately followed the definition of a measure. Though it is intended primarily for use in the chapter on product measure, it concerns a general construct of basic importance in measure theory.

Let I be a set of indices, fixed for the rest of the section. Let us assume that I is partially ordered, and is *directed to the right*, that is, for each pair of indices i and j, there exists an index k such that $i \leq k$ and $j \leq k$.

A family (α_i) of nonnegative extended real numbers indexed by I is said to be *increasingly directed* (by I) in case $\alpha_i \leq \alpha_j$ whenever $i \leq j$; in this case we shall write

$$\alpha_i \uparrow.$$

If moreover

$$\alpha = \text{LUB } \alpha_i,$$

we write

$$\alpha_i \uparrow \alpha.$$

We emphasize that only families of *nonnegative* extended real numbers are being considered.

Lemma 1. *If $\alpha_i \uparrow \alpha$ and $\beta_i \uparrow \beta$, then $\alpha_i + \beta_i \uparrow \alpha + \beta$.*

Proof. Observe that the indicated sums can always be formed. It is clear that $\alpha_i + \beta_i \uparrow$; let $\gamma = \text{LUB } (\alpha_i + \beta_i)$. For each i, $\alpha_i + \beta_i \leq \alpha + \beta$; hence $\gamma \leq \alpha + \beta$.

Let i, j be any pair of indices. If k is any index such that $i \leq k$ and $j \leq k$, then

$$\alpha_i + \beta_j \leq \alpha_k + \beta_k \leq \gamma,$$

thus $\alpha_i + \beta_j \le \gamma$. If $\beta_j = \infty$, then clearly $\gamma = \infty$, hence $\alpha + \beta_j \le \gamma$; if β_j is finite, then $\alpha_i \le \gamma - \beta_j$ for all i, hence $\alpha \le \gamma - \beta_j$. In either case, $\alpha + \beta_j \le \gamma$. Since j is arbitrary, a similar argument yields $\alpha + \beta \le \gamma$. ∎

A family (μ_i) of measures indexed by I, defined on the same ring $\mathscr{R}$, is said to be **increasingly directed** in case $\mu_i \le \mu_j$ whenever $i \le j$; in other words, for each E in $\mathscr{R}$, the family $(\mu_i(E))$ of nonnegative extended real numbers is increasingly directed.

Theorem 1. *If (μ_i) is an increasingly directed family of measures on a ring $\mathscr{R}$, and μ is the set function on $\mathscr{R}$ defined by the formula*

$$\mu(E) = \text{LUB}\, \mu_i(E),$$

then μ is a measure on $\mathscr{R}$. Notation: $\mu = \text{LUB}\, \mu_i$.

Proof. It is clear that μ is positive, and $\mu(\varnothing) = 0$. If E and F are disjoint sets in $\mathscr{R}$, then

$$\mu_i(E \cup F) = \mu_i(E) + \mu_i(F)$$

for each i, and hence

$$\mu_i(E \cup F) \uparrow \mu(E) + \mu(F)$$

by Lemma 1; thus

$$\mu(E \cup F) = \mu(E) + \mu(F),$$

that is, μ is additive. Finally, suppose $E_n \uparrow E$, where E and the E_n $(n = 1, 2, 3, \ldots)$ are sets in $\mathscr{R}$. For each i,

$$\mu_i(E_n) \uparrow \mu_i(E),$$

hence by the associativity of LUB's we have

$$\begin{aligned} \mu(E) &= \text{LUB}_i\, \mu_i(E) = \text{LUB}_i\, \text{LUB}_n\, \mu_i(E_n) \\ &= \text{LUB}_n\, \text{LUB}_i\, \mu_i(E_n) = \text{LUB}_n\, \mu(E_n). \end{aligned}$$ ∎

The following lemma will be used in Chapter 6 (in the proof of Lemma 5 to 39.2):

Lemma 2. *If $\alpha_i \uparrow \alpha$ and $\beta_i \uparrow \beta$, then $\alpha_i\beta_i \uparrow \alpha\beta$.*

Proof. Since the extended real numbers we are dealing with are nonnegative, it is clear that $\alpha_i\beta_i\uparrow$; let

$$\gamma = \text{LUB}\, \alpha_i\beta_i.$$

Since $0 \le \alpha_i \le \alpha$ and $0 \le \beta_i \le \beta$, we have

$$0 \le \alpha_i\beta_i \le \alpha\beta$$

for all i, hence $\gamma \le \alpha\beta$.

Observe that $\alpha_i\beta_j \leq \gamma$ for each pair of indices i, j; for, if k is an index such that $i \leq k$ and $j \leq k$, then $0 \leq \alpha_i \leq \alpha_k$ and $0 \leq \beta_j \leq \beta_k$, hence

$$\alpha_i\beta_j \leq \alpha_k\beta_k \leq \gamma.$$

Now, $\gamma \leq \alpha\beta$; the problem is to show that $\alpha\beta \leq \gamma$. This is clear if $\gamma = \infty$, or if $\alpha = 0$, or if $\beta = 0$. Hence we may assume $\alpha > 0$, $\beta > 0$, $\gamma < \infty$.

For each i, α_i is necessarily finite; for, if $\alpha_i = \infty$ for some i, then $\alpha_i\beta_j \leq \gamma < \infty$ shows that $\beta_j = 0$ for all j, contrary to the assumption that $\beta > 0$. Similarly, every β_j is finite.

Let i be any index for which $\alpha_i > 0$ (such an index must exist since $\alpha > 0$). For all indices j, we have $\alpha_i\beta_j \leq \gamma$, $\beta_j \leq (\alpha_i)^{-1}\gamma$, hence $\beta \leq (\alpha_i)^{-1}\gamma$. In particular, β is finite, $\alpha_i\beta \leq \gamma$, $\alpha_i \leq \beta^{-1}\gamma$. Since the last inequality holds trivially when $\alpha_i = 0$, clearly $\alpha \leq \beta^{-1}\gamma$, thus $\alpha\beta \leq \gamma$. ▮

Here is an elementary application of Theorem 1. Let us abandon the partial ordering on I, and consider an arbitrary family of measures (μ_i) on a ring $\mathscr{R}$, indexed by I. For each finite subset J of I, let us write

$$\mu_J = \sum_{i \in J} \mu_i;$$

it is clear from Lemma 1 that μ_J is a measure. The family (μ_J), indexed by the class of finite subsets J of I, is evidently increasingly directed. The measure $\mu = \text{LUB}_J\, \mu_J$ given by Theorem 1 is naturally denoted

$$\sum_{i \in I} \mu_i.$$

For each E in $\mathscr{R}$, $\mu(E)$ is the LUB of the finite sums

$$\sum_{i \in J} \mu_i(E).$$

EXERCISES

1. If μ is a measure defined on a ring $\mathscr{R}$, and $\mathscr{R}_\lambda$ is the algebra described in Exercise 1.7, then μ may be extended to a measure μ_λ on $\mathscr{R}_\lambda$ as follows. For each E in $\mathscr{R}$, the formula

$$\mu^E(A) = \mu(E \cap A)$$

defines a measure on $\mathscr{R}_\lambda$, and the family $\{\mu^E : E \in \mathscr{R}\}$ is increasingly directed; one defines

$$\mu_\lambda = \text{LUB}\, \mu^E.$$

2. If μ is a σ-finite measure, the measure μ_λ described in Exercise 1 need not be σ-finite.

***3.** Let μ be a measure on a ring $\mathscr{R}$, and suppose that

$$\mathfrak{S}(\mathscr{R}_\lambda) = (\mathfrak{S}(\mathscr{R}))_\lambda$$

(see Exercise 1.7). Let μ_λ be the measure on $\mathscr{R}_\lambda$ described in Exercise 1, and let $(\mu_\lambda)^-$ be the restriction of $(\mu_\lambda)^*$ to $\mathfrak{S}(\mathscr{R}_\lambda)$. On the other hand, let $\bar{\mu}$ be the restriction of μ^* to $\mathfrak{S}(\mathscr{R})$, and let $(\bar{\mu})_\lambda$ be the extension of $\bar{\mu}$ to $(\mathfrak{S}(\mathscr{R}))_\lambda$ via Exercise 1. Problem: compare $(\mu_\lambda)^-$ and $(\bar{\mu})_\lambda$.

4. The analogue of Lemma 2 for decreasingly directed families is false.

5. Suppose $\mathscr{R}$ and $\mathscr{S}$ are rings of subsets of X such that $\mathscr{R}$ is an *ideal* in $\mathscr{S}$; that is, $\mathscr{R} \subset \mathscr{S}$, and the relations $E \in \mathscr{S}$ and $F \in \mathscr{R}$ imply $E \cap F \in \mathscr{R}$. If μ is a measure on $\mathscr{R}$, then the formula

$$\rho(E) = \text{LUB}\,\{\mu(F) : F \subset E, F \in \mathscr{R}\}$$

defines a measure ρ on $\mathscr{S}$ which extends μ.

C H A P T E R 2

Measurable Functions

11. Measurable Spaces

A good deal of what has to be done in measure and integration theory is concerned only with the set theoretic concept of a σ-ring of sets; in some contexts, the presence of a measure on the σ-ring is unnecessary, and even confusing. The vehicle for this type of discussion is the concept of a **measurable space**; this is a pair $(X, \mathscr{S})$, where $\mathscr{S}$ is a σ-ring of subsets of X. Those subsets of X which belong to $\mathscr{S}$ are called **measurable** (with respect to $\mathscr{S}$). [It should be mentioned that in some expositions the definition of measurable space requires in addition that the union of the class $\mathscr{S}$ be equal to X, in other words, that every point of X belong to at least one measurable set E; we shall have no occasion to make use of this requirement.]

Of central importance is the measurable space $(R, \mathscr{B})$, where R is the set of all real numbers, and $\mathscr{B}$ is the class of all Borel sets, that is, the σ-ring generated by the class of all semiclosed finite intervals $[a, b)$. The notation $(R, \mathscr{B})$ is henceforth reserved for this example.

If $(X, \mathscr{S})$ is a measurable space, then the set X will be measurable if and only if $\mathscr{S}$ is a σ-algebra. This assumption on $\mathscr{S}$ will not be needed for any of our theorems. Our substitute is the concept of "locally measurable" set, a concept which seems to have considerable didactic value. We shall say that a subset A of X is **locally measurable** (with respect to $\mathscr{S}$) in case $A \cap E$ is measurable, for every measurable set E. The class of all locally measurable sets will be denoted $\mathscr{S}_\lambda$.

Theorem 1. *If $(X, \mathscr{S})$ is a measurable space, then the class $\mathscr{S}_\lambda$ of locally measurable sets is a σ-algebra, and $\mathscr{S} \subset \mathscr{S}_\lambda$.*

Proof. For each measurable set E, let $\mathscr{A}(E)$ be the class of all sets $A \subset X$ such that $A \cap E$ is measurable. By the corollary of 1.6, $\mathscr{A}(E)$ is a σ-ring; clearly X belongs to $\mathscr{A}(E)$, and $\mathscr{A}(E)$ contains $\mathscr{S}$. Briefly, $\mathscr{A}(E)$ is a σ-algebra containing $\mathscr{S}$. The theorem follows at once from the fact that $\mathscr{S}_\lambda$ is the intersection of all the classes $\mathscr{A}(E)$, as E varies over $\mathscr{S}$. ∎

Of course if $\mathscr{S}$ happens already to be a σ-algebra, then $\mathscr{S} = \mathscr{S}_\lambda$ (if A is locally measurable, then $A \cap X$ is measurable), and the concept of local measurability collapses to measurability.

The following remark is often useful: if E is measurable, $A \subset E$, and A is locally measurable, then $A = A \cap E$ shows that A is measurable. Briefly, a locally measurable subset of a measurable set is measurable.

EXERCISE

1. If $(X, \mathscr{S})$ is a measurable space, and A is the union of all the measurable sets, then A is locally measurable. Indeed, every subset of X which contains A is locally measurable. The measurable space $(X, \mathscr{S})$ can be replaced, for all practical purposes, by the measurable space $(A, \mathscr{S})$.

12. Measurable Functions

Let $(X, \mathscr{S})$ be a measurable space, fixed throughout the section. If f is a real-valued function defined on X, that is, if

$$f: X \to R,$$

we shall write $N(f)$ for the set of all points x such that $f(x) \neq 0$. Thus,

$$N(f) = \{x \in X: f(x) \neq 0\}.$$

A convenient verbalization of $N(f)$ is simply "N of f." (The term "nub of f" is not wholly inappropriate.)

We come now to a key definition. A function $f: X \to R$ is said to be **measurable** (with respect to $\mathscr{S}$), in case

$$N(f) \cap f^{-1}(M)$$

is a measurable set, for every Borel set M. We emphasize that the concept of measurability is defined here only for (finite) real-valued functions. [It should be mentioned that in many expositions a measurable function f is allowed to have extended real values, but it must then be stipulated that the sets $f^{-1}(\{\infty\})$ and $f^{-1}(\{-\infty\})$ are also measurable. However, we shall restrict the concept of measurability to real-valued functions; this makes the exposition simpler, without sacrificing any theorems.]

If A is any subset of X, the **characteristic function** of A, denoted χ_A, is the function on X whose values are 1 at the points of A, and 0 at the points of $X - A$. The simplest example of a measurable function is the characteristic function of a measurable set E. For, if $f = \chi_E$, we have $N(f) = E$. If M is any Borel set, then $f^{-1}(M)$ is either E, $X - E$, X, or $\varnothing$, according as M contains 1 but not 0, 0 but not 1, both 0 and 1, or neither 0 nor 1; thus $N(f) \cap f^{-1}(M)$ is either E, $\varnothing$, E, or $\varnothing$, respectively.

The function f which is identically 0 is measurable, since $f = \chi_\varnothing$. However, if c is a nonzero real number, and $g(x) = c$ for all x, then $N(g) = X$; since $g^{-1}(M)$ is either X or $\varnothing$, according as c does or does not belong to M, it is clear that g is measurable if and only if X is a measurable set.

A very useful characterization of measurability is as follows:

Theorem 1. *In order that a function f: $X \to R$ be measurable, it is necessary and sufficient that (i) $N(f)$ be measurable, and (ii) $f^{-1}(M)$ be locally measurable, for every Borel set M.*

Proof. If conditions (i) and (ii) hold, and M is a Borel set, then $N(f) \cap f^{-1}(M)$ is measurable by the definition of a locally measurable set, hence f is a measurable function.

On the other hand, suppose f is measurable. Since R is a Borel set, and $f^{-1}(R) = X$, it follows that the set

$$N(f) = N(f) \cap f^{-1}(R)$$

is measurable. Suppose M is any Borel set. If 0 belongs to M, then

$$f^{-1}(M) \supset f^{-1}(\{0\}) = \complement N(f),$$

hence

$$f^{-1}(M) - N(f) = \complement N(f);$$

if 0 does not belong to M, then $f^{-1}(M) \subset N(f)$, hence

$$f^{-1}(M) - N(f) = \varnothing.$$

In either case, it is clear that $f^{-1}(M) - N(f)$ is locally measurable; since $f^{-1}(M) \cap N(f)$ is measurable by hypothesis, it follows that the set

$$f^{-1}(M) = [f^{-1}(M) \cap N(f)] \cup [f^{-1}(M) - N(f)]$$

is locally measurable. ▌

The following test for measurability is in practice somewhat easier to apply:

Lemma. *Suppose $\mathscr{A}$ is a class of subsets of R such that the σ-ring generated by $\mathscr{A}$ is the class $\mathscr{B}$ of all Borel sets. In order that a function f: $X \to R$ be measurable, it is necessary and sufficient that (i) $N(f)$ be measurable, and (ii) $f^{-1}(M)$ be locally measurable, for every M in $\mathscr{A}$.*

Proof. Necessity is immediate from Theorem 1. On the other hand, suppose that f satisfies conditions (i) and (ii). The class of all sets $M \subset R$

such that $f^{-1}(M)$ belongs to the σ-ring $\mathscr{S}_\lambda$, is a σ-ring (1.6) containing $\mathscr{A}$, hence it contains $\mathfrak{S}(\mathscr{A}) = \mathscr{B}$. Thus $f^{-1}(M)$ is locally measurable for every Borel set M, hence f is measurable by Theorem 1. ▮

Theorem 2. *If $f\colon X \to R$ is a function such that $N(f)$ is measurable, then each of the following conditions is necessary and sufficient for the measurability of f:*

(a) *$\{x : f(x) < c\}$ is locally measurable, for each real number c.*

(b) *$\{x : f(x) \le c\}$ is locally measurable, for each real number c.*

(c) *$\{x : f(x) > c\}$ is locally measurable, for each real number c.*

(d) *$\{x : f(x) \ge c\}$ is locally measurable, for each real number c.*

Proof. We are looking at the inverse image, under f, of sets of the form $(-\infty, c)$, $(-\infty, c]$, (c, ∞), and $[c, \infty)$. That these are Borel sets is immediate from the formulas

$$(-\infty, c) = \bigcup_{n=1}^{\infty} [c - n, c)$$

$$(-\infty, c] = (-\infty, c) \cup \bigcap_{n=1}^{\infty} [c, c + 1/n)$$

$$(c, \infty) = \bigcup_{n=1}^{\infty} [c + 1/n, c + n)$$

$$[c, \infty) = \bigcup_{n=1}^{\infty} [c, c + n).$$

It then follows from Theorem 1 that each of the conditions (a) through (d) is necessary for the measurability of f.

On the other hand suppose condition (a) holds. Since

$$[a, b) = (-\infty, b) - (-\infty, a),$$

the intervals $(-\infty, c)$ generate $\mathscr{B}$, hence f is measurable by the lemma.

Suppose condition (b) holds. Since

$$(-\infty, c) = \bigcup_{n=1}^{\infty} (-\infty, c - 1/n],$$

and since the intervals $(-\infty, c)$ generate $\mathscr{B}$, so do the intervals $(-\infty, c]$, hence f is measurable by the lemma.

Since $[a, b) = [a, \infty) - [b, \infty)$, the intervals $[c, \infty)$ generate $\mathscr{B}$, hence condition (d) implies that f is measurable.

Finally,

$$[a, \infty) = \bigcap_{n=1}^{\infty} (a - 1/n, \infty)$$

shows that the intervals (c, ∞) also generate $\mathscr{B}$, hence condition (c) implies that f is measurable. ▮

The following result is the key to Sect. 13:

Theorem 3. *If f and g are measurable functions defined on X, and c is any real number, then each of the sets*

$$A = \{x: f(x) < g(x) + c\},$$

$$B = \{x: f(x) \leq g(x) + c\},$$

$$C = \{x: f(x) = g(x) + c\}$$

is locally measurable.

Proof. Let Q be the set of all rational numbers. For a point x in X, the relation $f(x) < g(x) + c$ holds if and only if there exists a rational number r such that both $f(x) < r$ and $r < g(x) + c$; hence

$$A = \bigcup_{r \in Q} \{x: f(x) < r\} \cap \{x: g(x) > r - c\}.$$

By Theorem 2, each of the sets $\{x: f(x) < r\} \cap \{x: g(x) > r - c\}$ is locally measurable, hence so is their countable union A.

Interchanging the roles of f and g, it follows that the set $\{x: g(x) < f(x) - c\}$ is locally measurable, hence so is its complement B. Finally, $C = B - A$ is locally measurable. ▮

The following remarks are frequently useful. Suppose f is a measurable function. If $c > 0$, then the set $\{x: f(x) \geq c\}$ is measurable; for, it is locally measurable, and is contained in the measurable set $N(f)$. Similarly, if $c \geq 0$, then $\{x: f(x) > c\}$ is measurable. If $c < 0$, then $\{x: f(x) \leq c\}$ is measurable. If $c \leq 0$, then $\{x: f(x) < c\}$ is measurable. If $c > 0$, then the set

$$\{x: |f(x)| \geq c\} = \{x: f(x) \leq -c\} \cup \{x: f(x) \geq c\}$$

is measurable. Similarly, if $c \geq 0$, then $\{x: |f(x)| > c\}$ is measurable.

EXERCISES

1. A complex-valued function is said to be *measurable* in case its real and imaginary parts are measurable. In order that a complex-valued function f be measurable, it is necessary and sufficient that

$$N(f) \cap f^{-1}(M)$$

be measurable for every Borel set M of complex numbers. (A set of complex numbers is called a Borel set if it belongs to the σ-ring generated by the "semiclosed" rectangles

$$[a, b) \times [c, d)$$

in the Gaussian plane.)

2. If $(X, \mathscr{S})$ is a measurable space, and A is the union of all the measurable sets, then every measurable function f vanishes on $X - A$.

13. Combinations of Measurable Functions

Let $(X, \mathscr{S})$ be a measurable space, and let f and g be real-valued functions defined on X. We wish to show that various combinations of f and g are measurable, provided f and g are measurable.

For each real number c, the function cf is defined by the formula

$$(cf)(x) = cf(x).$$

If $c = 0$, then cf is identically 0, hence $N(cf) = \varnothing$; if $c \neq 0$, clearly $N(cf) = N(f)$.

Theorem 1. *If f is measurable, and c is a real number, then cf is measurable.*

Proof. This is clear if $c = 0$. Assuming $c \neq 0$, let $h = cf$; then $N(h) = N(f)$ is by assumption measurable. Suppose first that $c > 0$. For each real number a, the set

$$\{x: h(x) < a\} = \{x: cf(x) < a\} = \{x: f(x) < a/c\}$$

is locally measurable by 12.2 applied to f, hence h is measurable by 12.2. If $c < 0$, then for each real number a, the set

$$\{x: h(x) < a\} = \{x: cf(x) < a\} = \{x: f(x) > a/c\}$$

is locally measurable, hence h is measurable. ▮

The function $f + g$ is defined by the formula

$$(f + g)(x) = f(x) + g(x).$$

Evidently $N(f + g) \subset N(f) \cup N(g)$.

Theorem 2. *If f and g are measurable, then $f + g$ is also measurable.*

Proof. By Theorem 1, $-g = (-1)g$ is measurable; it follows that for each real number c, the set

$$\{x: (f + g)(x) < c\} = \{x: f(x) < -g(x) + c\}$$

is locally measurable by 12.3.

It remains to show that $N(f + g)$ is measurable. Since

$$N(f + g) \subset N(f) \cup N(g),$$

and $N(f) \cup N(g)$ is measurable, it will suffice to show that $N(f + g)$ is locally measurable. Indeed,

$$\complement N(f + g) = \{x: f(x) + g(x) = 0\} = \{x: f(x) = -g(x) + 0\}$$

is a locally measurable set by 12.3. ▮

We may summarize by saying that the set of all measurable functions on X is a real vector space with respect to the pointwise linear operations.

We digress for a moment to discuss measurability with respect to a σ-algebra. Suppose $(Y, \mathscr{T})$ is a measurable space, where $\mathscr{T}$ is a σ-algebra, and consider a function $h: Y \to R$. In order that h be measurable, it is necessary and sufficient that $h^{-1}(M)$ be measurable for every Borel set M (or for M running over a system of generators for the σ-ring $\mathscr{B}$). Sufficiency is clear, since

$$N(h) = h^{-1}(R - \{0\})$$

will also be a measurable set; necessity results from the fact that local measurability collapses to measurability.

A function $\varphi: R \to R$ is said to be **Borel measurable** in case it is measurable with respect to $\mathscr{B}$. In view of the foregoing remarks, this means: $\varphi^{-1}(M)$ is a Borel set, for every Borel set M (or for M running over a system of generators for the σ-ring $\mathscr{B}$).

Theorem 3. *If f is measurable, and $\varphi: R \to R$ is a Borel measurable function such that $\varphi(0) = 0$, then the composite function $\varphi \circ f$ is also measurable.*

Proof. Let $h = \varphi \circ f$. If $f(x) = 0$, then $h(x) = \varphi(f(x)) = \varphi(0) = 0$, thus $\complement N(f) \subset \complement N(h)$; that is, $N(h) \subset N(f)$. Since $N(f)$ is measurable, it will suffice, in showing that $N(h)$ is measurable, to show that $N(h)$ is locally measurable.

If M is any Borel set, then $h^{-1}(M) = f^{-1}[\varphi^{-1}(M)]$; since $\varphi^{-1}(M)$ is a Borel set, $h^{-1}(M)$ is locally measurable by 12.1 applied to f. In particular, putting

$$M = R - \{0\} = (-\infty, 0) \cup (0, \infty),$$

it follows that $N(h) = f^{-1}[N(\varphi)]$ is locally measurable. ∎

Lemma. *If $\varphi: R \to R$ is continuous, then φ is Borel measurable.*

Proof. For any real number c, the set

$$\{a \in R: \varphi(a) < c\} = \varphi^{-1}((-\infty, c))$$

is the inverse image of an open set, hence is open by the continuity of f. In view of the remarks preceding Theorem 3, it will suffice to show that every open set U is a Borel set. Indeed, if x is any point of U, there exists an interval $[r, s)$ with rational end points (briefly, a "rational interval") such that

$$x \in [r, s) \subset U.$$

Thus, U is the union of all the rational intervals $[r, s)$ which it contains;

since there are at most countably many such rational intervals, and these are Borel sets, so is their union U a Borel set. ∎

If α is a real number >0, the function $|f|^\alpha$ is defined by the formula

$$|f|^\alpha(x) = |f(x)|^\alpha.$$

Since the function $\varphi(a) = |a|^\alpha$ is continuous, the lemma and Theorem 3 imply the following:

Theorem 4. *If f is measurable, and $\alpha > 0$, then the function $|f|^\alpha$ is also measurable. In particular, if f is measurable, then so is the function $|f|$.*

Incidentally, the measurability of $|f|$ follows more directly from the remarks at the end of the preceding section.

The functions $f \cup g$ and $f \cap g$ are defined by the formulas

$$(f \cup g)(x) = \max\{f(x), g(x)\},$$

$$(f \cap g)(x) = \min\{f(x), g(x)\}.$$

Since, for every pair of real numbers a and b, we have

$$\max\{a, b\} = (a + b + |a - b|)/2$$

and

$$\min\{a, b\} = (a + b - |a - b|)/2,$$

it follows that

$$f \cup g = (f + g + |f - g|)/2$$

and

$$f \cap g = (f + g - |f - g|)/2.$$

Quoting Theorems 1, 2, and 4, we have:

Theorem 5. *If f and g are measurable, then the functions $f \cup g$ and $f \cap g$ are also measurable.*

The function fg is defined by the formula

$$(fg)(x) = f(x)g(x).$$

Note, incidentally, that $N(fg) = N(f) \cap N(g)$. One writes f^2 for the function ff. Since $f^2 = |f|^2$, the measurability of f implies that of f^2, by Theorem 4. Since

$$fg = [(f + g)^2 - (f - g)^2]/4,$$

Theorems 1 and 2 then yield the following result:

Theorem 6. *If f and g are measurable, then the function fg is also measurable.*

The functions f^+ and f^- are defined by the formulas

$$f^+ = f \cup 0,$$
$$f^- = -(f \cap 0).$$

Clearly $f^+ \geq 0$, $f^- \geq 0$, and $f = f^+ - f^-$. Since 0 is a measurable function, the measurability of f implies that of f^+ and f^-, by Theorems 1 and 5. If conversely both f^+ and f^- are measurable, then $f = f^+ - f^-$ is measurable by Theorems 1 and 2. Summarizing:

Theorem 7. *In order that f be measurable, it is necessary and sufficient that both f^+ and f^- be measurable.*

Incidentally, it follows from the relations $f = f^+ - f^-$ and $f^+f^- = 0$, that $N(f) = N(f^+) \cup N(f^-)$.

It can happen that $|f|$ is measurable without f being measurable. For example, suppose there exists in $\mathscr{S}$ a set E not all of whose subsets are measurable. Say $S \subset E$, where S is not measurable. Defining

$$f = \chi_S - \chi_{E-S},$$

we have

$$|f| = \chi_S + \chi_{E-S} = \chi_E,$$

thus $|f|$ is measurable. However,

$$N(f) \cap f^{-1}(\{1\}) = E \cap S = S$$

shows that f is not measurable.

EXERCISES

1. If f is measurable, then so is the function $|f|/(1 + |f|)$.

2. If f is a measurable function, and g is the function such that $g(x) = 1/f(x)$ when $f(x) \neq 0$, and $g(x) = 0$ otherwise, then g is a measurable function.

14. Limits of Measurable Functions

Again we consider functions defined on a fixed measurable space $(X, \mathscr{S})$. In some of the definitions it will be convenient to admit functions with extended real values, but we maintain our convention that only real-valued functions are eligible to be measurable.

Suppose f_n is a sequence of extended real valued functions defined on X. Since the set of extended real numbers is a complete lattice, we may define a function g on X by the formula

$$g(x) = \text{GLB}\{f_n(x) : n = 1, 2, 3, \ldots\}.$$

Briefly, $g = \text{GLB} f_n$. Similarly, the function $h = \text{LUB} f_n$ is defined by the formula

$$h(x) = \text{LUB}\{f_n(x) \colon n = 1, 2, 3, \ldots\}.$$

Theorem 1. *Let f_n be a sequence of measurable functions. Suppose that for each point x in X, the sequence $f_n(x)$ is bounded below, so that $g = \text{GLB} f_n$ is real-valued. Then the function g is also measurable. Similarly, if the sequence $f_n(x)$ is bounded above, for each point x in X, then the function $h = \text{LUB} f_n$ is measurable.*

Proof. Our first task is to show that $N(g)$ is measurable. Since

$$N(g) = \{x \colon g(x) > 0\} \cup \{x \colon g(x) < 0\},$$

we may consider the summands separately. Now,

$$\begin{aligned} \{x \colon g(x) > 0\} &= \bigcup_{k=1}^{\infty} \{x \colon g(x) \geq 1/k\} \\ &= \bigcup_{k} \{x \colon f_n(x) \geq 1/k \text{ for all } n\} \\ &= \bigcup_{k} \bigcap_{n} \{x \colon f_n(x) \geq 1/k\}; \end{aligned}$$

since each of the sets $\{x \colon f_n(x) \geq 1/k\}$ is measurable by the remarks at the end of Sect. 12, it is clear that $\{x \colon g(x) > 0\}$ is measurable. Also,

$$\begin{aligned} \{x \colon g(x) < 0\} &= \bigcup_{k=1}^{\infty} \{x \colon g(x) < -1/k\} \\ &= \bigcup_{k} \{x \colon f_n(x) < -1/k \text{ for some } n\} \\ &= \bigcup_{k} \bigcup_{n} \{x \colon f_n(x) < -1/k\}, \end{aligned}$$

hence the measurability of $\{x \colon g(x) < 0\}$ results from the measurability of the sets $\{x \colon f_n(x) < -1/k\}$.

If c is any real number,

$$\begin{aligned} \{x \colon g(x) < c\} &= \{x \colon f_n(x) < c \text{ for some } n\} \\ &= \bigcup_{n} \{x \colon f_n(x) < c\}; \end{aligned}$$

since each set $\{x \colon f_n(x) < c\}$ is locally measurable, so is $\{x \colon g(x) < c\}$. It follows from 12.2 that g is measurable.

The last assertion then results from the evident relation

$$\text{LUB} f_n = -\text{GLB}(-f_n). \quad \blacksquare$$

Suppose again that f_n is a sequence of extended real valued functions defined on X. For each n, let g_n be the GLB of the sequence $f_n, f_{n+1}, f_{n+2}, \ldots$; briefly,

$$g_n = \underset{k \geq n}{\text{GLB}} f_k.$$

Evidently $g_n\uparrow$, in the sense that $g_n(x)\uparrow$ for each x in X. Define

$$f_* = \underset{n\geq 1}{\text{LUB}}\, g_n.$$

Thus,

$$f_* = \underset{n\geq 1}{\text{LUB}}\ \underset{k\geq n}{\text{GLB}}\, f_k.$$

The function f_* is called the **limit inferior** of the sequence f_n, and one writes

$$f_* = \liminf f_n.$$

Evidently $g_n \uparrow f_*$, in the sense that

$$g_n(x) \uparrow f_*(x)$$

for each x.

Similarly, define $h_n = \underset{k\geq n}{\text{LUB}}\, f_k$, and $f^* = \underset{n\geq 1}{\text{GLB}}\, h_n$. Thus,

$$f^* = \underset{n\geq 1}{\text{GLB}}\ \underset{k\geq n}{\text{LUB}}\, f_k,$$

and one has $h_n \downarrow f^*$, in the sense that

$$h_n(x) \downarrow f^*(x)$$

for each x. The function f^* is called the **limit superior** of the sequence f_n, and one writes

$$f^* = \limsup f_n.$$

One also defines lim inf and lim sup for a sequence α_n of extended real numbers, namely,

$$\liminf \alpha_n = \underset{n\geq 1}{\text{LUB}}\ \underset{k\geq n}{\text{GLB}}\, \alpha_k$$

and

$$\limsup \alpha_n = \underset{n\geq 1}{\text{GLB}}\ \underset{k\geq n}{\text{LUB}}\, \alpha_k.$$

If α_n is any sequence of extended real numbers, then

$$\liminf \alpha_n \leq \limsup \alpha_n.$$

Indeed, let us define

$$\beta_n = \underset{k\geq n}{\text{GLB}}\, \alpha_k, \qquad \beta = \underset{n\geq 1}{\text{LUB}}\, \beta_n,$$

and

$$\gamma_n = \underset{k\geq n}{\text{LUB}}\, \alpha_k, \qquad \gamma = \underset{n\geq 1}{\text{GLB}}\, \gamma_n.$$

Thus, $\beta = \liminf \alpha_n$ and $\gamma = \limsup \alpha_n$, and the problem is to show that $\beta \leq \gamma$. Given any pair of natural numbers m and n; if $k \geq \max\{m, n\}$, then $\beta_m \leq \alpha_k \leq \gamma_n$, thus $\beta_m \leq \gamma_n$. Since m is arbitrary, $\beta \leq \gamma_n$, and since

n is arbitrary, $\beta \le \gamma$. It is an easy exercise in elementary analysis that if α_n is a bounded sequence of real numbers, then α_n is convergent if and only if

$$\liminf \alpha_n = \limsup \alpha_n,$$

and in this case the common value is $\lim \alpha_n$.

It is clear from the foregoing definitions that

$$(\liminf f_n)(x) = \liminf f_n(x)$$

and

$$(\limsup f_n)(x) = \limsup f_n(x),$$

for each x. Moreover,

$$\liminf f_n \le \limsup f_n,$$

in the sense that

$$(\liminf f_n)(x) \le (\limsup f_n)(x)$$

for all x.

Theorem 2. *If f_n is a sequence of measurable functions such that $f_n(x)$ is a bounded sequence, for each x, then the functions* $\liminf f_n$ *and* $\limsup f_n$ *are also measurable.*

Proof. Let us adopt the notations g_n, f_* and h_n, f^*, described above. We are assuming that for each x, there is a finite interval $I_x = [a_x, b_x]$ such that $f_n(x) \in I_x$ for all n. Evidently $g_n(x), f_*(x), h_n(x)$, and $f^*(x)$ all belong to I_x, thus the functions g_n, f_*, h_n, and f^* are all real-valued. By Theorem 1,

$$g_n = \operatorname*{GLB}_{k \ge n} f_k$$

is measurable, hence so is

$$f_* = \operatorname*{LUB}_{n \ge 1} g_n.$$

Similarly for f^*. ▮

For our purposes, the most valuable result of this section is the following:

Theorem 3. *If f_n is a sequence of measurable functions, converging pointwise to the real-valued function f, then f is also measurable.*

Proof. With notations as in the proof of Theorem 2, we have $f_* = f^* = f$, hence f is measurable. ▮

15. Localization of Measurability

Let f be a real-valued function defined on a measurable space $(X, \mathscr{S})$.

Theorem 1. *If f is a measurable function, and A is a locally measurable set, then the function $\chi_A f$ is also measurable.*

Proof. Let $g = \chi_A f$. Then $N(g) = N(\chi_A) \cap N(f) = A \cap N(f)$; since A is locally measurable and $N(f)$ is measurable, it follows that $N(g)$ is measurable.

Let M be a Borel set; we know that $f^{-1}(M)$ is locally measurable (12.1). If 0 belongs to M, then the set

$$g^{-1}(M) = [A \cap f^{-1}(M)] \cup (X - A)$$

is locally measurable; if 0 is not in M, then the set

$$g^{-1}(M) = A \cap f^{-1}(M)$$

is again locally measurable (even measurable). Thus, g is measurable by 12.1. Incidentally, another proof results from the relation

$$\chi_A f = \chi_{A \cap N(f)} f$$

and 13.6. ▮

Theorem 2. *If A_n is a sequence of locally measurable sets such that*

$$\bigcup_1^\infty A_n = X,$$

and if $\chi_{A_n} f$ is measurable for each n, then f is a measurable function.

Proof. Let $g_n = \chi_{A_n} f$. We have $N(g_n) = A_n \cap N(f)$, and

$$N(f) = X \cap N(f) = \bigcup_1^\infty A_n \cap N(f) = \bigcup_1^\infty N(g_n);$$

since $N(g_n)$ is measurable for each n, so is $N(f)$. If M is any Borel set, then

$$\begin{aligned} f^{-1}(M) &= \{x \in X : f(x) \in M\} \\ &= \bigcup_1^\infty \{x \in A_n : f(x) \in M\} \\ &= \bigcup_1^\infty \{x \in A_n : g_n(x) \in M\} \\ &= \bigcup_1^\infty A_n \cap g_n^{-1}(M); \end{aligned}$$

since each $A_n \cap g_n^{-1}(M)$ is locally measurable, so is $f^{-1}(M)$. Thus, f is measurable by 12.1. ▮

Theorem 3. *If f is a measurable function, E is a measurable set, c is a real number, and g is the function on X defined by the formulas*

$$g(x) = \begin{cases} f(x) & \textit{for } x \in X - E, \\ c & \textit{for } x \in E, \end{cases}$$

then g is a measurable function.

Proof. We know from Theorem 1 that $\chi_{X-E}g = \chi_{X-E}f$ is measurable. Also, $\chi_E g = c\chi_E$ is clearly measurable. Since $X - E$ and E are both locally measurable, g is measurable by Theorem 2. Incidentally, another proof results from the relation $g = \chi_{X-E}f + c\chi_E$. ▮

EXERCISES

1. If f is measurable with respect to $\mathscr{S}_\lambda$, and $N(f) \in \mathscr{S}$, then f is measurable with respect to $\mathscr{S}$.

2. If f is measurable with respect to $\mathscr{S}$ and g is measurable with respect to $\mathscr{S}_\lambda$, then fg is measurable with respect to $\mathscr{S}$. Briefly, the product of a measurable function and a "locally measurable" function is measurable.

16. Simple Functions

A function f defined on a measurable space $(X, \mathscr{S})$ is said to be **simple** in case (i) f is measurable, and (ii) the range of f is a finite set of real numbers. Simple functions have an especially transparent structure (Theorem 1), and it is a fact of considerable technical importance that every measurable function can be approximated by a suitable sequence of simple functions (Theorems 3, 4).

Theorem 1. *Let f be a real-valued function defined on X. In order that f be simple, it is necessary and sufficient that there exist a finite number of measurable sets $E_1, \ldots, E_n$, and real numbers $c_1, \ldots, c_n$, such that*

$$f = \sum_1^n c_k \chi_{E_k}.$$

In this case, if $c_1, \ldots, c_n$ are taken to be the set of distinct nonzero values of f, one can take $E_k = f^{-1}(\{c_k\})$. The E_k are then nonempty, mutually disjoint, and the coefficients c_k are distinct and nonzero; a representation

$$f = \sum_1^n c_k \chi_{E_k}$$

with these properties is unique.

Proof. Suppose first that f is a simple function, and let $c_1, \ldots, c_n$ be the distinct nonzero values of f. Define $E_k = f^{-1}(\{c_k\})$; since E_k is locally measurable and is contained in $N(f)$, it is measurable. Evidently

$$f = \sum_1^n c_k \chi_{E_k}.$$

On the other hand, the characteristic function of a measurable set is measurable, and a linear combination of measurable functions is measurable. Then, if $E_1, \ldots, E_n$ are measurable sets, and $c_1, \ldots, c_n$ are real numbers, the function

$$f = \sum_1^n c_k \chi_{E_k}$$

is measurable. It is easy to see that f has at most 2^n values, and hence is a simple function. ▌

Theorem 2. *If f and g are simple functions, c is a real number, and A is a locally measurable set, then all of the following functions are simple: $cf, f + g, |f|, f \cup g, f \cap g, f^+, f^-, \chi_A f,$ and fg.*

Proof. The functions in question are measurable by Sect. 13 and by 15.1, and it is clear that their ranges are finite subsets of R. ▌

We now turn to the problem of approximating measurable functions by simple functions. We shall first consider the case of a bounded measurable function. If f_n is a sequence of bounded real-valued functions defined on X, and f is another such function, one says that f_n converges to f **uniformly** on X in case: given any $\varepsilon > 0$, there is an index N such that $n \geq N$ implies

$$|f_n(x) - f(x)| \leq \varepsilon$$

for *all* x in X. If g is a bounded function, let us write

$$\|g\|_\infty = \text{LUB}\{|g(x)| : x \in X\}.$$

The uniform convergence of f_n to f can then be expressed by the condition

$$\|f_n - f\|_\infty \to 0.$$

Our first result (Theorem 3) is that any *bounded* measurable function can be *uniformly* approximated by simple functions:

Lemma 1. *If f is a bounded measurable function, $f \geq 0$, then there exists a simple function g such that*

$$0 \leq g \leq f$$

and

$$\|f - g\|_\infty \leq \|f\|_\infty/2.$$

Proof. Write $c = \|f\|_\infty/2$, and define $E = \{x : f(x) > c\}$; the function $g = c\chi_E$ meets all requirements. ▌

Theorem 3. *If f is a bounded measurable function, there exists a sequence of simple functions f_n such that*

$$\|f_n - f\|_\infty \to 0,$$

that is, f_n converges to f uniformly on X. If moreover $f \geq 0$, one can further suppose that

$$0 \leq f_n \uparrow f.$$

Proof. Writing $f = f^+ - f^-$, we are reduced (13.7) to the case that $f \geq 0$. Let $c = \|f\|_\infty$.

By the lemma, there is a simple function g_1 such that

$$0 \leq g_1 \leq f$$

and

$$\|f - g_1\|_\infty \leq c/2.$$

Applying the lemma to $f - g_1$, there is a simple function g_2 such that

$$0 \leq g_2 \leq f - g_1,$$

and

$$\|(f - g_1) - g_2\|_\infty \leq \|f - g_1\|_\infty/2 \leq c/4.$$

Continuing inductively, we obtain a simple function g_n such that

$$0 \leq g_n \leq f - (g_1 + \cdots + g_{n-1})$$

and

$$\|f - (g_1 + \cdots + g_n)\|_\infty \leq c/2^n.$$

The functions $f_n = g_1 + \cdots + g_n$ evidently meet all requirements. ∎

The problem of approximating an arbitrary measurable function is reduced to the bounded case with the aid of the following:

Lemma 2. *If f is a measurable function, c is a real number, and $c > 0$, then $f \cap c$ is a measurable function.*

Proof. The function $f \cap c$ is of course defined by the formula

$$(f \cap c)(x) = \min\{f(x), c\};$$

so to speak, it is a "truncation" of f. Let

$$E = \{x : f(x) \geq c\};$$

then E is a measurable set, and

$$(f \cap c)(x) = \begin{cases} f(x) & \text{for } x \in X - E, \\ c & \text{for } x \in E; \end{cases}$$

hence $f \cap c$ is measurable by 15.3. ∎

Theorem 4. *If f is a measurable function, there exists a sequence of simple functions f_n such that f_n converges to f pointwise on X, that is,*

$$f_n(x) \to f(x)$$

for each x in X. If moreover $f \geq 0$, one can make

$$0 \leq f_n \uparrow f.$$

Proof. Writing $f = f^+ - f^-$, we are reduced to the case that $f \geq 0$. Let $g_n = f \cap n$; clearly $g_n \uparrow f$, and g_n is measurable by Lemma 2. Since g_n is bounded, measurable, and $g_n \geq 0$, by Theorem 3 there exists a simple function h_n such that $0 \leq h_n \leq g_n$ and

$$\|h_n - g_n\|_\infty \leq 1/n.$$

Since already $g_n \to f$ pointwise, it is clear that also $h_n \to f$ pointwise.

Define $f_n = h_1 \cup \cdots \cup h_n$. Then f_n is simple by Theorem 2, and evidently $f_n\uparrow$. Since

$$0 \leq h_n \leq g_n \leq f$$

for all n, clearly $f_n \leq f$. Thus, $0 \leq h_n \leq f_n \leq f$; since $h_n \to f$ pointwise, it is clear that $f_n \to f$ pointwise, and hence $f_n \uparrow f$. ∎

CHAPTER 3

Sequences of Measurable Functions

17. Measure Spaces

A **measure space** is a triple $(X, \mathcal{S}, \mu)$, where $(X, \mathcal{S})$ is a measurable space and μ is a measure whose domain of definition is the σ-ring $\mathcal{S}$. The principal objects of study from this point onward are measurable (with respect to $\mathcal{S}$) functions defined on a measure space $(X, \mathcal{S}, \mu)$. For instance, we shall assume for the rest of the chapter that we are working in the context of a given measure space $(X, \mathcal{S}, \mu)$. The measure space is said to be **finite** (respectively σ-**finite**) if the measure μ is finite (respectively σ-finite). There are obvious definitions for a totally finite or a totally σ-finite measure space, but we shall have no need for these concepts.

In this section we shall prove two simple and useful theorems about finite measure spaces.

Theorem 1. *In a finite measure space $(X, \mathcal{S}, \mu)$, the measure μ is necessarily bounded, that is, the number*

$$\text{LUB}\,\{\mu(E) : E \in \mathcal{S}\}$$

is finite.

Proof. Let M be the indicated LUB, and choose a sequence of measurable sets E_n such that LUB $\mu(E_n) = M$. Define $E = \bigcup_1^\infty E_n$; the set E is measurable, and $E_n \subset E$, hence $\mu(E_n) \leq \mu(E)$ for all n. Then

$$M = \text{LUB}\,\mu(E_n) \leq \mu(E) < \infty.$$

Incidentally, $\mu(E) = M$. ∎

If E_n is any sequence of sets, the **limit superior** of the sequence is defined to be the set of all points x such that $x \in E_n$ for infinitely many n. One writes lim sup E_n for this set; it is easy to see that

$$\limsup E_n = \bigcap_{n \geq 1} \bigcup_{k \geq n} E_k.$$

Dually, one defines the **limit inferior** of the sequence to be

$$\liminf E_n = \bigcup_{n\geq 1} \bigcap_{k\geq n} E_k;$$

this is easily seen to be the set of all points x for which an index n exists such that $x \in E_k$ for all $k \geq n$. It is clear that if the E_n are measurable sets, then so are $\limsup E_n$ and $\liminf E_n$. The next result is known as the Arzela-Young theorem:

Theorem 2. *If E_n is a sequence of measurable sets in a finite measure space, such that $\mu(E_n) \geq \varepsilon$ for all n, where $\varepsilon > 0$, then*

$$\mu(\limsup E_n) \geq \varepsilon.$$

Proof. Let $E = \limsup E_n$, and define

$$F_n = \bigcup_{k\geq n} E_k.$$

Evidently $F_n \downarrow E$; since μ is a finite measure, $\mu(F_n) \downarrow \mu(E)$ by 4.2. But $F_n \supset E_n$ shows that $\mu(F_n) \geq \mu(E_n) \geq \varepsilon$, hence

$$\mu(E) = \text{GLB}\, \mu(F_n) \geq \varepsilon. \quad \blacksquare$$

If A is a locally measurable set, the set function μ_A defined on $\mathscr{S}$ by the formula $\mu_A(E) = \mu(A \cap E)$ is called the **contraction** of μ **by** A; it is clearly a measure on $\mathscr{S}$. In particular, if F is a measurable set such that $\mu(F) < \infty$, then μ_F is a finite measure on $\mathscr{S}$; indeed, $\mu_F(E) = \mu(F \cap E) \leq \mu(F)$ for all E in $\mathscr{S}$. It is often convenient to pass to the measure space

$$(X, \mathscr{S}, \mu_A),$$

where A is an appropriately selected locally measurable set; so convenient, in fact, that contraction may be described as one of the basic constructs of measure theory.

EXERCISES

1. Let $(X, \mathscr{S}, \mu)$ be any measure space. For each E in $\mathscr{S}$, we may define a set function μ^E on the class $\mathscr{S}_\lambda$ of all locally measurable sets by the formula

$$\mu^E(A) = \mu(E \cap A).$$

The family $\{\mu^E : E \in \mathscr{S}\}$ is increasingly directed in the obvious sense, and so the formula

$$\mu_\lambda = \text{LUB}\, \{\mu^E : E \in \mathscr{S}\}$$

defines a measure μ_λ on the σ-algebra $\mathscr{S}_\lambda$ (10.1), and μ_λ is an extension of μ.

2. If $(X, \mathcal{S}, \mu)$ is a finite measure space, then the measure μ_λ is also finite.

3. If μ is discrete measure on the class $\mathcal{S}$ of all countable subsets of an uncountable set X, then μ is σ-finite, but μ_λ is not.

4. If $(X, \mathcal{S}, \mu)$ is a finite measure space, $\mathcal{R}$ is any ring such that $\mathfrak{S}(\mathcal{R}) = \mathcal{S}$, and if $E \in \mathcal{S}$, then:

(i) $\mu(E) = \text{GLB}\,\{\sum_1^\infty \mu(F_n) : E \subset \bigcup_1^\infty F_n, F_n \in \mathcal{R}\}$.

(ii) There exists a sequence $F_n \in \mathcal{R}$ such that $\mu(F_n \bigtriangleup E) \to 0$.

5. If E_n is a sequence of sets, and f_n is the characteristic function of E_n, then $\liminf f_n$ is the characteristic function of $\liminf E_n$, and $\limsup f_n$ is the characteristic function of $\limsup E_n$.

6. If E_n is any sequence of measurable sets in a measure space $(X, \mathcal{S}, \mu)$, then

$$\mu(\liminf E_n) \leq \liminf \mu(E_n).$$

If, moreover, μ is finite, then

$$\mu(\limsup E_n) \geq \limsup \mu(E_n).$$

***7.** Given a measure space $(X, \mathcal{S}, \mu)$, consider the measure space $(X, \mathcal{S}_\lambda, \mu_\lambda)$ constructed as in Exercise 1, and the complete measure space $(X, \hat{\mathcal{S}}, \hat{\mu})$ constructed as in Sect. 9. What is the relation between the measure spaces

$$(X, (\mathcal{S}_\lambda)\hat{}\,, (\mu_\lambda)\hat{}\,) \qquad \text{and} \qquad (X, (\hat{\mathcal{S}})_\lambda, (\hat{\mu})_\lambda)\,?$$

8. With notation as in Exercise 1, $(\mu_\lambda)_E = \mu^E$ for every E in $\mathcal{S}$.

9. If $(X, \mathcal{S}, \mu)$ is any measure space, and A is the union of all the measurable sets, then A is locally measurable, and $\mu_A = \mu$. The measure space $(X, \mathcal{S}, \mu)$ can be replaced, for all practical purposes, by the measure space $(A, \mathcal{S}, \mu)$.

10. The definition of contraction may be generalized in the following way. If μ is a measure on a ring $\mathcal{R}$ of subsets of X, and S is any subset of X, we may define a measure μ_S on $\mathcal{R}$ by the formula

$$\mu_S(E) = \mu^*(S \cap E).$$

If in particular $S \in \mathcal{R}_\lambda$, then

$$\mu_S(E) = \mu(S \cap E)$$

for all E in $\mathcal{R}$. If S is μ^*-measurable, then

$$\mu = \mu_S + \mu_{X-S}.$$

11. If E_n is a sequence of measurable sets in a finite measure space, such that $\mu(E_n) \geq \varepsilon$ for infinitely many n, where $\varepsilon > 0$, then

$$\mu(\limsup E_n) \geq \varepsilon.$$

12. If $(X, \mathcal{S}, \mu)$ is any measure space, one can extend μ to a measure ν on $\mathcal{S}_\lambda$ by defining $\nu(A) = \infty$ for every set A in $\mathcal{S}_\lambda$ which does not belong to $\mathcal{S}$.

18. The "Almost Everywhere" Concept

A measurable set of measure zero will be called a **null set**. Thus, a null set is a set E in $\mathscr{S}$ such that $\mu(E) = 0$. If, due to the presence of other measures, it is necessary to emphasize the given measure μ, we shall speak of E as a null set "with respect to μ," or briefly a **μ-null** set. We emphasize that a null set is required to be measurable.

If E is a null set, and A is a locally measurable set, then $A \cap E$ is a null set; for, $A \cap E$ is measurable and $\mu(A \cap E) \leq \mu(E) = 0$. In particular, every measurable subset of a null set is a null set. The union of any sequence of null sets is a null set; this follows at once from the countable subadditivity of μ (4.1). Since the difference of null sets is obviously a null set, the class of null sets is a σ-ring. (In the language of Boolean rings, the class of null sets is an *ideal* in $\mathscr{S}_\lambda$, and hence in $\mathscr{S}$.)

A measure is said to be **complete** if every subset of a null set is measurable (see the remarks after 5.2), and hence is a null set. None of our results will require that the measure in question be complete.

The first key definition was the concept of a measurable function (Sect. 12). The second is the "almost everywhere" concept. Suppose that for each point x in X, there is given a proposition $P(x)$. One says that $P(x)$ is true **almost everywhere** if there exists a null set E such that $x \in X - E$ implies that $P(x)$ is true. This does not mean that $\{x: P(x) \text{ false}\}$ is a null set, but it does mean that it is a subset of a null set. Other commonly used renditions of this concept are: "$P(x)$ is essentially true," "$P(x)$ is true for almost all x," "$P(x)$ is true a.e.," or briefly "$P(x)$ a.e."; when it is necessary to emphasize the measure μ, one adjoins phrases such as "with respect to μ," or "modulo μ," or the symbol $[\mu]$.

It should not be supposed that from this point onward we shall allow an "almost everywhere" vagueness to permeate everything we do; on the contrary, we shall be quite explicit in indicating the null set in question, because the vague approach, though it has the advantage of brevity, can easily lead to an "almost understanding" of the subject. Null sets are a natural convenience of the theory, not a way of sweeping difficulties under the rug.

We shall now consider some particular examples of the a.e. concept which are very frequently used. In these examples, f and g are extended real valued functions defined on X.

(1) $f = g$ *a.e.* means: there exists a null set E such that $x \notin E$ implies $f(x) = g(x)$. In other words, the restrictions of f and g to $X - E$ are

equal. Alternatively, $\chi_{X-E}f = \chi_{X-E}g$. In this example, the proposition $P(x)$ is "$f(x) = g(x)$."

(2) $f \le g$ *a.e.* means: there exists a null set E such that $x \notin E$ implies $f(x) \le g(x)$. Suppose that also $g \le f$ a.e., and let F be a null set such that $x \notin F$ implies $g(x) \le f(x)$. Then $G = E \cup F$ is a null set, and $x \notin G$ implies $f(x) = g(x)$. Summarizing: if $f \le g$ a.e. and $g \le f$ a.e., then $f = g$ a.e. The converse of this proposition is also true, and obvious.

(3) *f is constant a.e.* means: there exists an extended real number α, and a null set E, such that $x \notin E$ implies $f(x) = \alpha$. In other words, $f = \alpha$ a.e. in the sense of Example 1 (where α is allowed to stand for the constant function it determines).

(4) *f is essentially bounded* (or *f is bounded a.e.*) means: there exists a real number $M \ge 0$ and a null set E, such that $x \notin E$ implies $|f(x)| \le M$. In other words, $|f| \le M$ a.e. in the sense of Example 2 (incidentally, the convention is that $|\infty| = |-\infty| = \infty$).

(5) *f is finite a.e.* means: there exists a null set E such that $x \notin E$ implies $f(x) \in R$.

(6) Let f be a function which satisfies a set Γ of conditions. One says that f is *a.e. unique* or *essentially unique* (with respect to the conditions Γ) in case: if g is any other function satisfying the conditions Γ, necessarily $g = f$ a.e.

(7) Suppose S is any subset of X, and suppose that for each x in S there is defined a proposition $P(x)$. One says that *$P(x)$ is true a.e. on S* in case there exists a null set E such that $x \in S - E$ implies that $P(x)$ is true.

The relation " = a.e." is an equivalence relation in the set of all extended real valued functions defined on X:

Theorem 1. *Let f, g, h be extended real valued functions on X. Then:*

(1) $f = f$ a.e.

(2) *If $f = g$ a.e., then $g = f$ a.e.*

(3) *If $f = g$ a.e., and $g = h$ a.e., then $f = h$ a.e.*

Proof. (3) Let E be a null set such that $x \notin E$ implies $f(x) = g(x)$, and let F be a null set such that $x \notin F$ implies $g(x) = h(x)$. Then $G = E \cup F$ is a null set, and $f(x) = g(x) = h(x)$ on the complement of G. ∎

It is important to note that a measurable function may be equal a.e. to a real-valued function which is not measurable. For example, suppose there exists a null set E, one of whose subsets, say S, is nonmeasurable.

Then χ_E is measurable, χ_S is not, and $\chi_E = \chi_S$ a.e. Incidentally, if S and T are arbitrary subsets of X, it is easy to see that $\chi_S = \chi_T$ a.e. if and only if there exists a null set E such that $S \triangle T \subset E$.

19. Almost Everywhere Convergence

A sequence f_n of real-valued functions, defined on the measure space $(X, \mathcal{S}, \mu)$, is said to **converge almost everywhere** to the real-valued function f, defined on X, in case there exists a null set E such that $x \notin E$ implies $f_n(x) \to f(x)$. Briefly, $f_n \to f$ *a.e.* A sequence f_n of real-valued functions is said to be **fundamental almost everywhere** if there exists a null set E such that $x \notin E$ implies that $f_n(x)$ is a Cauchy sequence; briefly, f_n is *fundamental a.e.* We shall use the terms "$f_n \to f$ a.e." and "f_n is fundamental a.e." only with the tacit understanding that f_n is a sequence of real-valued functions, and f is a real-valued function (defined on X). This simplifies the statements of theorems such as the following:

Theorem 1. *If $f_n \to f$ a.e., then f_n is fundamental a.e.*

Proof. If E is a null set such that $f_n(x) \to f(x)$ for all x in the complement of E, then obviously $x \notin E$ implies that $f_n(x)$ is a Cauchy sequence. ∎

Almost everywhere limits are unique a.e.:

Theorem 2. *If $f_n \to f$ a.e. and $f_n \to g$ a.e., then $f = g$ a.e.*

Proof. If E is a null set on whose complement $f_n(x) \to f(x)$, and F is a null set on whose complement $f_n(x) \to g(x)$, then $E \cup F$ is a null set on whose complement $f(x) = \lim f_n(x) = g(x)$. ∎

If $f_n \to f$ a.e., the limit function f can be varied a.e. without essentially disturbing the convergence:

Theorem 3. *If $f_n \to f$ a.e., and g is a real-valued function such that $f = g$ a.e., then $f_n \to g$ a.e.*

Proof. If E is a null set on whose complement $f_n(x) \to f(x)$, and F is a null set on whose complement $f(x) = g(x)$, then $E \cup F$ is a null set on whose complement $f_n(x) \to f(x) = g(x)$. ∎

If $f_n \to f$ a.e., the terms of the sequence may be altered a.e.:

Theorem 4. *If $f_n \to f$ a.e., and g_n is a sequence of real-valued functions such that $f_n = g_n$ a.e. for each n, then $g_n \to f$ a.e.*

Proof. If E is a null set on whose complement $f_n(x) \to f(x)$, and E_n is a null set on whose complement $f_n(x) = g_n(x)$, then

$$E \cup \bigcup_1^{\infty} E_n$$

is a null set on whose complement $g_n(x) = f_n(x) \to f(x)$. ∎

Convergence a.e. behaves agreeably with respect to the various finite operations:

Theorem 5. *If $f_n \to f$ a.e., $g_n \to g$ a.e., c is a real number, and A is any subset of X, then:*

(1) $cf_n \to cf$ *a.e.*
(2) $f_n + g_n \to f + g$ *a.e.*
(3) $|f_n| \to |f|$ *a.e.*
(4) $f_n \cup g_n \to f \cup g$ *a.e., and* $f_n \cap g_n \to f \cap g$ *a.e.*
(5) $f_n^+ \to f^+$ *a.e., and* $f_n^- \to f^-$ *a.e.*
(6) $\chi_A f_n \to \chi_A f$ *a.e.,*
(7) $f_n g_n \to fg$ *a.e.*

Proof. Let E be a null set on the complement of which both $f_n(x) \to f(x)$ and $g_n(x) \to g(x)$; the theorem then follows at once from the corresponding properties of convergent sequences of real numbers. ∎

Similarly:

Theorem 6. *If f_n is fundamental a.e., g_n is fundamental a.e., c is a real number, and A is any subset of X, then all of the following sequences of functions are fundamental a.e.: cf_n, $f_n + g_n$, $|f_n|$, $f_n \cup g_n$, $f_n \cap g_n$, f_n^+, f_n^-, $\chi_A f_n$, and $f_n g_n$.*

We now prove a converse for Theorem 1; for our purposes, the second assertion of the theorem is the important one:

Theorem 7. *If f_n is fundamental a.e., there exists a real-valued function f such that $f_n \to f$ a.e. If moreover the f_n are measurable, we may take f to be measurable.*

Proof. Let E be a null set on whose complement $f_n(x)$ is a Cauchy sequence. Define $f(x) = \lim f_n(x)$ for $x \in X - E$, and $f(x) = 0$ for $x \in E$. Clearly $f_n \to f$ a.e. Suppose, moreover, that the f_n are measurable. Define $g_n = \chi_{X-E} f_n$. The g_n are measurable by 15.1, and $g_n(x) \to f(x)$ for all x in X, hence f is measurable by 14.3. ∎

The next group of results concerns the preservation of order properties on passage to limits a.e.:

Lemma. *If $f_n \to f$ a.e., and $f_n \geq 0$ a.e., then $f \geq 0$ a.e.*

Proof. Let E be a null set on whose complement $f_n(x) \to f(x)$, E_n a null set on whose complement $f_n(x) \geq 0$, and define

$$F = E \cup \bigcup_1^\infty E_n.$$

Then F is a null set, and $x \notin F$ implies $f(x) = \lim f_n(x) \geq 0$. ▮

Theorem 8. *If f_n, f, g are real-valued functions, $f_n \to f$ a.e., and $f_n \leq g$ a.e. for each n, then $f \leq g$ a.e.*

Proof. Since $g - f_n \geq 0$ a.e., and $g - f_n \to g - f$ a.e., we have $g - f \geq 0$ a.e. by the lemma. ▮

Corollary. *If f_n, f, g are real-valued functions, $f_n \to f$ a.e., and $|f_n| \leq |g|$ a.e., then $|f| \leq |g|$ a.e.*

Proof. Since $|f_n| \to |f|$ a.e., $|f| \leq |g|$ a.e. by the theorem. ▮

Theorem 9. *Suppose $f_n \to f$ a.e., where f_n is a sequence of real-valued functions such that $f_n \leq f_{n+1}$ a.e. for each n. Then $f_n \uparrow f$ a.e.*

Proof. Let E be a null set on whose complement $f_n(x) \to f(x)$, E_n a null set on whose complement $f_n(x) \leq f_{n+1}(x)$, and define

$$F = E \cup \bigcup_1^\infty E_n.$$

Then F is a null set, and $x \notin F$ implies both $f_n(x)\uparrow$ and $f_n(x) \to f(x)$, hence $f_n(x) \uparrow f(x)$. ▮

EXERCISES

1. If f_n is a sequence of measurable functions, and E is the set of all points x such that $f_n(x)$ is not a Cauchy sequence, then E is a measurable set.

2. If $f_n \to f$ a.e., $g_n \to g$ a.e., and $f_n = g_n$ a.e. for all n, then $f = g$ a.e.

3. If $f_n \to f$ a.e., $f_n = g_n$ a.e. for all n, and $f = g$ a.e., then $g_n \to g$ a.e.

4. Suppose the measure space $(X, \mathscr{S}, \mu)$ is not complete, and let E be a null set having a subset N which is not measurable. If f_n is identically zero for all n, then $f_n \to \chi_N$ a.e., the f_n are all measurable, but χ_N is not measurable.

20. Convergence in Measure

Our first theorem is intended to motivate the definition of "convergence in measure"; the result itself will later be superseded by Egoroff's theorem (21.1):

Theorem 1. *Let $(X, \mathscr{S}, \mu)$ be a finite measure space, f_n a sequence of measurable functions, f a measurable function, and suppose that $f_n \to f$ a.e. Then for each $\varepsilon > 0$,*

$$\mu(\{x: |f_n(x) - f(x)| \geq \varepsilon\}) \to 0$$

as $n \to \infty$.

Proof. Let $E_n = \{x: |f_n(x) - f(x)| \geq \varepsilon\}$, and assume to the contrary that $\mu(E_n)$ does not converge to 0. Then there exists a $\delta > 0$, and a subsequence E_{n_k}, such that $\mu(E_{n_k}) \geq \delta$ for all k. Define

$$E = \limsup E_{n_k}.$$

Since μ is finite, we have $\mu(E) \geq \delta$ by the Arzela-Young theorem (17.2). In particular, E is not a null set, hence the assumption $f_n \to f$ a.e. implies that there is at least one point x in E such that $f_n(x) \to f(x)$. But

$$|f_{n_k}(x) - f(x)| \geq \varepsilon$$

for infinitely many k, by the definition of E, and we have arrived at a contradiction. ∎

The condition in the conclusion of the theorem is meaningful in any measure space. Accordingly, if f_n $(n = 1, 2, 3, \ldots)$ and f are measurable functions defined on an arbitrary measure space, we shall say that f_n **converges** to f **in measure** if, for each $\varepsilon > 0$,

$$\mu(\{x: |f_n(x) - f(x)| \geq \varepsilon\}) \to 0$$

as $n \to \infty$. Briefly, $f_n \to f$ *in measure*; we shall employ a compact and suggestive notation due to Halmos: $f_n \to f$ in $\textcircled{m}$. The analogous Cauchy condition is defined as follows: a sequence f_n of measurable functions is said to be **fundamental in measure** if, for each $\varepsilon > 0$,

$$\mu(\{x: |f_m(x) - f_n(x)| \geq \varepsilon\}) \to 0$$

as $m, n \to \infty$. Briefly, *f_n is fundamental in $\textcircled{m}$*. In this terminology, Theorem 1 states that in a finite measure space, and for measurable functions, convergence a.e. implies convergence in $\textcircled{m}$. The terms "$f_n \to f$ in $\textcircled{m}$" and "f_n is fundamental in $\textcircled{m}$" will be used only with the tacit understanding that f_n is a sequence of measurable functions, and f is a measurable function. The statement of the next theorem illustrates the economy that results from this convention:

Theorem 2. *If $f_n \to f$ in $\textcircled{m}$, then f_n is fundamental in $\textcircled{m}$.*

Proof. Given any $\varepsilon > 0$, define

$$E_{mn} = \{x: |f_m(x) - f_n(x)| \geq \varepsilon\}$$

and

$$E_n = \{x: |f_n(x) - f(x)| \geq \varepsilon/2\}.$$

The relation

$$|f_m(x) - f_n(x)| \leq |f_m(x) - f(x)| + |f(x) - f_n(x)|$$

implies that $E_{mn} \subset E_m \cup E_n$, hence

$$\mu(E_{mn}) \leq \mu(E_m) + \mu(E_n) \to 0$$

as $m, n \to \infty$. ∎

Conversely, if a sequence f_n is fundamental in $\textcircled{m}$, it is shown in the next section (21.4) that there exists a measurable function f such that $f_n \to f$ in $\textcircled{m}$; no use will be made of this rather complicated result, but in any case, it is easy to see that such an f is necessarily unique a.e.:

Theorem 3. *If $f_n \to f$ in $\textcircled{m}$ and $f_n \to g$ in $\textcircled{m}$, then $f = g$ a.e.*

Proof. Given any $\varepsilon > 0$, we have

$$\{x: |f(x) - g(x)| \geq \varepsilon\} \subset \{x: |f(x) - f_n(x)| \geq \varepsilon/2\} \cup \{x: |f_n(x) - g(x)| \geq \varepsilon/2\},$$

hence it is clear that

$$\mu(\{x: |f(x) - g(x)| \geq \varepsilon\}) = 0.$$

Our assertion then follows from the relation

$$N(f - g) = \bigcup_{m=1}^{\infty} \{x: |f(x) - g(x)| \geq 1/m\}$$

and the fact that a countable union of null sets is a null set. ∎

Conversely, if $f_n \to f$ in $\textcircled{m}$, the limit function may be (measurably) altered a.e. without disturbing convergence:

Theorem 4. *If $f_n \to f$ in $\textcircled{m}$, and g is a measurable function such that $f = g$ a.e., then also $f_n \to g$ in $\textcircled{m}$.*

Proof. Let E be a null set such that $f(x) \neq g(x)$ implies $x \in E$. Then for any $\varepsilon > 0$,

$$\{x: |f_n(x) - g(x)| \geq \varepsilon\} \subset E \cup \{x: |f_n(x) - f(x)| \geq \varepsilon\}. \quad ∎$$

If $f_n \to f$ in $\textcircled{m}$, the terms of the sequence may be measurably altered a.e.:

Theorem 5. *If $f_n \to f$ in $\textcircled{m}$, and g_n is a sequence of measurable functions such that $f_n = g_n$ a.e., then also $g_n \to f$ in $\textcircled{m}$.*

Proof. Let E_n be a null set such that $f_n(x) = g_n(x)$ on the complement of E_n. Then

$$E = \bigcup_1^\infty E_n$$

is a null set, and for any $\varepsilon > 0$,

$$\{x: |g_n(x) - f(x)| \geq \varepsilon\} \subset E \cup \{x: |f_n(x) - f(x)| \geq \varepsilon\}$$

for all n. ∎

Convergence in $\mathcal{M}$ behaves agreeably with respect to the linear and lattice operations:

Theorem 6. *If $f_n \to f$ in $\mathcal{M}$, $g_n \to g$ in $\mathcal{M}$, c is a real number, and A is a locally measurable set, then:*

(1) $cf_n \to cf$ *in* $\mathcal{M}$.
(2) $f_n + g_n \to f + g$ *in* $\mathcal{M}$.
(3) $|f_n| \to |f|$ *in* $\mathcal{M}$.
(4) $f_n \cup g_n \to f \cup g$ *in* $\mathcal{M}$, *and* $f_n \cap g_n \to f \cap g$ *in* $\mathcal{M}$.
(5) $f_n^+ \to f^+$ *in* $\mathcal{M}$, *and* $f_n^- \to f^-$ *in* $\mathcal{M}$.
(6) $\chi_A f_n \to \chi_A f$ *in* $\mathcal{M}$.

Proof. (1) This is clear if $c = 0$; if $c \neq 0$, then

$$\{x: |cf_n(x) - cf(x)| \geq \varepsilon\} = \{x: |f_n(x) - f(x)| \geq \varepsilon/|c|\}.$$

(2) Let $h_n = f_n + g_n$, $h = f + g$. The relation

$$|h_n(x) - h(x)| \leq |f_n(x) - f(x)| + |g_n(x) - g(x)|$$

implies

$$\{x: |h_n(x) - h(x)| \geq \varepsilon\} \subset \{x: |f_n(x) - f(x)| \geq \varepsilon/2\} \cup \{x: |g_n(x) - g(x)| \geq \varepsilon/2\}.$$

(3) A consequence of $\big||f_n(x)| - |f(x)|\big| \leq |f_n(x) - f(x)|$.
(4) Immediate from (1) through (3) and the formulas preceding 13.5.
(5) Clear from (4).
(6) The functions in question are measurable by 15.1, and

$$|\chi_A f_n - \chi_A f| = \chi_A |f_n - f| \leq |f_n - f|. \quad ∎$$

We leave to the reader the similar proof for the Cauchy version of Theorem 6:

Theorem 7. *If f_n is fundamental in $\mathcal{M}$, g_n is fundamental in $\mathcal{M}$, c is a real number, and A is a locally measurable set, then each of the following sequences is fundamental in $\mathcal{M}$: $cf_n, f_n + g_n, |f_n|, f_n \cup g_n, f_n \cap g_n, f_n^+, f_n^-, \chi_A f_n$.*

The next group of results has to do with the preservation of order relations on passage to limits in $\textcircled{m}$:

Lemma. *If $f_n \to f$ in $\textcircled{m}$, where $f_n \geq 0$ a.e., then $f \geq 0$ a.e.*

Proof. Modifying on a null set, we may assume (in view of Theorem 5) that $f_n \geq 0$ (everywhere) for all n. Given $\varepsilon > 0$, define

$$E = \{x : f(x) \leq -\varepsilon\}$$

and

$$E_n = \{x : |f(x) - f_n(x)| \geq \varepsilon\};$$

we know that $\mu(E_n) \to 0$. If $x \in E$, then $f(x) \leq -\varepsilon$; since

$$f(x) = [f(x) - f_n(x)] + f_n(x) \geq f(x) - f_n(x),$$

we conclude that $f(x) - f_n(x) \leq -\varepsilon$, and hence

$$|f(x) - f_n(x)| \geq \varepsilon.$$

We have shown that $E \subset E_n$ for all n, hence $\mu(E) = 0$. It follows that

$$\{x : f(x) < 0\} = \bigcup_{m=1}^{\infty} \{x : f(x) \leq -1/m\}$$

is a null set, and so $f \geq 0$ a.e. ▮

Theorem 8. *If f_n, f, g are measurable, $f_n \to f$ in $\textcircled{m}$, and $f_n \leq g$ a.e. for all n, then $f \leq g$ a.e.*

Proof. We have $g - f_n \geq 0$ a.e., and $g - f_n \to g - f$ in $\textcircled{m}$ by Theorem 6, hence $g - f \geq 0$ a.e. by the lemma. ▮

Corollary. *If f_n, f, g are measurable, $f_n \to f$ in $\textcircled{m}$, and $|f_n| \leq |g|$ a.e. for all n, then $|f| \leq |g|$ a.e.*

Proof. Since $|f_n| \to |f|$ in $\textcircled{m}$ by Theorem 6, we have $|f| \leq |g|$ a.e. by Theorem 8. ▮

EXERCISES

1. In a finite measure space, if a sequence of measurable functions is fundamental a.e., then it is fundamental in $\textcircled{m}$. The converse of this proposition is false.

2. If $f_n \to f$ in $\textcircled{m}$, $g_n \to g$ in $\textcircled{m}$, and $f_n = g_n$ a.e. for all n, then $f = g$ a.e.

3. If a sequence f_n of measurable functions is fundamental in $\textcircled{m}$, and there exists a measurable function f to which some subsequence f_{n_k} converges in $\textcircled{m}$, then $f_n \to f$ in $\textcircled{m}$.

4. If $f_n \to f$ in $\textcircled{m}$, and g is an essentially bounded measurable function, then $f_n g \to fg$ in $\textcircled{m}$.

5. In a finite measure space, if $f_n \to f$ in $\textcircled{m}$ and $g_n \to g$ in $\textcircled{m}$, then $f_n g_n \to fg$ in $\textcircled{m}$.

6. In a finite measure space, if $f_n \to f$ in $\textcircled{m}$, and c_n is a sequence of real numbers such that $c_n \to c$, then $c_n f_n \to cf$ in $\textcircled{m}$.

7. Convergence a.e. does not always imply convergence in measure. For example, let μ be discrete measure on the class $\mathscr{S}$ of all subsets of the set $X = \{1, 2, 3, \ldots\}$. If f_n is the characteristic function of the singleton $\{n\}$, then $f_n \to 0$ pointwise on X, but f_n does not converge to 0 in measure.

8. Convergence in measure does not imply convergence a.e. For example, let μ be Lebesgue measure, and consider the sequence of intervals $[0, 1)$, $[0, \frac{1}{2})$, $[\frac{1}{2}, 1)$, $[0, \frac{1}{3})$, $[\frac{1}{3}, \frac{2}{3})$, $[\frac{2}{3}, 1)$, $[0, \frac{1}{4})$, $[\frac{1}{4}, \frac{2}{4})$, $[\frac{2}{4}, \frac{3}{4})$, $[\frac{3}{4}, 1), \ldots$. If f_n is the characteristic function of the nth term of this sequence, then $f_n \to 0$ in measure, but $f_n(x)$ is divergent for every x in $[0, 1)$.

*21. Almost Uniform Convergence, Egoroff's Theorem

Almost uniform convergence is a concept which is in a sense auxiliary to the concepts of convergence a.e. and convergence in $\textcircled{m}$; it will play no role in the rest of the book. Indeed, the entire section may be omitted without loss of continuity, and is presented simply for its intrinsic interest. Let us motivate the definition of almost uniform convergence by first proving *Egoroff's theorem*:

Theorem 1. *Suppose $f_n \to f$ a.e., where the f_n and f are measurable functions defined on a* **finite** *measure space $(X, \mathscr{S}, \mu)$. Then, given any $\delta > 0$, there exists a measurable set F such that $\mu(F) \le \delta$ and $f_n(x) \to f(x)$ uniformly on $X - F$.*

Proof. Let us assume first that $f_n(x) \to f(x)$ for *all* x; we shall see at the end of the proof that this entails no loss of generality.

For all $m, n = 1, 2, 3, \ldots$, define

$$\begin{aligned} F_n^m &= \{x\colon |f_k(x) - f(x)| \ge 1/m \text{ for some } k \ge n\} \\ &= \bigcup_{k \ge n} \{x\colon |f_k(x) - f(x)| \ge 1/m\}; \end{aligned}$$

since each $f_k - f$ is a measurable function, the sets F_n^m are measurable. It is clear that for fixed m, the sequence of sets F_n^m is decreasing. Indeed, we assert that $F_n^m \downarrow \varnothing$ as $n \to \infty$. For, given any positive integer m and any $x \in X$, the assumption $f_k(x) \to f(x)$ implies the existence of an index r such that

$$|f_k(x) - f(x)| < 1/m$$

whenever $k \geq r$, and so $x \notin F_r^m$; all the more,

$$x \notin \bigcap_{n=1}^{\infty} F_n^m,$$

and we have shown that the intersection of the F_n^m (for each fixed m) is empty.

For each m, we have $\mu(F_n^m) \downarrow 0$ by 4.2 (μ is a finite measure). Given any $\delta > 0$. For each m, $\mu(F_n^m) \to 0$ as $n \to \infty$, hence there exists an index $n(m)$ such that

$$\mu(F_{n(m)}^m) \leq \delta/2^m.$$

Defining

$$F = \bigcup_{m=1}^{\infty} F_{n(m)}^m,$$

F is a measurable set such that

$$\mu(F) \leq \sum_{m=1}^{\infty} \mu(F_{n(m)}^m) \leq \delta \sum_{1}^{\infty} 2^{-m} = \delta.$$

It remains to show that $f_n(x) \to f(x)$ uniformly on $X - F$. Given $\varepsilon > 0$, we seek an index r such that $k \geq r$ implies

$$|f_k(x) - f(x)| < \varepsilon$$

for all $x \in X - F$. Now,

$$\begin{aligned} X - F &= \bigcap_{m=1}^{\infty} \complement F_{n(m)}^m \\ &= \bigcap_{m} \bigcap_{k \geq n(m)} \complement\{x: |f_k(x) - f(x)| \geq 1/m\} \\ &= \bigcap_{m} \bigcap_{k \geq n(m)} \{x: |f_k(x) - f(x)| < 1/m\} \\ &= \bigcap_{m} \{x: |f_k(x) - f(x)| < 1/m \quad \text{for all } k \geq n(m)\}. \end{aligned}$$

Choose an m so that $1/m < \varepsilon$. Since

$$X - F \subset \{x: |f_k(x) - f(x)| < 1/m \quad \text{for all } k \geq n(m)\},$$

evidently $|f_k(x) - f(x)| < \varepsilon$ for all $x \in X - F$ provided $k \geq n(m)$.

Finally, consider the general case, namely, $f_n \to f$ a.e. Let E be a null set on whose complement $f_n(x) \to f(x)$, and define

$$g_n = \chi_{X-E} f_n, \qquad g = \chi_{X-E} f.$$

The functions g_n and g are measurable by 15.1, and $g_n(x) \to g(x)$ for all x. Given any $\delta > 0$; by the first part of the proof, there exists a measurable set G such that $\mu(G) \leq \delta$ and $g_n(x) \to g(x)$ uniformly on

$X - G$. Define $F = E \cup G$; then $\mu(F) \leq 0 + \mu(G) \leq \delta$, and on the complement of F we have

$$f_n(x) = g_n(x) \to g(x) = f(x)$$

uniformly. ▮

For the rest of the section, we consider a fixed measure space $(X, \mathscr{S}, \mu)$. A sequence f_n of real-valued functions defined on X is said to converge **almost uniformly** to the real-valued function f in case: given any $\delta > 0$, there exists a measurable set F such that $\mu(F) \leq \delta$ and $f_n(x) \to f(x)$ uniformly on $X - F$; briefly, $f_n \to f$ *a.u.* In this terminology, Egoroff's theorem reads as follows: in a finite measure space, if f_n is a sequence of measurable functions converging a.e. to a measurable function f, then $f_n \to f$ a.u. In general, the term "$f_n \to f$ a.u." will be used only with the tacit understanding that f and f_n $(n = 1, 2, 3, \ldots)$ are real-valued.

A sequence f_n of real-valued functions defined on a measure space is said to be **almost uniformly fundamental** in case: given any $\delta > 0$, there exists a measurable set F such that $\mu(F) \leq \delta$, and such that $f_n(x)$ is uniformly fundamental on $X - F$; this means that for each $\varepsilon > 0$, there is an index n_0 such that $m, n \geq n_0$ imply

$$|f_m(x) - f_n(x)| \leq \varepsilon$$

for all $x \in X - F$. Briefly, f_n is *fundamental a.u.* If $f_n \to f$ a.u., then f_n is fundamental a.u.; this is immediate from the relation

$$|f_m - f_n| \leq |f_m - f| + |f - f_n|.$$

Conversely:

Theorem 2. *If the sequence f_n of real-valued functions is fundamental a.u., there exists a real-valued function f such that $f_n \to f$ a.u. If moreover the f_n are measurable, one can take f to be measurable.*

Proof. For each positive integer m, let F_m be a measurable set such that $\mu(F_m) \leq 1/m$ and $f_n(x)$ is uniformly fundamental on $X - F_m$. Then the set

$$F = \bigcap_1^\infty F_m$$

is measurable, and $\mu(F) \leq \mu(F_m) \leq 1/m$ for all m, hence F is a null set. Moreover, for each x in the set

$$X - F = \bigcup_1^\infty X - F_m,$$

the sequence $f_n(x)$ is Cauchy. For each x in $X - F$, define $f(x) = \lim f_n(x)$, and define $f(x) = 0$ for $x \in F$. For each m, $f_n(x) \to f(x)$ on $X - F_m$, and $f_n(x)$ is uniformly fundamental on $X - F_m$, hence it is clear that $f_n(x) \to f(x)$ uniformly on $X - F_m$. It follows at once that $f_n \to f$ a.u.

If the f_n are also measurable, then so are the functions $g_n = \chi_{X-F} f_n$ by 15.1; since $g_n(x) \to f(x)$ for *all* x, f is measurable by 14.3. ▌

Almost uniform convergence is a "stronger" mode of convergence than convergence a.e. or convergence in ⓜ:

Theorem 3. *If $f_n \to f$ a.u., then $f_n \to f$ a.e. If moreover the f_n and f are measurable, then $f_n \to f$ in ⓜ.*

Proof. For each positive integer m, let F_m be a measurable set such that $\mu(F_m) \le 1/m$ and $f_n(x) \to f(x)$ uniformly on $X - F_m$. Then $F = \bigcap_1^\infty F_m$ is a null set, and $f_n(x) \to f(x)$ for each $x \in X - F$, thus $f_n \to f$ a.e.

Suppose, in addition, that the f_n and f are measurable. Given any $\varepsilon > 0$, define

$$E_n = \{x: |f_n(x) - f(x)| \ge \varepsilon\};$$

the problem is to show that $\mu(E_n) \to 0$. Given any $\delta > 0$, choose a positive integer m so that $1/m < \delta$, and consider the set F_m defined earlier. Since $f_n(x) \to f(x)$ uniformly on $X - F_m$, there exists an index n_0 such that $n \ge n_0$ implies that $|f_n(x) - f(x)| < \varepsilon$ for all $x \in X - F_m$, and hence $X - F_m \subset X - E_n$ whenever $n \ge n_0$. Then, $n \ge n_0$ implies $E_n \subset F_m$, and hence $\mu(E_n) \le \mu(F_m) \le 1/m < \delta$. ▌

Almost uniform convergence plays an auxiliary role in the following remarkable theorem:

Theorem 4 (Riesz-Weyl). *If the sequence f_n of measurable functions is fundamental in ⓜ, then there exists a measurable function to which f_n converges in ⓜ. Indeed, there exists a subsequence f_{n_k} and a measurable function f such that $f_{n_k} \to f$ a.u.; necessarily $f_n \to f$ in ⓜ, and this condition determines f uniquely a.e.*

Proof. For each positive integer k, we have

$$\mu(\{x: |f_m(x) - f_n(x)| \ge 2^{-k}\}) \to 0$$

as $m, n \to \infty$; hence there exists an index n_k such that

$$\mu(\{x: |f_m(x) - f_n(x)| \ge 2^{-k}\}) < 2^{-k}$$

whenever $m, n \ge n_k$. We may assume $n_1 < n_2 < n_3 < \cdots$. Write

$g_k = f_{n_k}$; it will eventually be shown that the sequence g_k is fundamental a.u. Define

$$E_k = \{x: |g_{k+1}(x) - g_k(x)| \geq 2^{-k}\};$$

since $n_{k+1} > n_k$, we have $\mu(E_k) < 2^{-k}$ by choice of n_k.

Defining

$$F_k = \bigcup_{j \geq k} E_j,$$

it follows from the countable subadditivity of μ that $\mu(F_k) < 2^{-(k-1)}$. What happens on the complement of F_k? If $x \in X - F_k$, then $x \in X - E_j$ for all $j \geq k$, hence

$$|g_{j+1}(x) - g_j(x)| < 2^{-j} \leq 2^{-k}$$

for all $j \geq k$; it follows that for any $j \geq k$ and for any positive integer p,

$$|g_{j+p}(x) - g_j(x)| \leq |g_{j+p}(x) - g_{j+p-1}(x)| + \cdots + |g_{j+1}(x) - g_j(x)|$$
$$< 2^{-(j+p-1)} + \cdots + 2^{-j} < 2^{-(j-1)}.$$

Summarizing,

$$(*) \qquad |g_{j+p}(x) - g_j(x)| < 2^{-(j-1)},$$

provided $x \in X - F_k$, $j \geq k$, $p \geq 1$. We assert that the sequence g_k is fundamental a.u. Given any $\delta > 0$, choose a positive integer r so that $2^{-(r-1)} < \delta$, and let $F = F_r$. Then $\mu(F) < 2^{-(r-1)} < \delta$, and it will suffice to show that the sequence $g_k(x)$ is uniformly fundamental on $X - F$. Given any $\varepsilon > 0$, choose an index k so large that both $k \geq r$ and $2^{-(k-1)} < \varepsilon$. If $j \geq k$, then also $j \geq r$, hence it follows from $(*)$ (with r in the role of k) that

$$|g_{j+p}(x) - g_j(x)| < 2^{-(j-1)} \leq 2^{-(k-1)} < \varepsilon$$

for all $x \in X - F_r$ and $p \geq 1$.

By Theorem 2, there exists a measurable function f such that $g_k \to f$ a.u., and hence $g_k \to f$ in $\mathscr{m}$ by Theorem 3. It then follows from the relation

$$f_n - f = (f_n - f_{n_k}) + (f_{n_k} - f) = (f_n - f_{n_k}) + (g_k - f)$$

that $f_n \to f$ in $\mathscr{m}$.

If also $f_n \to h$ in $\mathscr{m}$, then $f = h$ a.e. by 20.3. ∎

EXERCISES

1. In a finite measure space, if a sequence of measurable functions is fundamental a.e., then it is also fundamental a.u.

2. Deduce another proof of 20.1 from Egoroff's theorem.

3. If $f_n \to f$ in $\mathscr{m}$, and f_n is fundamental a.u., then $f_n \to f$ a.u.

4. If $f_n \to f$ in $\textcircled{m}$, and f_n is fundamental a.e., then $f_n \to f$ a.e.

5. If $f_n \to f$ in $\textcircled{m}$, and $f_n \le f_{n+1}$ a.e. for each n, then $f_n \uparrow f$ a.e.

6. Let $(X, \mathscr{S}, \mu)$ be an arbitrary measure space, and suppose A is a locally measurable set such that the contraction μ_A is a finite measure. If f and f_n $(n = 1, 2, 3, \ldots)$ are measurable functions such that $f_n \to f$ a.e. on A, then $f_n \to f$ a.u. on A (in the obvious sense).

7. If f and f_n $(n = 1, 2, 3, \ldots)$ are measurable functions on a finite measure space, then $f_n \to f$ a.u. if and only if for each $\varepsilon > 0$, the measure of the set

$$\bigcup_{k \ge n} \{x\colon |f_k(x) - f(x)| \ge \varepsilon\}$$

tends to zero as $n \to \infty$.

8. (*i*) If $f_n \to f$ a.u. and $f_n \to g$ a.u., then $f = g$ a.e.

(*ii*) If $f_n \to f$ a.u. and $f = g$ a.e., then $f_n \to g$ a.u.

(*iii*) If $f_n \to f$ a.u. and $g_n = f_n$ a.e., then $g_n \to f$ a.u.

(*iv*) If $f_n \to f$ a.u., $g_n = f_n$ a.e., and $g = f$ a.e., then $g_n \to g$ a.u.

9. If $f_n \to f$ in $\textcircled{m}$, then $f_{n_k} \to f$ a.u. for a suitable subsequence.

10. If f and f_n $(n = 1, 2, 3, \ldots)$ are measurable functions defined on a measure space, the following conditions are equivalent:

(*a*) $f_n \to f$ in $\textcircled{m}$.

(*b*) Every subsequence of f_n has a subsequence converging to f a.u.

If, moreover, the measure space is finite, these conditions are also equivalent to the following:

(*c*) Every subsequence of f_n has a subsequence converging to f a.e.

CHAPTER 4

Integrable Functions

22. Integrable Simple Functions

The context for the entire chapter is a fixed measure space $(X, \mathscr{S}, \mu)$. If f is a simple function (with respect to $\mathscr{S}$), we say that f is **integrable** (with respect to μ) in case $\mu(N(f)) < \infty$. Briefly, f is an ISF. Thus, an ISF is a measurable function which takes on at most a finite number of nonzero (real) values, each on a measurable set of finite measure; it may also take on the value 0, on a locally measurable set. In place of the phrase "measurable set of finite measure," it will be convenient to denote the class of all such sets by $\mathscr{S}_\varphi$, or, if there were any possible doubt as to the measure in question, $\mathscr{S}_\varphi(\mu)$. (On the other hand, the class $\mathscr{S}_\lambda$ of locally measurable sets depends only on the measurable space $(X, \mathscr{S})$.) It is clear that $\mathscr{S}_\varphi$ is a ring of sets. An ISF is a simple function f such that $N(f) \in \mathscr{S}_\varphi$.

If f and g are real-valued functions defined on X, c is a real number, and A is any subset of X, evidently $N(cf) \subset N(f)$, $N(f + g) \subset N(f) \cup N(g)$, $N(|f|) = N(f)$, $N(f \cup g) \subset N(f) \cup N(g)$, $N(f \cap g) \subset N(f) \cup N(g)$, $N(f^+) \cup N(f^-) = N(f)$, $N(\chi_A f) = A \cap N(f)$, and $N(fg) = N(f) \cap N(g)$; in view of 16.2, we have at once:

Theorem 1. *If f and g are* ISF, *c is a real number, and A is a locally measurable set, then all of the following functions are* ISF: *cf, $f + g$, $|f|$, $f \cup g$, $f \cap g$, f^+, f^-, $\chi_A f$, and fg.*

Suppose f is an ISF; equivalently, f can be written as a linear combination

$$f = \sum_1^n a_k \chi_{E_k},$$

where the a_k are real numbers and the E_k are in $\mathscr{S}_\varphi$. Our first task is to define the "integral" of f with respect to μ, and this should presumably be

$$\sum_1^n a_k \mu(E_k).$$

Such a definition must be justified by showing that this sum has the same value for every such representation of f. A way out is to assume that f is in its unique "disjoint form," where $a_1, \ldots, a_n$ are the distinct non-zero values of f and E_k is the set on which f takes the value a_k; however, this only parks the problem, for it reappears when we consider $f + g$ and ask that the integral of $f + g$ be equal to the sum of the integrals of f and g. Accordingly, the purpose of the next two lemmas is to show that

$$\sum_1^n a_k\mu(E_k)$$

is indeed independent of the particular representation of f.

Lemma 1. *If $A_1, \ldots, A_n$ are subsets of X, there exist sets $B_1, \ldots, B_m$ $(m = 2^n - 1)$ such that:*

(i) The B_k are mutually disjoint.

(ii) Each A_i is a union of certain of the B_k.

(iii) Each B_k is in the ring generated by $A_1, \ldots, A_n$, and is contained in some A_i.

Proof. For any subset A of X, let us write $A^1 = A$ and $A^{-1} = X - A$. For each n-ple $\varepsilon = (\varepsilon_1, \ldots, \varepsilon_n)$, where $\varepsilon_i = \pm 1$, define

$$A_\varepsilon = A_1^{\varepsilon_1} \cap \cdots \cap A_n^{\varepsilon_n}.$$

In particular, writing $\varepsilon_0 = (-1, -1, \ldots, -1)$, we have

$$A_{\varepsilon_0} = (X - A_1) \cap \cdots \cap (X - A_n) = X - (A_1 \cup \cdots \cup A_n);$$

for all other ε, it is clear that A_ε belongs to the ring generated by $A_1, \ldots, A_n$.

We agree to write $\varepsilon = \eta$ only when $\varepsilon_i = \eta_i$ for all i. It follows that if $\varepsilon \neq \eta$, then $A_\varepsilon \cap A_\eta = \varnothing$; indeed, if $\varepsilon_j \neq \eta_j$ for some coordinate j, say $\varepsilon_j = 1$ and $\eta_j = -1$, then $A_\varepsilon \subset A_j$ whereas $A_\eta \subset X - A_j$. Thus, the A_ε are mutually disjoint.

We assert that A_j is the union of those A_ε for which the jth coordinate of ε is $+1$; that is,

$$A_j = \bigcup_{\varepsilon_j = 1} A_\varepsilon.$$

Indeed, if ε is an n-ple for which $\varepsilon_j = 1$, then $A_\varepsilon \subset A_j^{\varepsilon_j} = A_j^1 = A_j$. On the other hand if $x \in A_j$, define an n-ple ε by setting $\varepsilon_i = 1$ if $x \in A_i$ and $\varepsilon_i = -1$ if $x \in X - A_i$; then $\varepsilon_j = 1$ and clearly $x \in A_\varepsilon$.

Take for $B_1, \ldots, B_m$ any listing of the $2^n - 1$ sets A_ε, where $\varepsilon \neq \varepsilon_0$. ∎

Lemma 2. *If $\sum_1^n c_i\chi_{E_i} = 0$, where the c_i are real numbers and the E_i are in $\mathscr{S}_\varphi$, then $\sum_1^n c_i\mu(E_i) = 0$.*

Proof. With notation as in the proof of Lemma 1, consider n-ples $\varepsilon = (\varepsilon_1, \ldots, \varepsilon_n)$, where $\varepsilon_i = \pm 1$, and define

$$F_\varepsilon = E_1^{\varepsilon_1} \cap \cdots \cap E_n^{\varepsilon_n}.$$

We know:

(*i*) the F_ε are mutually disjoint.

(*ii*) $E_i = \bigcup_{\varepsilon_i = 1} F_\varepsilon$,

(*iii*) $F_\varepsilon \in \mathscr{S}_\varphi$ whenever $\varepsilon \neq \varepsilon_0$.

By (*i*) and (*ii*), $\chi_{E_i} = \sum_{\varepsilon_i = 1} \chi_{F_\varepsilon}$, hence

$$(*) \qquad \sum_1^n c_i \chi_{E_i} = c_1 \sum_{\varepsilon_1 = 1} \chi_{F_\varepsilon} + \cdots + c_n \sum_{\varepsilon_n = 1} \chi_{F_\varepsilon}.$$

For a particular $\varepsilon \neq \varepsilon_0$, what is the total coefficient of χ_{F_ε} on the right side of $(*)$? Since χ_{F_ε} occurs precisely in those terms for which $\varepsilon_i = 1$, its total coefficient is evidently

$$\sum_{\varepsilon_i = 1} c_i,$$

hence the right side of $(*)$ may be reorganized as follows:

$$(**) \qquad \sum_1^n c_i \chi_{E_i} = \sum_{\varepsilon \neq \varepsilon_0} c(\varepsilon) \chi_{F_\varepsilon},$$

where $c(\varepsilon) = \sum_{\varepsilon_i = 1} c_i$. Since the left side of $(**)$ is assumed to be identically 0, we have

$$\sum_{\varepsilon \neq \varepsilon_0} c(\varepsilon) \chi_{F_\varepsilon} = 0;$$

the disjointness of the F_ε then implies that $c(\varepsilon) = 0$ whenever $F_\varepsilon \neq \varnothing$, and hence clearly $c(\varepsilon)\mu(F_\varepsilon) = 0$ for all $\varepsilon \neq \varepsilon_0$. Then

$$\sum_{\varepsilon \neq \varepsilon_0} c(\varepsilon) \mu(F_\varepsilon) = 0,$$

thus

$$(***) \qquad 0 = \sum_{\varepsilon \neq \varepsilon_0} \Big(\sum_{\varepsilon_i = 1} c_i \Big) \mu(F_\varepsilon).$$

What is the total coefficient of c_i on the right side of $(***)$? It is precisely the sum of all the $\mu(F_\varepsilon)$ for which $\varepsilon_i = 1$, that is,

$$\sum_{\varepsilon_i = 1} \mu(F_\varepsilon);$$

reorganizing the right side of $(***)$, we have

$$0 = \sum_1^n c_i \Big(\sum_{\varepsilon_i = 1} \mu(F_\varepsilon) \Big)$$

$$= \sum_1^n c_i \mu\left(\bigcup_{\varepsilon_i = 1} F_\varepsilon\right)$$

$$= \sum_1^n c_i \mu(E_i). \quad \blacksquare$$

Incidentally, it is clear from the proof of Lemma 2 that it is sufficient to assume

$$\sum_1^n c_i \chi_{E_i} = 0 \text{ a.e.};$$

for, one argues that $c(\varepsilon) = 0$ unless F_ε is a null set, and hence $c(\varepsilon)\mu(F_\varepsilon) = 0$.

We are now prepared to define the integral (with respect to μ) of an ISF f. Suppose

$$f = \sum_1^n a_i \chi_{E_i} \quad \text{and} \quad f = \sum_1^m b_j \chi_{F_j}$$

are any two representations of f, where the coefficients a_i, b_j are real, and the sets E_i, F_j are in $\mathscr{S}_\varphi$. Since

$$\sum_1^n a_i \chi_{E_i} + \sum_1^m (-b_j) \chi_{F_j} = 0,$$

it follows from Lemma 2 that

$$\sum_1^n a_i \mu(E_i) + \sum_1^m (-b_j) \mu(F_j) = 0,$$

and hence

$$\sum_1^n a_i \mu(E_i) = \sum_1^m b_j \mu(F_j);$$

thus we may unambiguously define the **integral** of f (with respect to μ) to be the real number

$$\sum_1^n a_i \mu(E_i).$$

We shall temporarily denote the integral of f by $I(f)$. In particular if E is a measurable set of finite measure, then $I(\chi_E) = \mu(E)$.

The rest of the section is devoted to proving the elementary properties of the passage from f to $I(f)$, where f varies over the class of all ISF.

Theorem 2. *If f and g are* ISF, *c is a real number, and E is a measurable set of finite measure, then:*

(1) $I(cf) = cI(f)$.

(2) $I(f + g) = I(f) + I(g)$.

(3) $I(\chi_E) = \mu(E)$.

These properties determine the integral uniquely on the class of all ISF.

Proof. Suppose

$$f = \sum_1^n a_i\chi_{E_i} \qquad \text{and} \qquad g = \sum_1^m b_j\chi_{F_j},$$

where the a_i, b_j are real and the sets E_i, F_j are in $\mathscr{S}_\varphi$. Since

$$f + g = \sum_1^n a_i\chi_{E_i} + \sum_1^m b_j\chi_{F_j},$$

it is immediate from the definition of integral that

$$I(f + g) = \sum_1^n a_i\mu(E_i) + \sum_1^m b_j\mu(F_j) = I(f) + I(g).$$

Similarly,

$$cf = \sum_1^n (ca_i)\chi_{E_i}$$

implies that

$$I(cf) = \sum_1^n ca_i\mu(E_i) = c\sum_1^n a_i\mu(E_i) = cI(f).$$

Property (3) is obvious.

Suppose now that J is any real-valued function defined on the class of all integrable simple functions, such that

$$J(cf) = cJ(f), \quad J(f + g) = J(f) + J(g), \quad \text{and} \quad J(\chi_E) = \mu(E)$$

for all f, g, c, and E. If $f = \sum_1^n a_i\chi_{E_i}$ is any ISF, then

$$J(f) = \sum_1^n J(a_i\chi_{E_i}) = \sum_1^n a_iJ(\chi_{E_i}) = \sum_1^n a_i\mu(E_i) = I(f). \quad \blacksquare$$

In the next theorem, we consider the order properties of the integral, and the behavior of the integral with respect to null sets:

Theorem 3. *Let f and g be* ISF.

(4) *If $f \geq 0$ a.e., then $I(f) \geq 0$.*
(5) *If $f \geq 0$ a.e. and $I(f) = 0$, then $f = 0$ a.e.*
(6) *If $f \leq g$ a.e. then $I(f) \leq I(g)$.*
(7) *If $f = g$ a.e., then $I(f) = I(g)$.*
(8) $|I(f)| \leq I(|f|)$.
(9) *$f = 0$ a.e. if and only if $I(|f|) = 0$.*
(10) *If $M = \max\{|f(x)| : x \in X\}$, then $|I(f)| \leq M\mu(N(f))$.*

Proof. (4) and (5): Suppose $f \geq 0$ a.e., and write

$$f = \sum_1^n a_i \chi_{E_i},$$

where $a_1, \ldots, a_n$ are the distinct nonzero values of f and E_i is the set on which f takes the value a_i. Since $f \geq 0$ a.e., necessarily $\mu(E_i) = 0$ whenever $a_i < 0$, hence $a_i\mu(E_i) \geq 0$ for all i, and

$$I(f) = \sum_1^n a_i\mu(E_i) \geq 0.$$

If moreover this sum is 0, then every term, being nonnegative, must be 0; thus $a_i\mu(E_i) = 0$, hence $\mu(E_i) = 0$ for all i, and so $f = 0$ a.e.

(6) If $f \leq g$ a.e., then $g - f \geq 0$ a.e., hence $I(g - f) \geq 0$, that is, $I(g) - I(f) \geq 0$.

(7) Immediate from (6).

(8) Since $|f|$ is an ISF, we may apply (6) to the relation $-|f| \leq f \leq |f|$.

(9) If $f = 0$ a.e., then $|f| = 0$ a.e., hence $I(|f|) = 0$ by (7). If conversely $I(|f|) = 0$, then $|f| = 0$ a.e. by (5).

(10) Apply (6) to the relation $-M\chi_{N(f)} \leq f \leq M\chi_{N(f)}$. ∎

EXERCISES

1. The ring $\mathscr{S}_\varphi$ is an ideal in the ring $\mathscr{S}_\lambda$.

2. The class of all ISF is an ideal in the algebra (pointwise operations) of simple functions with respect to $\mathscr{S}_\lambda$.

23. Heuristics

Our next task is to extend the notion of integral, already defined on the class of ISF, to a wider class of functions to be called "integrable" (with respect to μ).

What should an "integrable" function be? For a simple function f, the answer is easy: $N(f)$ should have finite measure. Just as every measurable function is the pointwise limit of a suitable sequence of simple functions, we shall wish every "integrable" function to be a limit (in a suitable sense) of a suitable sequence of ISF. However this is done, the following desiderata for the class of "integrable" functions seem uncontroversial:

(A) Every ISF should be "integrable."

(B) Every "integrable" function should be measurable.

(C) If f and g are "integrable" functions, and c is a real number, the functions $f + g$ and cf should also be "integrable."

We shall naturally want the "integral" of an "integrable" function f, to be denoted $\int f\,d\mu$, to have the following properties:

(D) $\int f\,d\mu$ is a real number.

(E) If in particular f is an ISF, then $\int f\,d\mu = I(f)$.

(F) If f and g are "integrable" functions, and c is a real number, then

$$\int (f + g)\,d\mu = \int f\,d\mu + \int g\,d\mu \qquad \text{and} \qquad \int cf\,d\mu = c \int f\,d\mu.$$

So far this is all very innocuous. The next desiderata are somewhat more subtle, and are concerned with order:

(G) If f and g are "integrable," and $f \leq g$, then $\int f\,d\mu \leq \int g\,d\mu$.

(H) If $0 \leq f \leq g$, where g is "integrable" and f is a simple function, then f is an ISF.

(I) If $f_n \uparrow f$, where the f_n $(n = 1, 2, 3, \ldots)$ and f are "integrable," then $\int f_n\,d\mu \uparrow \int f\,d\mu$.

We are still nowhere. All of the desiderata are fulfilled by the function $\int f\,d\mu = I(f)$ defined on the class of ISF f (the property (I) has not yet been verified, but this will be done in the next section). We must look further to get off the ground.

Suppose f is "integrable," and $f \geq 0$. Let f_n be any sequence of simple functions such that $0 \leq f_n \uparrow f$. The f_n are ISF by (H), and $\int f_n\,d\mu \uparrow \int f\,d\mu$ by (I). In particular, $I(f_n) = \int f_n\,d\mu$ is a bounded sequence. We are thus led to consider *nonnegative measurable functions* f, *with the property that there exists a sequence* f_n *of* ISF *such that* $0 \leq f_n \uparrow f$ *and* $I(f_n)$ *is bounded.* Indeed, this can be made the starting point for continuing the theory of integration beyond the class of ISF, and this we proceed to do in the next section.

24. Nonnegative Integrable Functions

We begin by defining the concept of integrability for nonnegative measurable functions. If f is a measurable function, $f \geq 0$, we shall say that f is **integrable** (with respect to μ) if there exists a sequence f_n of

ISF such that (*i*) $0 \le f_n \uparrow f$, and (*ii*) $I(f_n)$ is bounded. Now, since $I(f_n)$ is a bounded increasing sequence, it converges to its least upper bound. In view of Sect. 23, it is natural to define the "integral" of f to be LUB $I(f_n)$. We must verify that this LUB depends only on f, and not on the particular sequence f_n which has been chosen to exhibit the integrability of f. This is accomplished in the next three lemmas.

Lemma 1. *If f_n is a sequence of* ISF *such that $f_n \downarrow 0$, then $I(f_n) \downarrow 0$.*

Proof. Let $F = N(f_1)$, and $M = \max\{f_1(x): x \in X\}$. Evidently $f_n \le M\chi_F$ for all n.

Given any $\varepsilon > 0$, define $E_n = \{x: f_n(x) \ge \varepsilon\}$. We assert that $\mu(E_n) \to 0$. In any case it is clear that $E_n \subset F$, and $E_n \downarrow \varnothing$ results from the assumption that $f_n(x) \downarrow 0$ for each x. Since the E_n have finite measure, it follows from 4.2 that $\mu(E_n) \downarrow 0$.

Since $N(f_n) \subset N(f_1) = F$, we have

$$f_n = \chi_F f_n = \chi_{F-E_n} f_n + \chi_{E_n} f_n.$$

Since $f_n(x) < \varepsilon$ on $X - E_n$, and $f_n \le M\chi_F$, it follows that

$$\chi_{F-E_n} f_n \le \varepsilon \chi_{F-E_n} \le \varepsilon \chi_F$$

and

$$\chi_{E_n} f_n \le M\chi_{E_n},$$

hence

$$f_n \le \varepsilon\chi_F + M\chi_{E_n};$$

then

$$I(f_n) \le \varepsilon\mu(F) + M\mu(E_n)$$

by 22.2 and 22.3. Since $\mu(F)$ is a finite number independent of ε, and $\mu(E_n) \to 0$, it is clear that $I(f_n) \to 0$. ∎

Lemma 2. *If $f_n \uparrow f$, where f and the f_n $(n = 1, 2, 3, \ldots)$ are* ISF, *then $I(f_n) \uparrow I(f)$.*

Proof. Since $f - f_n \downarrow 0$, we have $I(f - f_n) \downarrow 0$ by Lemma 1. ∎

Lemma 3. *Suppose $0 \le f_n \uparrow f$ and $0 \le g_m \uparrow f$, where f is measurable, the f_n and g_m are sequences of* ISF, *and $I(f_n)$ is bounded. Then $I(g_m)$ is also bounded, and*

$$\text{LUB}_n\, I(f_n) = \text{LUB}_m\, I(g_m).$$

Proof. Let $M = \text{LUB}_n\, I(f_n)$; we are assuming $M < \infty$. For each m,

$$f_n \cap g_m \uparrow f \cap g_m = g_m$$

as $n \to \infty$, hence

$$I(f_n \cap g_m) \uparrow I(g_m)$$

by Lemma 2, that is,

$$I(g_m) = \text{LUB}_n\, I(f_n \cap g_m).$$

Since $I(f_n \cap g_m) \leq I(f_n) \leq M$ for all n, it follows that $I(g_m) \leq M$. Thus, $I(g_m)$ is also bounded.

Interchanging roles, we have $I(f_n) = \text{LUB}_m\, I(f_n \cap g_m)$, hence by the associativity of LUB's,

$$\begin{aligned}\text{LUB}_n\, I(f_n) &= \text{LUB}_n\, \text{LUB}_m\, I(f_n \cap g_m) \\ &= \text{LUB}_m\, \text{LUB}_n\, I(f_n \cap g_m) = \text{LUB}_m\, I(g_m). \quad \blacksquare\end{aligned}$$

Suppose now that f is any nonnegative integrable function. Choose any sequence f_n of ISF such that $0 \leq f_n \uparrow f$; by Lemma 3, $I(f_n)$ is bounded, and the LUB is independent of the particular sequence f_n, hence we may unambiguously define the **integral** of f to be the non-negative real number

$$\text{LUB}_n\, I(f_n).$$

The integral of f will be denoted

$$\int f\, d\mu.$$

If in particular f is a nonnegative ISF, we may take $f_n = f$ for all n, hence

$$\int f\, d\mu = I(f);$$

that is, the two definitions of "integral" for such an f coincide. Nevertheless, we shall continue to use the notation $I(f)$ for an ISF f whenever it seems to contribute to clarity.

The next theorem, though simple in appearance and proof, is remarkably fertile in applications:

Theorem 1. *If $0 \leq f \leq g$, where g is integrable and f is measurable, then f is also integrable, and*

$$\int f\, d\mu \leq \int g\, d\mu.$$

Proof. Let g_n be a sequence of ISF such that $0 \leq g_n \uparrow g$ and $I(g_n)$ is bounded. Say $I(g_n) \leq M < \infty$ for all n. Let f_n be any sequence of simple functions such that $0 \leq f_n \uparrow f$ (16.4). Then $h_n = f_n \cap g_n$ is simple, and $0 \leq h_n \leq g_n$ shows that $N(h_n) \subset N(g_n)$; it follows that h_n is an ISF, and $I(h_n) \leq I(g_n) \leq M$ for all n. Since

$$0 \leq h_n = f_n \cap g_n \uparrow f \cap g = f,$$

we conclude that f is integrable, and

$$\int f\,d\mu = \text{LUB}\,I(h_n) \le \text{LUB}\,I(g_n) = \int g\,d\mu. \quad \blacksquare$$

Corollary. *If $f \ge 0$ is integrable, and $g \ge 0$ is measurable, then $f \cap g$ is integrable.*

Proof. $0 \le f \cap g \le f$, and $f \cap g$ is measurable by 13.5. ∎

Theorem 2. *If $f \ge 0$ is integrable, then for each $\varepsilon > 0$, the set*

$$\{x : f(x) \ge \varepsilon\}$$

has finite measure. In particular, $N(f)$ is the union of a sequence of measurable sets of finite measure.

Proof. After shrinking ε, it is sufficient to prove that the set

$$\{x : f(x) > \varepsilon\}$$

has finite measure.

Let f_n be a sequence of ISF such that $0 \le f_n \uparrow f$ and $0 \le I(f_n) \le M < \infty$ for all n, and define

$$E_n = \{x : f_n(x) > \varepsilon\}, \qquad E = \{x : f(x) > \varepsilon\}.$$

Clearly $E_n \uparrow E$, hence $\mu(E_n) \uparrow \mu(E)$. For each n,

$$\varepsilon\chi_{E_n} \le \chi_{E_n} f_n \le f_n,$$

hence $\varepsilon\mu(E_n) \le I(f_n) \le M$. Thus, $\mu(E_n) \le M/\varepsilon$ for all n, hence $\mu(E) \le M/\varepsilon$.

The second assertion then follows from the relation

$$N(f) = \bigcup_{m=1}^{\infty} \{x : f(x) \ge 1/m\}. \quad \blacksquare$$

The following corollary clears up a possible confusion in our terminology:

Corollary 1. *If $f \ge 0$ is a simple function, then f is integrable if and only if f is an* ISF.

Proof. If f is integrable in the sense of the present section, then it is clear from Theorem 2 (and the fact that f has only finitely many values) that $N(f)$ has finite measure, and hence f is an ISF. The converse was noted earlier in the section. ∎

Corollary 2. *If $0 \le f \le g$, where g is integrable and f is simple, then f is an* ISF.

Proof. This is trivial if $f = 0$. Otherwise, let ε be the minimum nonzero value of f. Our assertion follows from the relation

$$N(f) = \{x: f(x) \geq \varepsilon\} \subset \{x: g(x) \geq \varepsilon\},$$

and the fact that the latter set has finite measure by Theorem 2. ▮

Theorem 3. *If $f \geq 0$ and $g \geq 0$ are integrable, $c \geq 0$ is a real number, and A is a locally measurable set, then:*

(1) *cf is integrable, and $\int cf\, d\mu = c \int f\, d\mu$.*

(2) *$f + g$ is integrable, and*

$$\int (f + g)\, d\mu = \int f\, d\mu + \int g\, d\mu.$$

(3) *$\chi_A f$ is integrable.*

Proof. Let f_n and g_n be sequences of ISF such that $0 \leq f_n \uparrow f$, $0 \leq g_n \uparrow g$, and $I(f_n)$, $I(g_n)$ are bounded.

(1) The cf_n are ISF, $0 \leq cf_n \uparrow cf$, and $I(cf_n) = cI(f_n)$ is bounded, hence cf is integrable and

$$\int cf\, d\mu = \text{LUB}\, I(cf_n) = c\, \text{LUB}\, I(f_n) = c \int f\, d\mu.$$

(2) The $f_n + g_n$ are ISF, $0 \leq f_n + g_n \uparrow f + g$, and

$$I(f_n + g_n) = I(f_n) + I(g_n)$$

is bounded, hence $f + g$ is integrable and

$$\int (f + g)\, d\mu = \text{LUB}\, I(f_n + g_n) = \lim I(f_n + g_n)$$
$$= \lim I(f_n) + \lim I(g_n) = \int f\, d\mu + \int g\, d\mu.$$

(3) The $\chi_A f_n$ are ISF, $0 \leq \chi_A f_n \uparrow \chi_A f$, and $I(\chi_A f_n) \leq I(f_n)$ is bounded. Incidentally, $0 \leq \chi_A f \leq f$; since $\chi_A f$ is measurable by 15.1, another proof results from Theorem 1. ▮

EXERCISES

1. If $f \geq 0$ and $g \geq 0$ are integrable, then $f \cup g$ is integrable.

2. If $g \geq 0$ is integrable, then

$$\int g\, d\mu = \text{LUB} \int f\, d\mu,$$

where f varies over the class of all simple functions such that $0 \leq f \leq g$.

3. Let f be a nonnegative measurable function. The *lower integral* of f, denoted $L(f)$, is defined by the formula

$$L(f) = \text{LUB}\,\{I(g): 0 \le g \le f,\ g \text{ ISF}\}.$$

(*i*) If $L(f) < \infty$, there exists an integrable function g such that $0 \le g \le f$ and $L(f - g) = 0$.

(*ii*) If f is integrable, then $L(f) < \infty$, and

$$\int f\,d\mu = L(f).$$

(*iii*) If μ is *semifinite* (cf. Exercise 25.9) and $L(f) = 0$, then $f = 0$ a.e.

(*iv*) If μ is semifinite, then f is integrable if and only if $L(f) < \infty$.

25. Integrable Functions

Let us write $\mathscr{P}^1(\mu)$, or briefly $\mathscr{P}^1$, for the class of all measurable functions f such that $f \ge 0$ and f is integrable in the sense of the preceding section.

A measurable function f is said to be **integrable** (with respect to μ) if there exist functions g and h in $\mathscr{P}^1$ such that $f = g - h$; briefly, f is **μ-integrable.** Suppose also $f = g' - h'$, where g' and h' are in $\mathscr{P}^1$. Then $g - h = g' - h'$, $g + h' = g' + h$, hence

$$\int g\,d\mu + \int h'\,d\mu = \int g'\,d\mu + \int h\,d\mu$$

by 24.3. Thus,

$$\int g\,d\mu - \int h\,d\mu = \int g'\,d\mu - \int h'\,d\mu,$$

and we may unambiguously define

$$\int g\,d\mu - \int h\,d\mu$$

to be the **integral** of f (with respect to μ); the integral of f will be denoted

$$\int f\,d\mu.$$

If in particular f is in $\mathscr{P}^1$, then $f = f - 0$ shows that the integral of f defined above coincides with the integral of f as defined in Sect. 24. It follows that

$$\int f\,d\mu = I(f)$$

for every ISF f.

We shall write $\mathscr{L}^1(\mu)$, or briefly $\mathscr{L}^1$, for the class of all functions which are integrable (with respect to μ). The nonnegative functions in $\mathscr{L}^1$ are precisely the functions in $\mathscr{P}^1$:

Theorem 1. *If f is integrable, and $f \geq 0$, then f is in $\mathscr{P}^1$. If, moreover, g is a measurable function such that $0 \leq g \leq f$, then g is also integrable, and*

$$\int g\,d\mu \leq \int f\,d\mu.$$

Proof. Suppose $f = h - k$, where h and k are in $\mathscr{P}^1$. Then $0 \leq f \leq h$, hence f is in $\mathscr{P}^1$ by 24.1. The second assertion is then simply 24.1. ▌

The analogue of 24.3 holds also for $\mathscr{L}^1$:

Theorem 2. *If f and g are integrable functions, c is a real number, and A is a locally measurable set, then:*

(1) *cf is integrable, and $\int cf\,d\mu = c\int f\,d\mu$.*

(2) *$f + g$ is integrable, and*

$$\int (f + g)\,d\mu = \int f\,d\mu + \int g\,d\mu.$$

(3) *$\chi_A f$ is integrable.*

Proof. Suppose $f = f_1 - f_2$ and $g = g_1 - g_2$, where the f_i and g_i are in $\mathscr{P}^1$.

(1) If $c \geq 0$, then $cf = cf_1 - cf_2$ shows (see 24.3) that cf is integrable, and

$$\int cf\,d\mu = \int cf_1\,d\mu - \int cf_2\,d\mu = c\int f_1\,d\mu - c\int f_2\,d\mu = c\int f\,d\mu.$$

If $c \leq 0$, consider $cf = (-c)f_2 - (-c)f_1$.

(2) The relation $f + g = (f_1 + g_1) - (f_2 + g_2)$ shows (see 24.3) that $f + g$ is integrable, and

$$\begin{aligned}\int (f + g)\,d\mu &= \int (f_1 + g_1)\,d\mu - \int (f_2 + g_2)\,d\mu\\ &= \int f_1\,d\mu + \int g_1\,d\mu - \int f_2\,d\mu - \int g_2\,d\mu\\ &= \int f\,d\mu + \int g\,d\mu.\end{aligned}$$

(3) The relation $\chi_A f = \chi_A f_1 - \chi_A f_2$ shows (see 24.3) that $\chi_A f$ is integrable. ▌

For a measurable function f, the integrability of f is equivalent to the integrability of $|f|$:

Theorem 3. *If f is a measurable function, the following conditions are equivalent:*

(a) *f is integrable.*

(b) $|f|$ *is integrable.*

(c) f^+ *and* f^- *are integrable.*

In this case

$$\left|\int f\,d\mu\right| \le \int |f|\,d\mu.$$

Proof. (a) *implies* (b): Write $f = g - h$, with g and h in $\mathscr{P}^1$. Since $0 \le |f| \le g + h$, and $g + h$ is integrable, it follows from Theorem 1 that $|f|$ is integrable.

(b) *implies* (c): Suppose $|f|$ is integrable. Since f^+ and f^- are measurable by 13.7, and $0 \le f^+ \le |f|$, $0 \le f^- \le |f|$, it follows from Theorem 1 that f^+ and f^- are integrable.

(c) *implies* (a): If f^+ and f^- are integrable, then so is $f = f^+ - f^-$ by Theorem 2, and

$$\left|\int f\,d\mu\right| = \left|\int f^+\,d\mu - \int f^-\,d\mu\right| \le \int f^+\,d\mu + \int f^-\,d\mu = \int |f|\,d\mu. \quad \blacksquare$$

In view of the formulas

$$f \cup g = (f + g + |f - g|)/2,$$

$$f \cap g = (f + g - |f - g|)/2,$$

Theorems 2 and 3 yield the following:

Theorem 4. *If f and g are integrable, then so are $f \cup g$ and $f \cap g$.*

Integrability, and the value of the integral, are not disturbed by measurable alterations on null sets:

Theorem 5. *If f is integrable, g is measurable, and $f = g$ a.e., then g is also integrable, and*

$$\int f\,d\mu = \int g\,d\mu.$$

Proof. Let $h = f - g$; h is measurable, and $h = 0$ a.e. Writing $h = h^+ - h^-$, we have $h^+ = 0$ a.e. and $h^- = 0$ a.e. Moreover, $h^+ \ge 0$ and $h^- \ge 0$.

We assert that h^+ is integrable, with integral 0. Indeed, let h_n be a sequence of simple functions such that $0 \le h_n \uparrow h^+$. Since $h^+ = 0$ a.e., and since $N(h_n) \subset N(h^+)$, it follows that $\mu(N(h_n)) = 0$. Thus, h_n is an ISF, and $h_n = 0$ a.e., hence $I(h_n) = 0$ by 22.3. In particular $I(h_n)$ is bounded, hence by the definition in Sect. 24, h^+ is integrable, and

$$\int h^+\,d\mu = \text{LUB}\, I(h_n) = 0.$$

Similarly h^- is integrable, and $\int h^- \, d\mu = 0$. Then $h = h^+ - h^-$ is integrable, and $\int h \, d\mu = 0 - 0$.

Thus, $f - g$ is integrable, and $\int (f - g) \, d\mu = 0$. Since $g = f - (f - g)$, and f is also integrable, it follows from Theorem 2 that g is integrable, and

$$\int g \, d\mu = \int f \, d\mu - \int (f - g) \, d\mu = \int f \, d\mu - 0. \quad \blacksquare$$

The process of integration is monotone, that is, it preserves order:

Theorem 6. *If f and g are integrable, and $f \leq g$ a.e., then*

$$\int f \, d\mu \leq \int g \, d\mu.$$

Proof. Let $h = g - f$; then h is integrable by Theorem 2, and $h \geq 0$ a.e. Let E be a null set on whose complement $h(x) \geq 0$, and define $k = \chi_{X-E} h$. Then k is integrable by Theorem 2, and $h = k$ a.e., hence

$$\int h \, d\mu = \int k \, d\mu$$

by Theorem 5. Since $k \geq 0$, k is in $\mathscr{P}^1$ by Theorem 1, hence $\int k \, d\mu \geq 0$ by the definition of integral in Sect. 24. Thus,

$$0 \leq \int k \, d\mu = \int h \, d\mu = \int (g - f) \, d\mu = \int g \, d\mu - \int f \, d\mu. \quad \blacksquare$$

A measurable function which is "absolutely dominated" by an integrable function is necessarily integrable:

Theorem 7. *If f is measurable, g is integrable, and $|f| \leq |g|$ a.e., then f is integrable.*

Proof. Let E be a null set on whose complement $|f(x)| \leq |g(x)|$, and define $f' = \chi_{X-E} f$. Then f' is measurable by 15.1, and $|f'| \leq |g|$. Since $|g|$ is integrable by Theorem 3, and $|f'|$ is measurable by 13.4, it follows from Theorem 1 that $|f'|$ is integrable. Then f' is integrable by Theorem 3, and $f = f'$ a.e., hence f is integrable by Theorem 5. $\blacksquare$

Theorem 8. *If f is integrable, then for each $\varepsilon > 0$, the set*

$$\{x \colon |f(x)| \geq \varepsilon\}$$

has finite measure. In particular, $N(f)$ is the union of a sequence of measurable sets of finite measure; indeed, there exists an increasing sequence of measurable sets F_n such that $F_n \uparrow N(f)$, $\mu(F_n) < \infty$, and $\chi_{F_n} f$ is bounded for each n.

Proof. Since $|f|$ is integrable by Theorem 3, and belongs to $\mathscr{P}^1$ by Theorem 1, our first assertion is immediate from 24.2. In particular, for each positive integer n, the set

$$F_n = \{x: n \geq |f(x)| \geq 1/n\}$$

has finite measure, $F_n \uparrow N(f)$, and

$$n^{-1}\chi_{F_n} \leq |f|\chi_{F_n} \leq n\chi_{F_n}. \quad \blacksquare$$

The process of integration is "strictly positive":

Theorem 9. *If f is integrable, $f \geq 0$ a.e., and $\int f\,d\mu = 0$, then $f = 0$ a.e.*

Proof. Defining $E_n = \{x: |f(x)| \geq 1/n\}$, we have

$$N(f) = \bigcup_1^\infty E_n,$$

and $\mu(E_n) < \infty$ by Theorem 8. Since

$$n^{-1}\chi_{E_n} \leq |f|\chi_{E_n} \leq |f| = f \text{ a.e.},$$

Theorem 6 yields

$$n^{-1}\mu(E_n) \leq \int f\,d\mu = 0,$$

hence $\mu(E_n) = 0$ for all n. It follows that $N(f)$ is a null set, and so $f = 0$ a.e. $\blacksquare$

If f is an integrable function, and A is a locally measurable set, the **integral** of f **over** A is defined to be

$$\int \chi_A f\,d\mu;$$

this is permissible by Theorem 2. The integral of f over A will be denoted

$$\int_A f\,d\mu.$$

It is merely a matter of notational flexibility to allow A to be *locally* measurable; for, if $E = N(f)$, then

$$\chi_A f = \chi_A \chi_E f = \chi_{A\cap E} f,$$

and so

$$\int_A f\,d\mu = \int_{A\cap E} f\,d\mu.$$

Thus, since $A \cap E$ is measurable, no essential generality would be lost by integrating f only over measurable sets.

Theorem 10. *If f is integrable, E is a measurable set such that $f(x) > 0$ a.e on E, and $\int_E f\,d\mu = 0$, then $\mu(E) = 0$.*

Proof. We are assuming (see Sect. 18) that $\chi_E f \geq 0$ a.e., and $\int \chi_E f \, d\mu = 0$, hence $\chi_E f = 0$ a.e. by Theorem 9. Since $f(x) > 0$ a.e. on E, it follows that $\mu(E) = 0$. ▌

The conclusion of Theorem 10 holds also if one alters the hypothesis to read $f(x) < 0$ a.e. on E (consider $-f$).

An integrable function whose integral over every measurable set is zero must vanish almost everywhere:

Theorem 11. *If f is an integrable function such that $\int_E f \, d\mu = 0$ for every measurable set E, then $f = 0$ a.e.*

Proof. Define

$$E = \{x : f(x) > 0\} \quad \text{and} \quad F = \{x : f(x) < 0\} = \{x : -f(x) > 0\}.$$

Since $\int_E f \, d\mu = 0$, and $f(x) > 0$ on E, it follows from Theorem 10 that $\mu(E) = 0$. Similarly $\mu(F) = 0$. Thus, $N(f) = E \cup F$ is a null set, and so $f = 0$ a.e. ▌

EXERCISES

1. Let $(X, \mathscr{S})$ be a measurable space, let μ_i be an increasingly directed family of measures on $\mathscr{S}$, and define $\mu = \text{LUB } \mu_i$ as in Sect. 10. Then, a measurable function f is μ-integrable if and only if it is μ_i-integrable for all i and $\int |f| \, d\mu_i$ is bounded. In this case if $f \geq 0$, then

$$\int f \, d\mu = \text{LUB}_i \int f \, d\mu_i.$$

2. If f is integrable, a and b are real numbers, and E is a measurable set such that $a \leq f(x) \leq b$ a.e. on E, then

$$a\mu(E) \leq \int_E f \, d\mu \leq b\mu(E).$$

3. If f is integrable, and g is an essentially bounded measurable function, then fg is integrable.

4. In a finite measure space, if f is any measurable function, then the function

$$f/(1 + |f|)$$

is integrable.

5. A complex-valued function f on X is said to be *integrable* if its real and imaginary parts are integrable; if $f = g + ih$, one then defines

$$\int f \, d\mu = \int g \, d\mu + i \int h \, d\mu.$$

If f is a complex-valued measurable function (Exercise 12.1), then f is integrable if and only if $|f|$ is integrable, and in this case

$$\left| \int f \, d\mu \right| \leq \int |f| \, d\mu.$$

6. If g, f, and f_n $(n = 1, 2, 3, \cdots)$ are integrable functions such that $|f_n| \leq |g|$ a.e. and $f_n \to f$ a.e., then $f_n \to f$ a.u.

7. If f: $[a, b] \to R$ is integrable in the sense of Riemann, then the function g, defined by

$$g(x) = \begin{cases} f(x) & \text{for } a \leq x \leq b, \\ 0 & \text{otherwise,} \end{cases}$$

is integrable with respect to Lebesgue measure μ, and

$$\int g \, d\mu = \int_a^b f(x) \, dx.$$

8. Let E be a measurable set, and μ_E the contraction of μ by E. In order that a measurable function f be μ_E-integrable, it is necessary and sufficient that $\chi_E f$ be μ-integrable, and in this case

$$\int f \, d\mu_E = \int \chi_E f \, d\mu.$$

9. Let us say that μ is **semifinite** in case: for each measurable set E, $\mu(E)$ is the least upper bound of $\mu(F)$, as F varies over the class of all measurable subsets of E of finite measure; that is,

$$\mu(E) = \text{LUB} \{\mu(F) : F \subset E, F \in \mathscr{S}_\varphi\}.$$

We then speak of $(X, \mathscr{S}, \mu)$ as a **semifinite measure space**. An equivalent formulation of semifiniteness is

$$\mu(E) = \text{LUB} \{\mu(F \cap E) : F \in \mathscr{S}_\varphi\}$$

for all measurable sets E; that is,

$$\mu = \text{LUB} \{\mu_F : F \in \mathscr{S}_\varphi\}$$

in the sense of Sect 10. Every σ-finite measure is semifinite.

10. If μ is semifinite, then a measurable function f is integrable if and only if $\chi_F f$ is integrable for each F in $\mathscr{S}_\varphi$, and

$$\text{LUB} \left\{ \int \chi_F |f| \, d\mu : F \in \mathscr{S}_\varphi \right\} < \infty.$$

See also Exercise 27.2.

11. If μ is semifinite, then so is the measure μ_λ constructed in Exercise 17.1.

12. Let $(X, \mathscr{S}_\lambda, \mu_\lambda)$ be the measure space constructed in Exercise 17.1. If f is measurable (with respect to $\mathscr{S}$), then f is μ_λ-integrable if and only if it is μ-integrable, and in this case

$$\int f \, d\mu_\lambda = \int f \, d\mu.$$

13. Let $(X, \mathscr{S}_\lambda, \mu_\lambda)$ be the measure space constructed in Exercise 17.1. If g is measurable with respect to $\mathscr{S}_\lambda$, then g is μ_λ-integrable if and only if $\chi_E g$ is μ-integrable for each E in $\mathscr{S}$, and

$$\text{LUB} \left\{ \int \chi_E |g| \, d\mu : E \in \mathscr{S} \right\} < \infty.$$

26. Indefinite Integrals

If f is an integrable function, the **indefinite integral** of f is the real-valued set function μ_f defined on $\mathscr{S}$ by the formula

$$\mu_f(E) = \int_E f \, d\mu$$

for every measurable set E. (Though it is clear that the domain of μ_f could be expanded to the class $\mathscr{S}_\lambda$ of locally measurable sets, we shall have no occasion to do so.) If f and g are integrable functions, c is a real number, and E is a measurable set, it is clear from 25.2 that

$$\mu_{f+g}(E) = \mu_f(E) + \mu_g(E)$$

and

$$\mu_{cf}(E) = c\mu_f(E);$$

briefly, $\mu_{f+g} = \mu_f + \mu_g$ and $\mu_{cf} = c\mu_f$, as real-valued functions on $\mathscr{S}$.

In this section we develop the simplest properties of indefinite integrals. Absolute continuity, perhaps the deepest property, is deferred until Sect. 43.

Theorem 1. *Let f and g be integrable functions. Then:*

(i) $\mu_f = \mu_g$ *if and only if* $f = g$ *a.e.* $[\mu]$.

(ii) $\mu_f \geq 0$ *if and only if* $f \geq 0$ *a.e.* $[\mu]$.

Proof. (*i*) Let $h = f - g$. Then $\mu_h = \mu_f - \mu_g$, hence the condition $\mu_f = \mu_g$ is equivalent to $h = 0$ a.e. by 25.11 and 25.5.

(*ii*) Suppose $\mu_f \geq 0$, that is, $\mu_f(E) \geq 0$ for every measurable set E. In particular for the set

$$F = \{x : f(x) < 0\} = \{x : -f(x) > 0\},$$

we have $\mu_f(F) \geq 0$; but $\chi_F f \leq 0$, hence

$$\mu_f(F) = \int \chi_F f \, d\mu \leq 0$$

by 25.6. Thus $\int_F f \, d\mu = 0$, hence $\int_F (-f) \, d\mu = 0$; since $-f(x) > 0$ on F, it follows from 25.10 that $\mu(F) = 0$, and so $f \geq 0$ a.e.

If conversely $f \geq 0$ a.e., then $\chi_E f \geq 0$ a.e. for every measurable set E, hence

$$\mu_f(E) = \int \chi_E f \, d\mu \geq 0$$

by 25.6. ∎

Our next theorem asserts that μ_f is a countably additive set function on $\mathscr{S}$, for every integrable function f. We begin with an easy special case:

Lemma. *If f is an* ISF, *and $f \geq 0$, then μ_f is a (finite) measure on $\mathscr{S}$.*

Proof. Write

$$f = \sum_1^n a_i \chi_{E_i},$$

where the a_i are real numbers >0, and the E_i are measurable sets of finite measure (see Sect. 22). If E is any measurable set, then

$$\chi_E f = \sum_1^n a_i \chi_{E_i \cap E},$$

hence

$$\mu_f(E) = \sum_1^n a_i \mu(E_i \cap E) = \sum_1^n a_i \mu_{E_i}(E),$$

where μ_{E_i} is the contraction of μ by E_i (see Sect. 17). Since each μ_{E_i} is a finite measure, the a_i are >0, and

$$\mu_f = \sum_1^n a_i \mu_{E_i},$$

it is clear that μ_f is also a finite measure. ∎

Theorem 2. *If f is any integrable function, then μ_f is a countably additive set function on $\mathscr{S}$. Moreover, μ_f is a (finite) measure if and only if $f \geq 0$ a.e. $[\mu]$.*

Proof. Writing $f = g - h$, with g and h in $\mathscr{P}^1$, we have $\mu_f = \mu_g - \mu_h$. Changing notation, it clearly suffices to show that μ_f is countably additive when f is in $\mathscr{P}^1$. Assuming f is in $\mathscr{P}^1$, choose any sequence f_n of ISF such that $0 \leq f_n \uparrow f$. For any measurable set E, we have

$$0 \leq \chi_E f_n \uparrow \chi_E f,$$

hence by the definition of integral given in Sect. 24,

$$\int \chi_E f \, d\mu = \text{LUB}\, I(\chi_E f_n).$$

That is,

$$\mu_f(E) = \text{LUB}\, \mu_{f_n}(E)$$

for every measurable set E; briefly,

$$\mu_f = \text{LUB}\, \mu_{f_n}.$$

Since each μ_{f_n} is a measure by the lemma, it follows from 10.1 that μ_f is a measure on $\mathscr{S}$, hence μ_f is countably additive by 3.1. The last assertion is then immediate from Theorem 1. ∎

EXERCISES

1. Suppose $(X, \mathscr{S}, \mu)$ is a σ-finite measure space, and f is an essentially bounded measurable function, $f \geq 0$ a.e. The formula

$$\nu(P) = \int \chi_P f \, d\mu$$

defines a finite measure ν on the ring $\mathscr{R}$ of all measurable sets P such that $\mu(P) < \infty$. It follows that there exists a unique measure $\bar{\nu}$ on $\mathscr{S}$ which extends ν. It is natural to write $\int_E f \, d\mu$ in place of $\bar{\nu}(E)$, for every E in $\mathscr{S}$.

2. If $(X, \mathscr{S}, \mu)$ is any measure space, $\mathscr{R}$ is a ring such that $\mathfrak{S}(\mathscr{R}) = \mathscr{S}$, and f is an integrable function such that $\int_E f \, d\mu = 0$ for all E in $\mathscr{R}$, then $f = 0$ a.e.

3. If m is the Borel restriction of Lebesgue measure, and f is a function in $\mathscr{L}^1(m)$ such that $m_f([a, b)) = 0$ for all semiclosed intervals $[a, b)$, then $f = 0$ a.e. $[m]$.

4. If $(X, \mathscr{S}, \mu)$ is any measure space, $\mathscr{R}$ is a ring such that $\mathfrak{S}(\mathscr{R}) = \mathscr{S}$, and f is an integrable function such that $\int_E f \, d\mu \geq 0$ for all E in $\mathscr{R}$, then $f \geq 0$ a.e.

5. Suppose g, f, and f_n $(n = 1, 2, 3, \ldots)$ are integrable functions such that $|f_n| \leq |g|$ a.e. and $f_n \to f$ a.e. One knows from Exercise 25.6 that $f_n \to f$ a.u. and hence $f_n \to f$ in $\textcircled{m}$. Show directly that $f_n \to f$ in $\textcircled{m}$, by applying 20.1 to the indefinite integral of $|g|$.

27. The Monotone Convergence Theorem

Of the various classical "convergence theorems," the monotone convergence theorem lies nearest in spirit and proof to our definitions. We shall arrange the proof in a series of lemmas each of which is itself a special case of the monotone convergence theorem.

Lemma 1. *If f_n is a sequence of integrable functions such that $f_n \downarrow 0$, and if f_1 is bounded and $N(f_1)$ has finite measure, then $\int f_n \, d\mu \downarrow 0$.*

Proof. Let $F = N(f_1)$, and $M = \text{LUB}\,\{f_1(x) \colon x \in X\}$. Clearly $0 \leq f_n \leq M\chi_F$ for all n. The proof proceeds exactly as in Lemma 1 to 24.1. ∎

Lemma 2. *If $f_n \uparrow f$, where f and the f_n $(n = 1, 2, 3, \ldots)$ are integrable, then*

$$\int f_n \, d\mu \uparrow \int f \, d\mu.$$

Proof. Setting $g_n = f - f_n$, we have $g_n \downarrow 0$, and it will suffice to show that $\int g_n \, d\mu \downarrow 0$. Let $E = N(g_1)$. By 25.8, there exists a sequence of measurable sets G_k such that $G_k \uparrow E$, $\mu(G_k) < \infty$, and $\chi_{G_k} g_1$ is bounded.

Given any $\varepsilon > 0$; since μ_{g_1} is a finite measure by 26.2, the relation $E - G_k \downarrow \varnothing$ implies $\mu_{g_1}(E - G_k) \downarrow 0$, hence we may choose an index k_0 such that $\mu_{g_1}(E - G_{k_0}) \leq \varepsilon$. Writing $G = G_{k_0}$, we have the following properties: G is a measurable set, $\mu(G) < \infty$, $\mu_{g_1}(E - G) \leq \varepsilon$, and $\chi_G g_1$ is bounded.

Now

$$\begin{aligned}\int g_n\, d\mu &= \int_E g_n\, d\mu \\ &= \int_{E-G} g_n\, d\mu + \int_G g_n\, d\mu \leq \int_{E-G} g_1\, d\mu + \int_G g_n\, d\mu \\ &= \mu_{g_1}(E - G) + \int_G g_n\, d\mu \leq \varepsilon + \int_G g_n\, d\mu.\end{aligned}$$

Since $\chi_G\, g_n \downarrow 0$, it follows from Lemma 1 that $\int \chi_G\, g_n\, d\mu \downarrow 0$, hence there is an index n_0 such that

$$\int_G g_n\, d\mu \leq \varepsilon$$

whenever $n \geq n_0$; then

$$\int g_n\, d\mu \leq \varepsilon + \varepsilon$$

whenever $n \geq n_0$. ∎

Lemma 3. *If $0 \leq f_n \uparrow g$, where g is simple, the f_n $(n = 1, 2, 3, \ldots)$ are integrable, and $\int f_n\, d\mu$ is bounded, then g is integrable and*

$$\int f_n\, d\mu \uparrow \int g\, d\mu.$$

Proof. We may suppose $g \neq 0$. Given any $\varepsilon > 0$, consider the sets

$$E = \{x: g(x) > \varepsilon\} \qquad \text{and} \qquad E_n = \{x: f_n(x) > \varepsilon\}.$$

Evidently $E_n \uparrow E$, hence $\mu(E_n) \uparrow \mu(E)$. Since f_n is integrable, $\mu(E_n) < \infty$ by 25.8. Suppose

$$\int f_n\, d\mu \leq M < \infty$$

for all n. Since

$$\varepsilon\chi_{E_n} \leq \chi_{E_n} f_n \leq f_n,$$

we have $\varepsilon\mu(E_n) \leq \int f_n\, d\mu \leq M$, hence $\mu(E_n) \leq M/\varepsilon$; taking LUB over n, we have $\mu(E) \leq M/\varepsilon$. Thus, $\mu(E) < \infty$. If in particular ε is chosen so that $0 < \varepsilon < m$, where m is the minimum nonzero value of g, then $E = N(g)$; thus $\mu(N(g)) < \infty$, g is an ISF, and

$$\int f_n\, d\mu \uparrow \int g\, d\mu$$

follows from Lemma 2. ∎

Lemma 4. *If* $0 \le f_n \uparrow f$, *where the* f_n $(n = 1, 2, 3, \ldots)$ *are integrable,* f *is measurable, and* $\int f_n \, d\mu$ *is bounded, then* f *is integrable and*

$$\int f_n \, d\mu \uparrow \int f \, d\mu.$$

Proof. Say $\int f_n \, d\mu \le M < \infty$ for all n. Let g_m be a sequence of simple functions such that $0 \le g_m \uparrow f$. Fixing m, we have

$$0 \le f_n \cap g_m \uparrow f \cap g_m = g_m$$

as $n \to \infty$. Since $f_n \cap g_m$ is integrable by the corollary of 24.1, and $f_n \cap g_m \le f_n$, we have

$$\int f_n \cap g_m \, d\mu \le \int f_n \, d\mu \le M$$

for all n; it follows from Lemma 3 that g_m is integrable, and

$$\int g_m \, d\mu = \text{LUB}_n \int f_n \cap g_m \, d\mu \le M.$$

Thus, $0 \le g_m \uparrow f$, where g_m is an ISF by Corollary 1 of 24.2, and $I(g_m) \le M < \infty$ for all m; it follows from the definition given in Sect. 24 that f is integrable, hence

$$\int f_n \, d\mu \uparrow \int f \, d\mu$$

by Lemma 2. ∎

Lemma 5. *If* $f_n \uparrow f$ *a.e., where the* f_n $(n = 1, 2, 3, \ldots)$ *are integrable,* f *is measurable, and* $\int f_n \, d\mu$ *is bounded, then* f *is integrable and*

$$\int f_n \, d\mu \uparrow \int f \, d\mu.$$

Proof. Let E be a null set on whose complement $f_n(x) \uparrow f(x)$, and define

$$f_n' = \chi_{X-E} f_n \qquad \text{and} \qquad f' = \chi_{X-E} f.$$

Then $f_n' = f_n$ a.e., $f' = f$ a.e., and $f_n' \uparrow f'$; in view of 25.5, we may assume after a change of notation that $f_n \uparrow f$.

Define $g_n = f_n - f_1$; then g_n is integrable, and $0 \le g_n \uparrow f - f_1$, where $f - f_1$ is measurable. Moreover,

$$\int g_n \, d\mu = \int f_n \, d\mu - \int f_1 \, d\mu$$

is bounded; it follows from Lemma 4 that $f - f_1$ is integrable, and

$$\int g_n \, d\mu \uparrow \int (f - f_1) \, d\mu,$$

that is,

$$\int (f_n - f_1)\, d\mu \uparrow \int (f - f_1)\, d\mu.$$

Then $f = (f - f_1) + f_1$ is also integrable, and

$$\int f_n\, d\mu - \int f_1\, d\mu \uparrow \int f\, d\mu - \int f_1\, d\mu,$$

hence

$$\int f_n\, d\mu \uparrow \int f\, d\mu. \quad \blacksquare$$

Lemma 6. *If $0 \le f_n\uparrow$, where the f_n $(n = 1, 2, 3, \ldots)$ are integrable and $\int f_n\, d\mu$ is bounded, then there exists an integrable function f such that $f_n \uparrow f$ a.e. Necessarily*

$$\int f_n\, d\mu \uparrow \int f\, d\mu.$$

Proof. Say $\int f_n\, d\mu \le M < \infty$ for all n. Our first task is to show that the sequence $f_n(x)$ is bounded for almost all x.

For each $\varepsilon > 0$, let us write $E_n(\varepsilon) = \{x : f_n(x) \ge \varepsilon\}$. In order that the sequence $f_n(x)$ be unbounded, it is necessary and sufficient that for every positive integer m, there exist an index n such that $f_n(x) \ge m$. Thus, the sequence $f_n(x)$ is unbounded if and only if for every positive integer m, x belongs to the set

$$E(m) = \bigcup_{n=1}^{\infty} E_n(m).$$

Thus, writing $E = \bigcap_{m=1}^{\infty} E(m)$, we have

$$E = \{x : f_n(x) \text{ is unbounded}\},$$

and the problem is to show that E is a null set.

For each m, it is clear from the monotonicity of the sequence f_n that

$$E_n(m) \uparrow E(m),$$

hence

$$\mu(E_n(m)) \uparrow \mu(E(m)).$$

Since

$$m\chi_{E_n(m)} \le \chi_{E_n(m)} f_n \le f_n,$$

and $E_n(m)$ has finite measure by 25.8, we have

$$m\mu(E_n(m)) \le \int f_n\, d\mu \le M$$

for all n. Taking LUB over n, this yields $m\mu(E(m)) \le M$, thus $\mu(E(m)) \le M/m$. Now,

$$\mu(E) \le \mu(E(m)) \le M/m$$

for every positive integer m, hence $\mu(E) = 0$.

Define $f'_n = \chi_{X-E} f_n$; then $0 \leq f'_n \uparrow$, and $f'_n(x)$ is a bounded sequence for each x, hence we may define

$$f(x) = \lim f'_n(x)$$

for all x in X. Since f'_n is measurable by 15.1, f is measurable by 14.3. Since $f'_n = f_n$ a.e., we have $f_n \to f$ a.e. by 19.4, hence $f_n \uparrow f$ a.e. by 19.9. Finally, by Lemma 5 f is integrable and

$$\int f_n \, d\mu \uparrow \int f \, d\mu. \quad \blacksquare$$

Theorem 1 (**Monotone Convergence Theorem**). *If f_n is a sequence of integrable functions such that*

$$f_n \leq f_{n+1} \text{ a.e.}$$

$(n = 1, 2, 3, \ldots)$, *and*

$$\int f_n \, d\mu$$

is bounded, then there exists an integrable function f such that

$$f_n \uparrow f \text{ a.e.}$$

Necessarily

$$\int f_n \, d\mu \uparrow \int f \, d\mu;$$

indeed, if h is any measurable function such that $f_n \to h$ a.e., then $h = f$ a.e., h is integrable, and

$$\int f_n \, d\mu \uparrow \int h \, d\mu.$$

Proof. Let E be a null set on whose complement $f_n(x) \uparrow$ (see the proof of 19.9), and consider $\chi_{X-E} f_n$; in view of 25.5, we may assume after a change of notation that $f_n \uparrow$. Define $g_n = f_n - f_1$; then g_n is integrable, $0 \leq g_n \uparrow$, and

$$\int g_n \, d\mu = \int f_n \, d\mu - \int f_1 \, d\mu$$

is bounded. It follows from Lemma 6 that there exists an integrable function g such that $g_n \uparrow g$ a.e., that is, $f_n - f_1 \uparrow g$ a.e. Defining $f = g + f_1$, we have $f_n \uparrow f$ a.e., where f is an integrable function.

Suppose now that h is any measurable function such that $f_n \to h$ a.e. Since already $f_n \to f$ a.e., we have $h = f$ a.e. by 19.2, hence h is integrable by 25.5. Of course $f_n \uparrow h$ a.e., hence

$$\int f_n \, d\mu \uparrow \int h \, d\mu$$

by Lemma 5. $\blacksquare$

The monotone convergence theorem will be referred to in the sequel by the symbol (MCT). The motive is not abbreviation, but to attract attention to those crucial points where the theorem plays a role. This completes our list of "symbol theorems": (LMC), (UET), (MCT). They are the technical cornerstones, respectively, of our future work with σ-rings, measures, and integrals.

EXERCISES

1. Suppose f_n is a sequence of integrable functions such that $f_n \le f_{n+1}$ a.e., and $\int f_n \, d\mu$ is bounded. If h is a measurable function such that $f_n \to h$ in (m), then h is integrable, $f_n \uparrow h$ a.e., and

$$\int f_n \, d\mu \uparrow \int h \, d\mu.$$

2. If f is a nonnegative integrable function, then

$$\int f \, d\mu = \text{LUB}\left\{\int_F f \, d\mu : F \in \mathscr{S}_\varphi\right\}.$$

3. A measurable function f is integrable under either of the following hypotheses:

(*i*) There exists a sequence of measurable sets E_n such that $E_n \uparrow N(f)$, each $\chi_{E_n} f$ is integrable, and $\int \chi_{E_n} |f| \, d\mu$ is bounded.

(*ii*) There exists a sequence of measurable sets F_n such that $N(f) = \bigcup_1^\infty F_n$, each $\chi_{F_n} f$ is integrable, and

$$\sum_1^\infty \int \chi_{F_n} |f| \, d\mu < \infty.$$

4. Let f be any nonnegative measurable function on the measure space $(X, \mathscr{S}, \mu)$. For any locally measurable set A, define $\nu(A)$ to be $\int \chi_A f \, d\mu$ when $\chi_A f$ is integrable, and $\nu(A) = \infty$ otherwise. Then ν is a measure on $\mathscr{S}_\lambda$.

5. The sequential aspect of the (MCT) is essential, as is shown by the following example. Let μ be Lebesgue measure, write f for the characteristic function of $[0, 1]$, and f_A for the characteristic function of an arbitrary finite subset A of $[0, 1]$. Then the family (f_A) is increasingly directed (by the relation of inclusion for finite subsets), and $f_A \uparrow f$, but one does *not* have

$$\int f_A \, d\mu \uparrow \int f \, d\mu.$$

6. If f is measurable, $f \ge 0$ a.e., and f is not integrable, it is customary to formally define $\int f \, d\mu = +\infty$. If f and f_n $(n = 1, 2, 3, \ldots)$ are measurable functions such that

$$0 \le f_n \uparrow f \text{ a.e.},$$

it is a trivial consequence of 25.7 and the (MCT) that

$$\int f_n \, d\mu \uparrow \int f \, d\mu.$$

28. Mean Convergence

If f and g are integrable functions, we shall write

$$\rho(f, g) = \int |f - g| \, d\mu.$$

The function ρ has the following "distance-like" properties:

(1) $\rho(f, g) = 0$ if and only if $f = g$ a.e.

(2) $\rho(f, g) = \rho(g, f)$.

(3) $\rho(f, g) \leq \rho(f, h) + \rho(h, g)$.

Indeed, (1) is immediate from 25.5 and 25.9, (2) is trivial, and (3) results from integrating the relation

$$|f - g| \leq |f - h| + |h - g|.$$

Observe that

$$\rho(f, g) = \rho(f - g, 0), \qquad \rho(f + h, g + h) = \rho(f, g).$$

A sequence f_n of integrable functions is said to be **mean fundamental** (or *fundamental in mean*) if

$$\rho(f_m, f_n) \to 0$$

as $m, n \to \infty$; briefly, f_n is MF. A sequence f_n of integrable functions is said to **converge in mean** to the integrable function f in case

$$\rho(f_n, f) \to 0;$$

briefly, $f_n \to f$ *in mean.* If also $f_n \to g$ in mean, then $f = g$ a.e. results from the inequality

$$\rho(f, g) \leq \rho(f, f_n) + \rho(f_n, g).$$

If $f_n \to f$ in mean, then f_n is MF; this is immediate from

$$\rho(f_m, f_n) \leq \rho(f_m, f) + \rho(f, f_n).$$

Conversely, if f_n is a MF sequence of integrable functions, it will be shown in Sect. 31 that there exists an integrable function f such that $f_n \to f$ in mean (this is the "$\mathscr{L}^1$ completeness theorem"). Observe that if f_n is a MF sequence, and f_{n_k} is a subsequence which converges in mean to some integrable function f, then $f_n \to f$ in mean; this is immediate from

$$\rho(f_n, f) \leq \rho(f_n, f_{n_k}) + \rho(f_{n_k}, f).$$

In this section we shall consider some of the preliminary properties of mean convergence. Deeper properties are deferred to the next chapter.

Theorem 1. *Let f_n, f, g_n, g be integrable functions, c a real number, and A a locally measurable set. Assume that $f_n \to f$ in mean and $g_n \to g$ in mean. Then:*

(1) *$cf_n \to cf$ in mean.*

(2) *$f_n + g_n \to f + g$ in mean.*

(3) *$|f_n| \to |f|$ in mean.*

(4) *$f_n \cup g_n \to f \cup g$ in mean, and $f_n \cap g_n \to f \cap g$ in mean.*

(5) *$f_n^+ \to f^+$ in mean, and $f_n^- \to f^-$ in mean.*

(6) *$\chi_A f_n \to \chi_A f$ in mean.*

Proof. All of the functions in question are integrable by Sect. 25.

(1) $\rho(cf_n, cf) = \int |cf_n - cf| \, d\mu = |c| \rho(f_n, f)$.

(2)
$$\begin{aligned}\rho(f_n + g_n, f + g) &\leq \rho(f_n + g_n, f + g_n) + \rho(f + g_n, f + g) \\ &= \rho(f_n, f) + \rho(g_n, g).\end{aligned}$$

(3) Integrating the inequality

$$\big||f_n| - |f|\big| \leq |f_n - f|,$$

one has

$$\rho(|f_n|, |f|) \leq \rho(f_n, f).$$

(4) This is immediate from (1) through (3).

(5) This is a special case of (4).

(6) Integrating the relation

$$|\chi_A f_n - \chi_A f| = \chi_A |f_n - f| \leq |f_n - f|,$$

one has

$$\rho(\chi_A f_n, \chi_A f) \leq \rho(f_n, f). \quad \blacksquare$$

The same inequalities yield the analogous results for MF sequences:

Theorem 2. *If f_n and g_n are* MF *sequences of integrable functions, c is a real number, and A is a locally measurable set, then each of the following sequences is* MF: *$cf_n, f_n + g_n, |f_n|, f_n \cup g_n, f_n \cap g_n, f_n^+, f_n^-$, and $\chi_A f_n$.*

For sequences of integrable functions, convergence in mean is a "stronger" mode of convergence than convergence in measure:

Theorem 3. *Let f and f_n $(n = 1, 2, 3, \ldots)$ be integrable functions.*

(*i*) *If f_n is* MF, *then f_n is fundamental in* ⓜ.

(*ii*) *If $f_n \to f$ in mean, then $f_n \to f$ in* ⓜ.

Proof. (*i*) Given any $\varepsilon > 0$, let

$$E_{mn} = \{x: |f_m(x) - f_n(x)| \geq \varepsilon\};$$

the problem is to show that $\mu(E_{mn}) \to 0$ as $m, n \to \infty$. Indeed, $\mu(E_{mn}) < \infty$ by 25.8, and

$$\varepsilon\chi_{E_{mn}} \leq \chi_{E_{mn}}|f_m - f_n| \leq |f_m - f_n|,$$

hence $\varepsilon\mu(E_{mn}) \leq \rho(f_m, f_n)$ by 25.6. Since $\rho(f_m, f_n) \to 0$, clearly $\mu(E_{mn}) \to 0$.

(*ii*) Similarly, consider the sets

$$E_n = \{x: |f_n(x) - f(x)| \geq \varepsilon\}. \quad ▮$$

The $\mathscr{L}^1$ completeness theorem may be approached in the following way. If f_n is a MF sequence of integrable functions, then f_n is fundamental in ⓜ, hence by 21.4 there exists a measurable function f such that $f_n \to f$ in ⓜ. It would then remain to show that f is integrable and $\int |f_n - f| \, d\mu \to 0$. We shall use a different approach in Sect. 31.

Every integrable function is the limit in mean of a suitable sequence of integrable simple functions:

Theorem 4. *If f is any integrable function, there exists a sequence f_n of* ISF *such that $f_n \to f$ in mean, and $f_n \to f$ pointwise. If moreover $f \geq 0$, we may assume that $0 \leq f_n \uparrow f$.*

Proof. Writing $f = f^+ - f^-$, it suffices by Theorem 1 to consider the case that $f \geq 0$. Since $f \in \mathscr{P}^1$ by 25.1, by the definition of integrability in Sect. 24 there exists a sequence f_n of ISF such that $0 \leq f_n \uparrow f$, and $\int f \, d\mu$ is defined to be

$$\text{LUB}\, I(f_n) = \text{LUB} \int f_n \, d\mu.$$

Thus, $\int f_n \, d\mu \uparrow \int f \, d\mu$; since $f - f_n \geq 0$, we have

$$\rho(f, f_n) = \int (f - f_n) \, d\mu = \int f \, d\mu - \int f_n \, d\mu \to 0. \quad ▮$$

EXERCISES

1. If $f_n \to f$ in mean, and g is a measurable function such that $f_n \leq g$ a.e. for all n, then also $f \leq g$ a.e.

2. If f_n is an increasing sequence of integrable functions such that $\int f_n \, d\mu$ is bounded, there exists an integrable function f such that $f_n \to f$ in mean.

3. Convergence in measure does not imply convergence in mean. For example, let μ be Lebesgue measure, and define

$$f_n = n\chi_{[0,1/n)}.$$

Then $f_n \to 0$ in measure, but $\int f_n \, d\mu = 1$ for all n.

4. If $f_n \to f$ in mean, and $c_n \to c$, then $c_n f_n \to cf$ in mean. (Cf. Exercise 20.6.)

CHAPTER 5

Convergence Theorems

The first three sections are the heart of the chapter, culminating in the $\mathscr{L}^1$ completeness theorem, which is of fundamental importance in applications. This theorem is cited only in one of the starred sections (Sect. 86); aside from this, no reference is made, in later chapters, to the theorems of this chapter.

Throughout the chapter, we work in the context of a fixed measure space $(X, \mathscr{S}, \mu)$.

29. Dominated Convergence in Measure

The dominated convergence theorems to be presented in this and the next section will play a key role in the proof of the $\mathscr{L}^1$ completeness theorem to be given in Sect. 31. They are, of course, of independent interest, and are basic for questions of integration term by term.

Lemma. *If $0 \leq f_n \leq g$ a.e., where g is integrable, the f_n $(n = 1, 2, 3, \ldots)$ are measurable, and $f_n \to 0$ in $\mathscr{M}$, then the f_n are integrable and $f_n \to 0$ in mean.*

Proof. Since $|f_n| \leq |g|$ a.e., the f_n are integrable by 25.7. Replacing f_n and g by $\chi_{X-E}f_n$ and $\chi_{X-E}g$, where E is a suitable null set, we may assume (in view of 25.5) that $0 \leq f_n \leq g$. Given $\varepsilon > 0$, it will suffice to find an index n_0 such that

$$\int f_n \, d\mu \leq 3\varepsilon$$

whenever $n \geq n_0$.

Let $E = N(g)$; since $N(f_n) \subset N(g)$, we have $\chi_E f_n = f_n$ for all n. By 25.8, there exists a sequence F_k of measurable sets such that $\mu(F_k) < \infty$, $F_k \uparrow E$, and $\chi_{F_k} g$ is bounded; then $E - F_k \downarrow \varnothing$, hence $\mu_g(E - F_k) \downarrow 0$ because μ_g is a finite measure by 26.2. We extract from this the following fact: there exists a measurable set F such that $F \subset E$, $\mu(F) < \infty$, $\mu_g(E - F) \leq \varepsilon$, and $\chi_F g$ is bounded. Say $\chi_F g \leq M\chi_F$, where $0 \leq M < \infty$. It follows from $f_n \leq g$ that

$$\mu_{f_n}(E - F) \leq \mu_g(E - F) \leq \varepsilon;$$

summarizing, we have found a measurable set F, and a nonnegative real number M, such that

(1) $F \subset E$, $\mu(F) < \infty$, $\chi_F g \leq M\chi_F$, and $\int_{E-F} f_n \, d\mu \leq \varepsilon$ for all n.

Now,

$$\int f_n \, d\mu = \int_E f_n \, d\mu = \int_{E-F} f_n \, d\mu + \int_F f_n \, d\mu \leq \varepsilon + \int_F f_n \, d\mu,$$

hence it will suffice to find an index n_0 such that

$$\int_F f_n \, d\mu \leq 2\varepsilon$$

whenever $n \geq n_0$.

Since $\mu(F)$ is finite, we may choose a $\delta > 0$ such that $\delta\mu(F) \leq \varepsilon$. Define

$$E_n = \{x : f_n(x) \geq \delta\};$$

by assumption, $\mu(E_n) \to 0$. Since $f_n(x) < \delta$ on $X - E_n$, $f_n \leq g$, and $\chi_F g \leq M\chi_F$, we have

$$\begin{aligned} \chi_F f_n &= \chi_{F-E_n} f_n + \chi_{F \cap E_n} f_n \leq \delta\chi_{F-E_n} + \chi_{F \cap E_n} g \leq \delta\chi_F + \chi_{F \cap E_n}(M\chi_F) \\ &= \delta\chi_F + M\chi_{F \cap E_n} \leq \delta\chi_F + M\chi_{E_n}, \end{aligned}$$

hence

$$\int_F f_n \, d\mu \leq \delta\mu(F) + M\mu(E_n) \leq \varepsilon + M\mu(E_n).$$

Summarizing,

$$\text{(2)} \qquad \int_F f_n \, d\mu \leq \varepsilon + M\mu(E_n)$$

for all n. Since $\mu(E_n) \to 0$, there is an index n_0 such that $n \geq n_0$ implies $M\,\mu\,(E_n) \leq \varepsilon$, and hence (2) implies that

$$\int_F f_n \, d\mu \leq 2\varepsilon$$

whenever $n \geq n_0$. ∎

Theorem 1 (Dominated Convergence in Measure). *Suppose*

$$|f_n| \leq |g| \ a.e.$$

($n = 1, 2, 3, \ldots$), *where g is integrable and the f_n are measurable. If f is a measurable function such that*

$$f_n \to f \text{ in } \mathcal{m},$$

then $|f| \leq |g|$ a.e., the f_n and f are integrable, and $f_n \to f$ in mean.

Proof. We have $|f| \le |g|$ a.e. by the corollary of 20.8; it follows from 25.7 that f and the f_n are integrable. Now,

$$|f_n - f| \le |f_n| + |f| \le 2|g| \text{ a.e.},$$

and $|f_n - f| \to 0$ in $\mathcal{m}$, hence $\int |f_n - f| \, d\mu \to 0$ by the lemma. ∎

EXERCISE

1. In a finite measure space, we may define

$$d(f, g) = \int \frac{|f - g|}{1 + |f - g|} \, d\mu$$

for every pair of measurable functions f and g. Then: (*i*) $d(f, g) \ge 0$, and $d(f, g) = 0$ if and only if $f = g$ a.e., (*ii*) $d(f, g) = d(g, f)$, and (*iii*) $d(f, g) \le d(f, h) + d(h, g)$. In order that $f_n \to f$ in $\mathcal{m}$, it is necessary and sufficient that $d(f_n, f) \to 0$ as $n \to \infty$. A sequence f_n is fundamental in $\mathcal{m}$ if and only if $d(f_m, f_n) \to 0$ as $m, n \to \infty$.

30. Dominated Convergence Almost Everywhere

We now come to Lebesgue's dominated convergence (a.e.) theorem:

Theorem 1. *Suppose*

$$|f_n| \le |g| \ a.e.$$

($n = 1, 2, 3, \ldots$), *where the* f_n *and* g *are integrable. If* f_n *is fundamental a.e., then* f_n *is* MF. *More precisely, if* f *is a measurable function such that*

$$f_n \to f \ a.e.,$$

then f *is integrable and* $f_n \to f$ *in mean.*

Proof. Assuming f_n is fundamental a.e., there exists, by 19.7, a measurable function f such that $f_n \to f$ a.e.; since $|f| \le |g|$ a.e. by the corollary of 19.8, it follows from 25.7 that f is integrable.

Define $g_n = |f - f_n|$ and $h = 2|g|$; we have $g_n \to 0$ a.e., and

$$0 \le g_n \le |f| + |f_n| \le 2|g| = h \text{ a.e.}$$

Multiplying through by χ_{X-E} for a suitable null set E, we may assume that $0 \le g_n \le h$ and $g_n \to 0$ pointwise. Let us show that $g_n \to 0$ in $\mathcal{m}$.

Consider the finite measure $\nu = \mu_h$ (see 26.2). Applying 20.1 to the finite measure space $(X, \mathcal{S}, \nu)$, we conclude that $g_n \to 0$ in $\mathcal{m}$ with respect to ν. That is, given $\varepsilon > 0$, and setting

$$E_n = \{x: g_n(x) \ge \varepsilon\},$$

we have $\nu(E_n) \to 0$; since

$$\varepsilon\chi_{E_n} \le \chi_{E_n} g_n \le \chi_{E_n} h,$$

it follows that

$$\varepsilon\mu(E_n) \le \int_{E_n} h \, d\mu = \nu(E_n),$$

hence clearly $\mu(E_n) \to 0$. We have shown that $g_n \to 0$ in ⓜ with respect to μ, where $0 \le g_n \le h$, hence $\int g_n \, d\mu \to 0$ by the lemma to 29.1. That is,

$$\int |f_n - f| \, d\mu \to 0. \quad \blacksquare$$

Briefly, we have deduced the "dominated convergence a.e. theorem" from the "dominated convergence in ⓜ theorem," by a suitable application of 20.1.

EXERCISE

1. There is a "dominated" variant of Egoroff's theorem. Suppose g and f_n $(n = 1, 2, 3, \ldots)$ are integrable functions such that $|f_n| \le |g|$ a.e. If f is a measurable function such that $f_n \to f$ a.e., then $f_n \to f$ a.u.

31. The $\mathscr{L}^1$ Completeness Theorem

The $\mathscr{L}^1$ *completeness theorem* follows easily from the convergence theorems already proved:

Theorem 1. *If f_n is a* MF *sequence of integrable functions, there exists an integrable function f such that $f_n \to f$ in mean.*

Proof. We are assuming that $\rho(f_m, f_n) \to 0$. Passing to a subsequence, we may assume without loss of generality that

$$\rho(f_{n+1}, f_n) \le 2^{-n}$$

for all n. The first step is to construct an integrable function g which "dominates" the f_n, preparatory to applying the dominated convergence theory. Define $f_0 = 0$, and

$$g_n = \sum_1^n |f_k - f_{k-1}| = |f_1| + |f_2 - f_1| + \cdots + |f_n - f_{n-1}|,$$

for $n = 1, 2, 3, \ldots$. The g_n are integrable, and $0 \le g_n \uparrow$. Moreover,

$$\int g_n \, d\mu = \int |f_1| \, d\mu + \sum_2^n \rho(f_k, f_{k-1}) < 1 + \int |f_1| \, d\mu$$

for all n; hence by the (MCT), there exists an integrable function g such that $g_n \uparrow g$ a.e. Say E is a null set on whose complement $g_n(x) \uparrow g(x)$; multiplying through the functions f_n, g_n, g by χ_{X-E}, we may assume, without loss of generality (25.5), that $0 \le g_n \uparrow g$. It is clear from the definition of g_n that $|f_n| \le g_n$, and hence $|f_n| \le g$ for all n.

Next, we show that there exists a measurable function f such that $f_n \to f$ (pointwise). Define $h_n = f_n - f_{n-1}$ $(n = 1, 2, 3, \ldots)$. For each $x \in X$, we have

$$\sum_1^n |h_k(x)| = g_n(x) \le g(x)$$

for all n, hence the series $\sum_1^\infty |h_k(x)|$ is convergent. So, therefore, is the series $\sum_1^\infty h_k(x)$, and we may define

$$f(x) = \sum_1^\infty h_k(x).$$

But

$$\sum_1^n h_k(x) = f_n(x),$$

hence

$$f(x) = \lim_n \sum_1^n h_k(x) = \lim f_n(x);$$

that is, $f_n(x) \to f(x)$. Then f is measurable by 14.3; since $|f_n| \le g$ for all n, it follows that $|f| \le g$, hence f is integrable by 25.7. Finally, $f_n \to f$ in mean by the dominated convergence a.e. theorem (30.1). ∎

Corollary. *If f_n is a* MF *sequence of integrable functions, and f is a measurable function such that $f_n \to f$ in $\mathscr{m}$, then f is integrable and $f_n \to f$ in mean.*

Proof. By the $\mathscr{L}^1$ completeness theorem, there exists an integrable function g such that $f_n \to g$ in mean; since $f_n \to g$ in $\mathscr{m}$ by 28.3, it follows from 20.3 that $f = g$ a.e. Then by 25.5, f is integrable, and

$$\int |f_n - f| \, d\mu = \int |f_n - g| \, d\mu \to 0. \quad ∎$$

[It may be observed here that in some expositions, a measurable function f is *defined* to be integrable if there exists a MF sequence f_n of ISF such that $f_n \to f$ in $\mathscr{m}$. See also 28.4.]

EXERCISE

1. If f_n is a MF sequence of integrable functions, and f is a measurable function such that $f_n \to f$ a.e., then f is integrable, and $f_n \to f$ in mean.

32. Fatou's Lemma

We have bypassed a result known as *Fatou's Lemma*, which in many expositions is made the basis for the dominated convergence theorems; it will play an important role in the next section. This result is an easy consequence of the monotone convergence theorem:

Theorem 1. *If f_n is a sequence of integrable functions such that $f_n \geq 0$ a.e. and*

$$\liminf \int f_n \, d\mu < \infty,$$

then there exists an integrable function f such that

$$f = \liminf f_n \quad a.e.,$$

and one has

$$\int f \, d\mu \leq \liminf \int f_n \, d\mu.$$

Proof. Replacing the f_n by $\chi_{X-E} f_n$, where E is a suitable null set, we may assume without loss of generality that $f_n \geq 0$, for all n. For each n, define

$$g_n = \operatorname*{GLB}_{k \geq n} f_k.$$

Then $0 \leq g_n \leq f_n$; since g_n is measurable by 14.1, and f_n is integrable, g_n is integrable by 25.7. Moreover, $g_n \uparrow$.

For each n, $k \geq n$ implies $g_n \leq f_k$, and hence

$$\int g_n \, d\mu \leq \int f_k \, d\mu;$$

taking GLB over $k \geq n$, we have

$$\int g_n \, d\mu \leq \operatorname*{GLB}_{k \geq n} \int f_k \, d\mu.$$

Now, if

$$M = \liminf \int f_n \, d\mu = \operatorname*{LUB}_{n \geq 1} \operatorname*{GLB}_{k \geq n} \int f_k \, d\mu,$$

we have shown that $\int g_n \, d\mu \leq M$ for all n. It follows from the (MCT) that there exists an integrable function f such that $g_n \uparrow f$ a.e., and one has

$$\int g_n \, d\mu \uparrow \int f \, d\mu.$$

Since $\int g_n \, d\mu \leq M$ for all n, we have $\int f \, d\mu \leq M$. Finally, for almost all x,

$$f(x) = \operatorname{LUB}_n g_n(x) = \operatorname*{LUB}_{n \geq 1} \operatorname*{GLB}_{k \geq n} f_k(x) = \liminf f_n(x);$$

that is, $f = \liminf f_n$ a.e. ∎

There is a "dominated" variant of Fatou's Lemma:

Theorem 2. *Suppose* $|f_n| \le |g|$ *a.e.* $(n = 1, 2, 3, \ldots)$, *where the* f_n *and* g *are integrable functions. Then, there exists an integrable function* f *such that*

$$f = \liminf f_n \quad a.e.$$

Moreover, $|f| \le |g|$ *a.e., and*

$$\int f \, d\mu \le \liminf \int f_n \, d\mu.$$

Proof. We may assume, without loss of generality, that $|f_n| \le |g|$, for all n. Since $|g|$ is integrable, replacing g by $|g|$ we may assume that $|f_n| \le g$, for all n. Then,

$$(*) \qquad -g(x) \le f_n(x) \le g(x)$$

for all n and x, hence $f = \liminf f_n$ is real-valued, and $-g \le f \le g$. Since f is measurable by 14.2, and $|f| \le g$, it follows from 25.7 that f is integrable.

Define $h_n = g + f_n$. Then h_n is integrable, and it is clear from $(*)$ that $0 \le h_n \le 2g$. Let $h = \liminf h_n$; clearly $0 \le h \le 2g$, and h is integrable by a repetition of the above argument. Moreover,

$$0 \le \int h_n \, d\mu \le 2 \int g \, d\mu,$$

hence we have

$$(**) \qquad \int h \, d\mu \le \liminf \int h_n \, d\mu$$

by Theorem 1.

Since

$$\begin{aligned} h(x) &= \liminf h_n(x) = \liminf \,[g(x) + f_n(x)] \\ &= g(x) + \liminf f_n(x) = g(x) + f(x), \end{aligned}$$

we have $h = g + f$. Similarly,

$$\begin{aligned} \liminf \int h_n \, d\mu &= \liminf \int (g + f_n) \, d\mu \\ &= \liminf \left(\int g \, d\mu + \int f_n \, d\mu \right) = \int g \, d\mu + \liminf \int f_n \, d\mu. \end{aligned}$$

Substituting in $(**)$,

$$\int g \, d\mu + \int f \, d\mu \le \int g \, d\mu + \liminf \int f_n \, d\mu,$$

thus

$$\int f \, d\mu \le \liminf \int f_n \, d\mu. \quad \blacksquare$$

EXERCISES

1. If E_n is any sequence of measurable sets, then

$$\mu(\liminf E_n) \leq \liminf \mu(E_n.)$$

2. If g and f_n ($n = 1, 2, 3, \cdots$) are integrable functions such that $|f_n| \leq |g|$ a.e., then there exists an integrable function f such that $f = \limsup f_n$ a.e. Moreover, $|f| \leq |g|$ a.e., and

$$\limsup \int f_n \, d\mu \leq \int f \, d\mu.$$

3. Let f_n be a sequence of integrable functions such that $f_n \geq 0$ a.e. and $\int f_n \, d\mu \leq 1$ for all n. If f is a measurable function such that $f_n \to f$ a.e. or $f_n \to f$ in $\mathscr{m}$, then f is integrable, $f \geq 0$ a.e., and $\int f \, d\mu \leq 1$.

4. The proof of the $\mathscr{L}^1$ completeness theorem may be based on Fatou's Lemma, as follows. Suppose f_n is a MF sequence of integrable functions. Since f_n is fundamental in $\mathscr{m}$ by 28.3, by the Riesz-Weyl theorem there exists a measurable function f such that $f_n \to f$ in $\mathscr{m}$. For each fixed index m, the sequence

$$\int |f_m - f_n| \, d\mu$$

is bounded (even convergent); since

$$|f_m - f_n| \to |f_m - f| \qquad \text{in } \mathscr{m},$$

it follows from Exercise 3 that $|f_m - f|$ is integrable, and

$$\int |f_m - f| \, d\mu \leq \lim_n \int |f_m - f_n| \, d\mu.$$

Then $f = f_m - (f_m - f)$ is integrable, etc.

33. The Space $\mathscr{L}^2$, Riesz-Fischer Theorem

We shall write $\mathscr{L}^2(\mu)$, or briefly $\mathscr{L}^2$, for the class of all measurable functions f such that f^2 is integrable. In other words, $\mathscr{L}^2$ is the class of all "square-integrable" measurable functions.

Theorem 1. *If $f, g \in \mathscr{L}^2$, c is a real number, and A is a locally measurable set, then the following functions all belong to $\mathscr{L}^2$: $cf, f + g, |f|, f \cup g, f \cap g, f^+, f^-$, and $\chi_A f$.*

Proof. In any case these functions are all measurable (Sect. 13, and 15.1). The assertions concerning cf, $|f|$, and $\chi_A f$ follow at once from 25.2 and 13.4. In the elementary inequality

$$(f + g)^2 \leq 2(f^2 + g^2),$$

the left member is measurable, and the right member is integrable (25.2), and so $(f + g)^2$ is integrable by 25.7; thus, $f + g \in \mathscr{L}^2$. The assertions concerning $f \cup g$, $f \cap g$, f^+, and f^- now follow at once from the usual formulas (cf. 25.4). ▮

The product of two functions in $\mathscr{L}^2$ is integrable:

Theorem 2. *If $f, g \in \mathscr{L}^2$, then $fg \in \mathscr{L}^1$.*

Proof. In the elementary inequality

$$|fg| \leq (f^2 + g^2)/2,$$

fg is measurable (13.6), and $f^2 + g^2$ is integrable (25.2), and so fg is integrable by 25.7. ▮

In view of Theorem 2, we may define the **scalar product** of two functions f and g in $\mathscr{L}^2$ by the formula

$$(f|g) = \int fg \, d\mu.$$

The algebraic properties of scalar products are as follows:

Theorem 3. *If $f, g, h \in \mathscr{L}^2$ and c is a real number, then:*
(1) $(f|f) \geq 0$
(2) $(f|f) = 0$ *if and only if* $f = 0$ *a.e.*
(3) $(f|g) = (g|f)$
(4) $(f|g + h) = (f|g) + (f|h)$
(5) $(cf|g) = c(f|g)$
(6) $(f + g|h) = (f|h) + (g|h)$
(7) $(f|cg) = c(f|g)$.

Proof. Property (3) is trivial, and properties (4) through (7) are immediate from the linearity of integration (25.2). Property (1) follows from the definitions in Sects. 24 and 25, while property (2) follows from 25.9. ▮

If $f \in \mathscr{L}^2$, the *$\mathscr{L}^2$-norm* of f is the real number $\|f\|_2$ defined by the formula

$$\|f\|_2 = (f|f)^{1/2} = \left(\int |f|^2 \, d\mu\right)^{1/2}.$$

The scalar product of two functions in $\mathscr{L}^2$ is dominated by the product of their $\mathscr{L}^2$-norms:

Theorem 4 (Cauchy-Schwarz inequality). *If $f, g \in \mathscr{L}^2$, then*

$$|(f|g)| \le \|f\|_2\|g\|_2.$$

Proof. It is to be shown that

$$\left(\int fg\, d\mu\right)^2 \le \left(\int f^2\, d\mu\right)\left(\int g^2\, d\mu\right).$$

If $f = 0$ a.e. or $g = 0$ a.e., then $fg = 0$ a.e., and the assertion is trivial. Thus we may suppose that

$$\int f^2\, d\mu > 0 \qquad \text{and} \qquad \int g^2\, d\mu > 0.$$

For all real numbers $\alpha > 0$, the inequality $(\alpha f - g)^2 \ge 0$ leads to

$$2fg \le \alpha f^2 + g^2/\alpha;$$

integrating (as we may by Theorem 2), we have

$$2(f|g) \le \alpha(f|f) + (g|g)/\alpha.$$

In particular, setting $\alpha = \|g\|_2/\|f\|_2$, we obtain

$$(f|g) \le \|f\|_2\|g\|_2;$$

replacing f by $-f$, we have also

$$-(f|g) \le \|f\|_2\|g\|_2,$$

and so

$$|(f|g)| \le \|f\|_2\|g\|_2. \quad \blacksquare$$

Theorem 5 (Triangle inequality). *If $f, g \in \mathscr{L}^2$, then*

$$\|f + g\|_2 \le \|f\|_2 + \|g\|_2.$$

Proof. Invoking the Cauchy-Schwarz inequality at the appropriate step, we have

$$\begin{aligned}(\|f + g\|_2)^2 &= (f + g|f + g) = (f|f) + 2(f|g) + (g|g) \\ &\le (f|f) + 2\|f\|_2\|g\|_2 + (g|g) = (\|f\|_2 + \|g\|_2)^2. \quad \blacksquare\end{aligned}$$

If $f, g \in \mathscr{L}^2$, let us write

$$\rho_2(f, g) = \|f - g\|_2.$$

Analogous to the properties of the function ρ defined in Sect. 28, we have:

(1) $\rho_2(f, g) = 0$ if and only if $f = g$ a.e.,

(2) $\rho_2(f, g) = \rho_2(g, f)$,

(3) $\rho_2(f, g) \le \rho_2(f, h) + \rho_2(h, g)$,

where the latter inequality is proved by applying Theorem 5 to the relation

$$f - g = (f - h) + (h - g).$$

Thus ρ_2 is also a "pseudometric."

A sequence f_n in $\mathscr{L}^2$ is said to be **fundamental in mean of order 2** in case

$$\rho_2(f_m, f_n) \to 0,$$

equivalently,

$$\int |f_m - f_n|^2 \, d\mu \to 0,$$

as $m, n \to \infty$; briefly, the sequence is **2-mean fundamental.**

A sequence f_n in $\mathscr{L}^2$ is said to **converge in mean of order 2** to a function f in $\mathscr{L}^2$ in case

$$\rho_2(f_n, f) \to 0,$$

equivalently,

$$\int |f_n - f|^2 \, d\mu \to 0,$$

as $n \to \infty$; briefly, $f_n \to f$ **in 2-mean.**

For the 2-mean convergence of functions in $\mathscr{L}^2$, we have the analogues of the results in Sect. 28 concerning the mean convergence of functions in $\mathscr{L}^1$; we suppress both the statements and the easy proofs of these results. In the next theorem we shall prove the $\mathscr{L}^2$ analogue of the $\mathscr{L}^1$ completeness theorem (31.1).

Lemma. *If f_n is a 2-mean fundamental sequence in $\mathscr{L}^2$, there exists a subsequence f_{n_k} and a measurable function f such that $f_{n_k} \to f$ a.e.*

Proof. Passing to a subsequence, we may assume that

$$\|f_{n+1} - f_n\|_2 \le 2^{-n}.$$

If F is any measurable set of finite measure, we assert that the sequence $\chi_F f_n$ is fundamental a.e. Indeed, since $\chi_F \in \mathscr{L}^2$, we have $\chi_F f_n \in \mathscr{L}^1$ by Theorem 2, and

$$\int |\chi_F f_m - \chi_F f_n| \, d\mu = \int \chi_F |f_m - f_n| \, d\mu \le (\mu(F))^{1/2} \|f_m - f_n\|_2$$

by the Cauchy-Schwarz inequality, and this clearly implies that the sequence $\chi_F f_n$ is mean fundamental. Since, moreover,

$$\int |\chi_F f_{n+1} - \chi_F f_n| \, d\mu \le (\mu(F))^{1/2} \|f_{n+1} - f_n\|_2 \le 2^{-n}(\mu(F))^{1/2},$$

it follows from the proof of 31.1 that the sequence $\chi_F f_n$ is fundamental a.e.

Defining $E = \bigcup_1^\infty N(f_n)$, we may choose a sequence F_k of measurable sets of finite measure, such that $E = \bigcup_1^\infty F_k$ (25.8). By the foregoing argument, there exists, for each k, a measurable set G_k of measure zero such that $f_n(x)$ is convergent for each x in $F_k - G_k$. Defining $G = \bigcup_1^\infty G_k$, we have $\mu(G) = 0$; since

$$E - G \subset \bigcup_1^\infty F_k - G_k,$$

we conclude that $f_n(x)$ converges for each x in $E - G$. On the other hand, every f_n vanishes identically on $X - E$; it follows that $f_n(x)$ converges for all x in $X - G$. Thus, f_n is fundamental a.e., and the required f exists by 19.7. ▌

Theorem 6 (Riesz-Fischer). *If f_n is a 2-mean fundamental sequence in $\mathscr{L}^2$, there exists a function f in $\mathscr{L}^2$ such that $f_n \to f$ in 2-mean.*

Proof. By Theorem 5 it is sufficient to show that some subsequence of f_n possesses a 2-mean limit. Passing to a subsequence, we may assume, by the lemma, that $f_n \to f$ a.e. for a suitable measurable function f.

Fix any index m. It is clear from the hypothesis that the numbers

$$\int |f_m - f_n|^2 \, d\mu$$

are bounded, and so by Fatou's Lemma (32.1) there exists an integrable function $g_m \geq 0$ such that

$$g_m = \liminf_n |f_m - f_n|^2 \text{ a.e.,}$$

and

$$(*) \qquad \int g_m \, d\mu \leq \liminf_n \int |f_m - f_n|^2 \, d\mu.$$

Since $|f_m - f_n|^2 \to |f_m - f|^2$ a.e. as $n \to \infty$, it follows that

$$g_m = |f_m - f|^2 \text{ a.e.}$$

Since $f_m - f$ is measurable and g_m is integrable, it follows from 25.5 that $|f_m - f|^2$ is integrable; thus $f_m - f \in \mathscr{L}^2$, and relation $(*)$ yields

$$(**) \qquad \int |f_m - f|^2 \, d\mu \leq \liminf_n \int |f_m - f_n|^2 \, d\mu.$$

Since $f_m \in \mathscr{L}^2$ and $f_m - f \in \mathscr{L}^2$, we conclude from the relation $f = f_m - (f_m - f)$ that $f \in \mathscr{L}^2$ (Theorem 1).

It remains to show that $f_n \to f$ in 2-mean. Given any $\varepsilon > 0$, choose an index N such that

$$\int |f_m - f_n|^2 \, d\mu \leq \varepsilon$$

whenever $m, n \geq N$. It is then clear from (**) that

$$\int |f_m - f|^2 \, d\mu \leq \varepsilon$$

whenever $m \geq N$. ∎

The following alternative proof of the Riesz-Fischer theorem is of some interest. Suppose f_n is a 2-mean fundamental sequence in $\mathscr{L}^2$. Observe that the sequence f_n is fundamental in measure; for, the measure of the set

$$\{x\colon |f_m(x) - f_n(x)| \geq \varepsilon\} = \{x\colon |f_m(x) - f_n(x)|^2 \geq \varepsilon^2\}$$

has the limit 0, for each $\varepsilon > 0$ (see the proof of 28.3). According to the Riesz-Weyl theorem (21.4), there exists a measurable function f such that $f_n \to f$ in measure, and passing to a subsequence we may suppose that $f_n \to f$ a.e. The proof now proceeds as in Theorem 6.

EXERCISES

1. Strictly speaking, the classical Riesz-Fischer theorem is an application of Theorem 6 to Fourier series.

2. A *Hilbert space* is constructed from $\mathscr{L}^2$ in the following way. Let $\mathscr{N}$ be the class of all f in $\mathscr{L}^2$ such that $\|f\|_2 = 0$ (equivalently, f is a measurable function such that $f = 0$ a.e.). Evidently $\mathscr{N}$ is a linear subspace of the real vector space $\mathscr{L}^2$, and we may form the quotient vector space $\mathscr{L}^2/\mathscr{N}$, which we denote by $L^2(\mu)$ (or briefly L^2). The elements of L^2 are the cosets $u = f + \mathscr{N}$, $v = g + \mathscr{N}, \ldots$, with linear operations defined by

$$u + v = (f + g) + \mathscr{N} \qquad \text{and} \qquad cu = cf + \mathscr{N}.$$

One may unambiguously define scalar products in L^2 by

$$(u|v) = (f|g),$$

and it is easy to check that $(u|v)$ satisfies the axioms for a pre-Hilbert space. Defining

$$\|u\|_2 = (u|u)^{1/2} = \|f\|_2,$$

it is immediate from the Riesz-Fischer theorem that L^2 is a Hilbert space; that is, if u_n is a sequence in L^2 such that

$$\|u_m - u_n\|_2 \to 0$$

as $m, n \to \infty$, there exists an element u of L^2 such that

$$\|u_n - u\|_2 \to 0.$$

3. If p is a fixed real number, $p > 1$, the class of all measurable functions f such that $|f|^p$ is integrable is denoted $\mathscr{L}^p$; the number

$$\|f\|_p = \left(\int |f|^p \, d\mu\right)^{1/p}$$

is called the $\mathscr{L}^p$-*norm* of f. Let q be the unique real number, necessarily >1, such that $p + q = pq$.

(*i*) If $f \in \mathscr{L}^p$ and $g \in \mathscr{L}^q$, then $fg \in \mathscr{L}^1$, and

$$\left| \int fg \, d\mu \right| \leq \|f\|_p \|g\|_q \qquad \text{(Hölder's inequality).}$$

(*ii*) If $f, g \in \mathscr{L}^p$, then also $f + g \in \mathscr{L}^p$, and

$$\|f + g\|_p \leq \|f\|_p + \|g\|_p \qquad \text{(Minkowski's inequality).}$$

(*iii*) The analogue of the Riesz-Fischer theorem holds for $\mathscr{L}^p$.

4. The analogues of the dominated convergence theorems hold also for $\mathscr{L}^p$. For example, suppose $f_n \to f$ a.e., where the f_n and f are measurable, and $|f_n| \leq |g|$ a.e. for suitable g in $\mathscr{L}^p$. Then the f_n and f also belong to $\mathscr{L}^p$, and $f_n \to f$ in mean of order p.

5. With notation as in Exercise 3, suppose $f_n \to f$ in p-mean, and $g_n \to g$ in q-mean. Then $f_n g_n \to fg$ in mean.

CHAPTER 6

Product Measures

In the earlier sections of this chapter, we are concerned with the following problem: given two measure spaces $(X, \mathscr{S}, \mu)$ and $(Y, \mathscr{T}, \nu)$, how to define an appropriate measure for suitable subsets of $X \times Y$? The entire chapter rests, in an almost axiomatic fashion, on three key results:

(LMC) *The lemma on monotone classes* (see Sect. 2). If a monotone class contains a ring $\mathscr{R}$, then it also contains the generated σ-ring $\mathfrak{S}(\mathscr{R})$.

(UET) *The unique extension theorem* (see Sect. 6). If $\mathscr{R}$ is a ring, and μ_1, μ_2 are two measures on $\mathfrak{S}(\mathscr{R})$ whose restrictions to $\mathscr{R}$ are equal and σ-finite, then $\mu_1 = \mu_2$ on $\mathfrak{S}(\mathscr{R})$.

(MCT) *The monotone convergence theorem* (see Sect. 27). Suppose f_n is a sequence of integrable functions such that $f_n \le f_{n+1}$ a.e. $(n = 1, 2, 3, \ldots)$. In order that there exist an integrable function f such that $f_n \uparrow f$ a.e., it is necessary and sufficient that $\int f_n$ be bounded. In this case if g is any measurable function such that $f_n \to g$ a.e., then g is integrable, $f_n \uparrow g$ a.e., and $\int f_n \uparrow \int g$.

34. Rectangles

Let X and Y be arbitrary sets. If $A \subset X$ and $B \subset Y$, the set $A \times B$ will be called a **rectangle** in $X \times Y$, with **sides** A and B. In the theorem of this section, we shall suppose that we are given a ring $\mathscr{S}$ of subsets of X, and a ring $\mathscr{T}$ of subsets of Y; the class of all rectangles $E \times F$, with $E \in \mathscr{S}$ and $F \in \mathscr{T}$, generates a ring $\mathscr{R}$, and it is the purpose of the theorem to show that each set in $\mathscr{R}$ can be decomposed into a finite disjoint union of such rectangles $E \times F$. The preliminary lemmas also contain material which will be useful throughout the chapter.

Lemma 1. *If A and A_n $(n = 1, 2, 3, \ldots)$ are subsets of X, and if B and B_m $(m = 1, 2, 3, \ldots)$ are subsets of Y, then:*

(1) $A \times (\bigcup_1^\infty B_m) = \bigcup_1^\infty A \times B_m$

(2) $(\bigcup_1^\infty A_n) \times B = \bigcup_1^\infty A_n \times B$

(3) $(\bigcup_1^\infty A_n) \times (\bigcup_1^\infty B_m) = \bigcup_{n,m} A_n \times B_m$

(4) *If* $A_n \uparrow A$ *and* $B_n \uparrow B$, *then* $A_n \times B_n \uparrow A \times B$.

Proof. (3) A point (x, y) will belong to the left side if and only if x belongs to some A_n and y belongs to some B_m, in other words (x, y) belongs to some $A_n \times B_m$. This proves (3), and it is clear that (1) and (2) are special cases of (3).

(4) Obviously $A_n \times B_n \uparrow$. If $k = \max\{n, m\}$, then

$$A_n \times B_m \subset A_k \times B_k;$$

it follows that

$$A \times B = \bigcup_{n,m} A_n \times B_m \subset \bigcup_k A_k \times B_k \subset A \times B.$$

Thus $A \times B = \bigcup_k A_k \times B_k$. ▮

A similar (but not dual) result holds for intersections; the proof is left to the reader:

Lemma 2. *If* A *and* A_n $(n = 1, 2, 3, \ldots)$ *are subsets of* X, *and if* B *and* B_m $(m = 1, 2, 3, \ldots)$ *are subsets of* Y, *then:*

(1) $A \times (\bigcap_1^\infty B_m) = \bigcap_1^\infty A \times B_m$

(2) $(\bigcap_1^\infty A_n) \times B = \bigcap_1^\infty A_n \times B$

(3) $(\bigcap_1^\infty A_n) \times (\bigcap_1^\infty B_m) = \bigcap_{n,m} A_n \times B_m$

(4) *If* $A_n \downarrow A$ *and* $B_n \downarrow B$, *then* $A_n \times B_n \downarrow A \times B$.

The intersection of two rectangles is a rectangle:

Lemma 3. *If* A_1, A_2 *are subsets of* X, *and* B_1, B_2 *are subsets of* Y, *then*

$$(A_1 \times B_1) \cap (A_2 \times B_2) = (A_1 \cap A_2) \times (B_1 \cap B_2).$$

Proof. In order that (x, y) belong to both rectangles $A_i \times B_i$, it is necessary and sufficient that x belong to both of the A_i and y belong to both of the B_i. ▮

It is easy to see that

$$A \times B_1 - A \times B_2 = A \times (B_1 - B_2);$$

in general, a difference of two rectangles can be expressed as a disjoint union of two rectangles:

Lemma 4. *If* A_1, A_2 *are subsets of* X, *and* B_1, B_2 *are subsets of* Y, *then* $A_1 \times B_1 - A_2 \times B_2$ *is the union of the disjoint rectangles* $(A_1 - A_2) \times B_1$ *and* $(A_1 \cap A_2) \times (B_1 - B_2)$.

Proof. Evidently

$$\begin{aligned} A_1 \times B_1 &= [(A_1 - A_2) \cup (A_1 \cap A_2)] \times B_1 \\ &= [(A_1 - A_2) \times B_1] \cup [(A_1 \cap A_2) \times B_1] \\ &= [(A_1 - A_2) \times B_1] \\ &\qquad \cup [(A_1 \cap A_2) \times (B_1 - B_2)] \cup [(A_1 \cap A_2) \times (B_1 \cap B_2)], \end{aligned}$$

and it is clear from Lemma 3 that the three rectangles in the rightmost member of this equation are mutually disjoint. Since

$$\begin{aligned} A_1 \times B_1 - A_2 \times B_2 &= A_1 \times B_1 - (A_1 \times B_1) \cap (A_2 \times B_2) \\ &= A_1 \times B_1 - (A_1 \cap A_2) \times (B_1 \cap B_2), \end{aligned}$$

our assertion is clear. ▌

Theorem 1. *Let $\mathscr{S}$ be a ring of subsets of X, $\mathscr{T}$ a ring of subsets of Y, and let $\mathscr{R}$ be the ring generated by the class of all rectangles $E \times F$, where $E \in \mathscr{S}$ and $F \in \mathscr{T}$. Then, $\mathscr{R}$ coincides with the class of all finite disjoint unions*

$$M = \bigcup_1^n E_i \times F_i,$$

where $E_i \in \mathscr{S}$ and $F_i \in \mathscr{T}$.

Proof. Let $\mathscr{E}$ be the class of all such sets M. Evidently $\mathscr{E} \subset \mathscr{R}$, and $\mathscr{E}$ contains every $E \times F$. To show that $\mathscr{R} \subset \mathscr{E}$, it will suffice by 1.1 to show that $\mathscr{E}$ is a ring. The proof is organized into a series of remarks.

(*i*) *$\mathscr{E}$ is closed under finite intersections.* Suppose

$$M = \bigcup_1^n E_i \times F_i \qquad \text{and} \qquad M^* = \bigcup_1^m E_j^* \times F_j^*$$

are two sets in $\mathscr{E}$ of the indicated sort. Then,

$$\begin{aligned} M \cap M^* &= \left(\bigcup_1^n E_i \times F_i\right) \cap \left(\bigcup_1^m E_j^* \times F_j^*\right) \\ &= \bigcup_{i,j} (E_i \times F_i) \cap (E_j^* \times F_j^*) \\ &= \bigcup_{i,j} (E_i \cap E_j^*) \times (F_i \cap F_j^*) \end{aligned}$$

by Lemma 3. The terms of the latter union are disjoint. For, suppose $(i, j) \neq (i_0, j_0)$. Say $i \neq i_0$; then the disjointness of $E_i \times F_i$ and $E_{i_0} \times F_{i_0}$ guarantees that the sets

$$(E_i \cap E_j^*) \times (F_i \cap F_j^*) \qquad \text{and} \qquad (E_{i_0} \cap E_{j_0}^*) \times (F_{i_0} \cap F_{j_0}^*)$$

are also disjoint. Similarly if $j \neq j_0$. Thus, $M \cap M^*$ is in $\mathscr{E}$.

(*ii*) *$\mathscr{E}$ is closed under finite disjoint unions.* Obvious.

(*iii*) *$\mathscr{E}$ is closed under differences.* Let M and M^* be notated as in (*i*). Then,

$$\begin{aligned} M - M^* &= \Big(\bigcup_i E_i \times F_i\Big) \cap \complement\Big(\bigcup_j E_j^* \times F_j^*\Big) \\ &= \Big(\bigcup_i E_i \times F_i\Big) \cap \Big[\bigcap_j \complement(E_j^* \times F_j^*)\Big] \\ &= \bigcup_i \Big((E_i \times F_i) \cap \Big[\bigcap_j \complement(E_j^* \times F_j^*)\Big]\Big) \\ &= \bigcup_i \Big[\bigcap_j (E_i \times F_i) \cap \complement(E_j^* \times F_j^*)\Big] \\ &= \bigcup_i \Big[\bigcap_j (E_i \times F_i - E_j^* \times F_j^*)\Big]. \end{aligned}$$

Each $E_i \times F_i - E_j^* \times F_j^*$ belongs to $\mathscr{E}$ by Lemma 4, hence the set

$$M_i = \bigcap_j (E_i \times F_i - E_j^* \times F_j^*)$$

belongs to $\mathscr{E}$ by (*i*). Since $M_i \subset E_i \times F_i$, the M_i are mutually disjoint, hence

$$M - M^* = \bigcup_i M_i$$

belongs to $\mathscr{E}$ by (*ii*).

(*iv*) *$\mathscr{E}$ is closed under finite unions.* For, if M and M^* belong to $\mathscr{E}$, consider

$$M \cup M^* = (M - M^*) \cup (M^* - M) \cup (M \cap M^*);$$

then $\mathscr{E}$ contains the first two terms by (*iii*), the third by (*i*), and hence their (disjoint) union by (*ii*).

By (*iii*) and (*iv*), $\mathscr{E}$ is a ring. ∎

With notation as in the theorem, it follows that $\mathscr{R}$ coincides with the class of all finite unions $\bigcup_i E_i \times F_i$, where $E_i \in \mathscr{S}$ and $F_i \in \mathscr{T}$. For, if $\mathscr{F}$ denotes this class, it is clear that

$$\mathscr{R} = \mathscr{E} \subset \mathscr{F} \subset \mathscr{R}.$$

35. Cartesian Product of Two Measurable Spaces

For the rest of the chapter $(X, \mathscr{S})$ and $(Y, \mathscr{T})$ are fixed measurable spaces. The letters E and F will be used for generic elements of $\mathscr{S}$ and $\mathscr{T}$, respectively. The use of these letters in any context (in this chapter) is meant to indicate the σ-rings from which the sets are drawn. A set of the form $E \times F$ will be called a **measurable rectangle.** The σ-ring generated by the class of all measurable rectangles $E \times F$ is denoted $\mathscr{S} \times \mathscr{T}$, and the measurable space

$$(X \times Y, \mathscr{S} \times \mathscr{T})$$

is called the **Cartesian product** of the given measurable spaces. The subsets of $X \times Y$ which belong to $\mathscr{S} \times \mathscr{T}$ are of course called measurable (with respect to $\mathscr{S} \times \mathscr{T}$). Our first result is that each measurable subset of $X \times Y$ can be contained in some measurable rectangle:

Theorem 1. *If $M \in \mathscr{S} \times \mathscr{T}$, there exists a measurable rectangle $E \times F$ such that $M \subset E \times F$.*

Proof. By 1.4, there exists a sequence of measurable rectangles $E_n \times F_n$ such that $M \subset \bigcup_1^\infty E_n \times F_n$. Defining

$$E = \bigcup_1^\infty E_n \qquad \text{and} \qquad F = \bigcup_1^\infty F_n,$$

our assertion is immediate. ∎

The next theorem asserts that the "Cartesian product" of a system of generators for $\mathscr{S}$ with a system of generators for $\mathscr{T}$, is a system of generators for $\mathscr{S} \times \mathscr{T}$:

Theorem 2. *If $\mathscr{E}$ is a class of (measurable) subsets of X such that*

$$\mathfrak{S}(\mathscr{E}) = \mathscr{S},$$

and $\mathscr{F}$ is a class of (measurable) subsets of Y such that

$$\mathfrak{S}(\mathscr{F}) = \mathscr{T},$$

and if

$$\mathscr{G} = \{E \times F: E \in \mathscr{E}, F \in \mathscr{F}\},$$

then

$$\mathfrak{S}(\mathscr{G}) = \mathscr{S} \times \mathscr{T}.$$

Proof. Since $\mathscr{S} \times \mathscr{T}$ contains $\mathscr{G}$, we have

$$\mathfrak{S}(\mathscr{G}) \subset \mathscr{S} \times \mathscr{T}$$

by 1.3.

Fix $E \in \mathscr{E}$, and consider the class

$$\mathscr{D} = \{T \subset Y: E \times T \in \mathfrak{S}(\mathscr{G})\}.$$

Since

$$E \times \left(\bigcup_1^\infty T_n\right) = \bigcup_1^\infty E \times T_n$$

and

$$E \times (T_1 - T_2) = E \times T_1 - E \times T_2,$$

it is clear that $\mathscr{D}$ is a σ-ring. Since $\mathscr{D}$ evidently contains $\mathscr{F}$, we conclude that $\mathscr{D}$ contains $\mathfrak{S}(\mathscr{F}) = \mathscr{T}$. Summarizing: if $E \in \mathscr{E}$ and $F \in \mathscr{T}$, then

$$E \times F \in \mathfrak{S}(\mathscr{G}).$$

Similarly, fix $F \in \mathscr{T}$, and consider the class of all sets $S \subset X$ such that

$$S \times F \in \mathfrak{S}(\mathscr{G});$$

this class is a σ-ring containing $\mathscr{E}$, hence it contains $\mathfrak{S}(\mathscr{E}) = \mathscr{S}$. Summarizing: if $E \in \mathscr{S}$ and $F \in \mathscr{T}$, then

$$E \times F \in \mathfrak{S}(\mathscr{G}),$$

and from this it follows that

$$\mathscr{S} \times \mathscr{T} \subset \mathfrak{S}(\mathscr{G}). \quad \blacksquare$$

EXERCISES

***1.** Let $\mathscr{S}$ and $\mathscr{T}$ be σ-rings of subsets of X and Y, respectively.

(*i*) It is easy to see that $\mathscr{S}_\lambda \times \mathscr{T}_\lambda \subset (\mathscr{S} \times \mathscr{T})_\lambda$.

(*ii*) If $\mathscr{S}$ is the class of all countable subsets of X, and $\mathscr{T}$ is the class of all countable subsets of Y, then $\mathscr{S} \times \mathscr{T}$ is the class of all countable subsets of $X \times Y$. In this case, we have $\mathscr{S}_\lambda = \mathscr{P}(X)$, $\mathscr{T}_\lambda = \mathscr{P}(Y)$, and

$$(\mathscr{S} \times \mathscr{T})_\lambda = \mathscr{P}(X \times Y).$$

(*iii*) Does $\mathscr{S}_\lambda \times \mathscr{T}_\lambda = (\mathscr{S} \times \mathscr{T})_\lambda$?

2. If $\mathscr{S}$, $\mathscr{T}$, $\mathscr{U}$ are σ-rings of subsets of X, Y, Z, respectively, then

$$\mathscr{S} \times (\mathscr{T} \times \mathscr{U}) = (\mathscr{S} \times \mathscr{T}) \times \mathscr{U},$$

provided $X \times (Y \times Z)$ and $(X \times Y) \times Z$ are identified in the natural way.

3. If $\mathscr{S}_1$, $\mathscr{S}_2$, $\mathscr{T}_1$, $\mathscr{T}_2$ are σ-rings such that $\mathscr{S}_1$ is an ideal in $\mathscr{S}_2$ and $\mathscr{T}_1$ is an ideal in $\mathscr{T}_2$, then $\mathscr{S}_1 \times \mathscr{T}_1$ is an ideal in $\mathscr{S}_2 \times \mathscr{T}_2$.

4. If f is measurable with respect to $\mathscr{S}$, and g is measurable with respect to $\mathscr{T}$, then the function

$$h(x, y) = f(x) g(y)$$

is measurable with respect to $\mathscr{S} \times \mathscr{T}$.

36. Sections

Suppose M is a subset of $X \times Y$. For each $x \in X$, define

$$M_x = \{y \in Y: (x, y) \in M\};$$

M_x is called the x-**section** of M. Similarly, if $y \in Y$, the y-**section** of M is the set

$$M^y = \{x \in X: (x, y) \in M\}.$$

The following theorem is immediate from the definitions:

Theorem 1. *If $A \subset X$ and $B \subset Y$, then:*

$$(1) \qquad (A \times B)_x = \begin{cases} B & \text{when } x \in A, \\ \varnothing & \text{when } x \notin A. \end{cases}$$

$$(2) \qquad (A \times B)^y = \begin{cases} A & \text{when } y \in B, \\ \varnothing & \text{when } y \notin B. \end{cases}$$

Suppose $x \in X$. Define a mapping $g: Y \to X \times Y$ by the formula $g(y) = (x, y)$. If M is any subset of $X \times Y$, evidently $M_x = g^{-1}(M)$. Since g^{-1} preserves the usual set theoretic operations, we have at once:

Theorem 2. *Suppose M, N, and M_n $(n = 1, 2, 3, \ldots)$ are subsets of $X \times Y$, and $x \in X$. Then:*

(*i*) $(\bigcup_1^\infty M_n)_x = \bigcup_1^\infty (M_n)_x$

(*ii*) $(\bigcap_1^\infty M_n)_x = \bigcap_1^\infty (M_n)_x$

(*iii*) $(M - N)_x = M_x - N_x$

(*iv*) *If $M \subset N$, then $M_x \subset N_x$.*

(*v*) *If $M_n \uparrow M$, then $(M_n)_x \uparrow M_x$.*

(*vi*) *If $M_n \downarrow M$, then $(M_n)_x \downarrow M_x$.*

Similarly for y-sections.

Recall that we are assuming given measurable spaces $(X, \mathscr{S})$ and $(Y, \mathscr{T})$. Let $(X \times Y, \mathscr{S} \times \mathscr{T})$ be the product measurable space described in the preceding section. Every section of a measurable subset of $X \times Y$ is measurable:

Theorem 3. *If $M \in \mathscr{S} \times \mathscr{T}$, then $M_x \in \mathscr{T}$ and $M^y \in \mathscr{S}$, for every $x \in X$ and $y \in Y$.*

Proof. Fix $x \in X$, and define $g: Y \to X \times Y$ by the formula $g(y) = (x, y)$. Thus, $g^{-1}(U) = U_x$, for all $U \subset X \times Y$. Consider the class

$$\mathscr{E} = \{U \subset X \times Y: g^{-1}(U) \in \mathscr{T}\};$$

by 1.6, $\mathscr{E}$ is a σ-ring. If $E \in \mathscr{S}$ and $F \in \mathscr{T}$, then $g^{-1}(E \times F) = F$ or $\varnothing$ by Theorem 1, and in either case

$$g^{-1}(E \times F) \in \mathscr{T}.$$

Thus, $\mathscr{E}$ is a σ-ring containing every measurable rectangle, and so $\mathscr{S} \times \mathscr{T} \subset \mathscr{E}$. In particular $M \in \mathscr{E}$, that is, $M_x \in \mathscr{T}$. Similarly for y-sections. ▮

Corollary. *If $A \times B$ is a nonempty rectangle such that*

$$A \times B \in \mathscr{S} \times \mathscr{T},$$

then $A \in \mathscr{S}$ and $B \in \mathscr{T}$.

Proof. Fix $a \in A$ and $b \in B$. By Theorem 3,

$$A = (A \times B)^b \in \mathscr{S} \qquad \text{and} \qquad B = (A \times B)_a \in \mathscr{T}. \quad ▮$$

Thus, the class of "measurable rectangles," as defined in Sect. 35, coincides with the class of all rectangles $A \times B$ which are measurable with respect to $\mathscr{S} \times \mathscr{T}$. It should be noted that the empty rectangle can be written in the form $A \times \varnothing$, where A is an arbitrary subset of X. This need not cause any confusion, for we have agreed that when we say that $E \times F$ is a measurable rectangle, it is implied by the choice of letters that we are assuming $E \in \mathscr{S}$ and $F \in \mathscr{T}$.

Suppose $h: X \times Y \to Z$ (for our purposes it will always be the case that $Z = R_e$ or $Z = R$). For each $x \in X$, the x-**section** of h is the mapping $h_x: Y \to Z$ defined by the formula

$$h_x(y) = h(x, y).$$

Similarly, for each $y \in Y$, the y-**section** of h is the function $h^y: X \to Z$ defined by

$$h^y(x) = h(x, y).$$

The definition of sections of a function is compatible with the definition of sections of a set:

Theorem 4. *If M is any subset of $X \times Y$, then $(\chi_M)_x = \chi_{M_x}$ and $(\chi_M)^y = \chi_{M^y}$, for all $x \in X$ and $y \in Y$.*

Proof. Let $h = \chi_M$, and $x \in X$. Then $h_x(y) = h(x, y) = 1$ if and only if $(x, y) \in M$, that is, $y \in M_x$; otherwise, $h_x(y) = 0$. Thus, h_x is the characteristic function of M_x. Similarly for y-sections. ▮

The process of taking sections is a linear operation; the proof is obvious:

Theorem 5. *If h and k are real-valued functions defined on $X \times Y$, and if a and b are real numbers, then*

$$(ah + bk)_x = ah_x + bk_x,$$

$$(ah + bk)^y = ah^y + bk^y,$$

for all $x \in X$ and $y \in Y$.

The process of taking sections preserves pointwise limits (proof obvious):

Theorem 6. *If h and h_n $(n = 1, 2, 3, \ldots)$ are real-valued functions defined on $X \times Y$, such that $h_n \to h$ pointwise on $X \times Y$, then for each $x \in X$ and $y \in Y$, we have $(h_n)_x \to h_x$ pointwise on Y and $(h_n)^y \to h^y$ pointwise on X.*

Similarly for the monotone lattice operations:

Theorem 7. *Suppose h and h_n $(n = 1, 2, 3, \ldots)$ are extended real valued functions defined on $X \times Y$. If $h_n \uparrow h$ pointwise on $X \times Y$, then for each $x \in X$ and $y \in Y$, we have $(h_n)_x \uparrow h_x$ pointwise on Y and $(h_n)^y \uparrow h^y$ pointwise on X. Similarly if $h_n \downarrow h$.*

Every section of a measurable function on $X \times Y$ is measurable:

Theorem 8. *If h is measurable with respect to $\mathscr{S} \times \mathscr{T}$, then for each $x \in X$ the function h_x is measurable with respect to $\mathscr{T}$, and for each $y \in Y$ the function h^y is measurable with respect to $\mathscr{S}$.*

Proof. Fix $x \in X$. One has $N(h_x) = (N(h))_x$; for, the relation $y \in N(h_x)$ is equivalent to each of the following relations: $h_x(y) \neq 0$, $h(x, y) \neq 0$, $(x, y) \in N(h)$, $y \in (N(h))_x$. If B is any Borel set of real numbers, we have $h_x^{-1}(B) = (h^{-1}(B))_x$ since the following relations are equivalent: $h_x(y) \in B$, $h(x, y) \in B$, $(x, y) \in h^{-1}(B)$, $y \in (h^{-1}(B))_x$. Since $N(h) \cap h^{-1}(B)$ is by assumption measurable (see Sect. 12), by Theorem 3 so is the set

$$(N(h) \cap h^{-1}(B))_x = (N(h))_x \cap (h^{-1}(B))_x = N(h_x) \cap h_x^{-1}(B),$$

hence h_x is measurable. ∎

EXERCISES

1. Suppose $\bigcup \mathscr{S} = X$ and $\bigcup \mathscr{T} = Y$. If $M \in (\mathscr{S} \times \mathscr{T})_\lambda$, then every section of M is locally measurable. In particular, $\mathscr{S}_\lambda \times \mathscr{T}_\lambda$ and $(\mathscr{S} \times \mathscr{T})_\lambda$ contain the same rectangles.

2. If $\bigcup \mathscr{S} = X$ and $\bigcup \mathscr{T} = Y$, and h is measurable with respect to the σ-ring $(\mathscr{S} \times \mathscr{T})_\lambda$, then every x-section of h is measurable with respect to $\mathscr{T}_\lambda$, and every y-section of h is measurable with respect to $\mathscr{S}_\lambda$.

37. Preliminaries

The notations to be used for the rest of the chapter will now be fixed:

(1) $(X, \mathscr{S}, \mu)$ and $(Y, \mathscr{T}, \nu)$ are given measure spaces.

(2) $(X \times Y, \mathscr{S} \times \mathscr{T})$ is the product measurable space; *our problem is to define a suitable measure on* $\mathscr{S} \times \mathscr{T}$.

(3) The letters E, F, and M (or N) are reserved for sets in $\mathscr{S}$, $\mathscr{T}$, and $\mathscr{S} \times \mathscr{T}$, respectively. These are the measurable subsets of X, Y, and $X \times Y$, respectively.

(4) The letters f, g, and h (or k) are reserved for functions defined on X, Y, and $X \times Y$, respectively.

(5) We agree to take sections only of measurable subsets of $X \times Y$, so that these sections will always be measurable by 36.3.

(6) For each M in $\mathscr{S} \times \mathscr{T}$, the functions $f_M \colon X \to R_e$ and $g^M \colon Y \to R_e$ are defined by the formulas

$$f_M(x) = \nu(M_x),$$
$$g^M(y) = \mu(M^y).$$

Quoting 36.1, we have at once:

Theorem 1. *If $E \times F$ is a measurable rectangle such that $\mu(E) < \infty$ and $\nu(F) < \infty$, then*

$$f_{E \times F} = \nu(F)\chi_E,$$
$$g^{E \times F} = \mu(E)\chi_F;$$

in particular, these functions are integrable, and

$$\int f_{E \times F}\, d\mu = \int g^{E \times F}\, d\nu = \mu(E)\nu(F).$$

Suppose M and N are disjoint sets in $\mathscr{S} \times \mathscr{T}$. By 36.2,

$$(M \cup N)_x = M_x \cup N_x$$

and

$$M_x \cap N_x = (M \cap N)_x = (\varnothing)_x = \varnothing,$$

hence

$$\nu[(M \cup N)_x] = \nu(M_x) + \nu(N_x).$$

Similarly for y-sections. Summarizing:

Theorem 2. *If M and N are disjoint sets in $\mathscr{S} \times \mathscr{T}$, then*

$$f_{M \cup N} = f_M + f_N,$$
$$g^{M \cup N} = g^M + g^N.$$

If M_n is a sequence of sets in $\mathscr{S} \times \mathscr{T}$, and if $M_n \uparrow M$, then

$$(M_n)_x \uparrow M_x$$

by 36.2, hence

$$\nu[(M_n)_x] \uparrow \nu(M_x).$$

Similarly for y-sections. Evidently:

Theorem 3. *If $M_n \uparrow M$, then $f_{M_n} \uparrow f_M$ and $g^{M_n} \uparrow g^M$.*

38. The Product of Two Finite Measure Spaces

Assuming now that $(X, \mathscr{S}, \mu)$ and $(Y, \mathscr{T}, \nu)$ are *finite* measure spaces, we shall show in this section how to define a natural measure on $\mathscr{S} \times \mathscr{T}$; the general case, to be presented in the next section, will be based on the finite case.

Since μ is a finite measure, its values are bounded (17.1); let us write $\|\mu\|_\infty$ for the (finite) real number

$$\text{LUB}\,\{\mu(E) : E \in \mathscr{S}\}.$$

Similarly,

$$\|\nu\|_\infty = \text{LUB}\,\{\nu(F) : F \in \mathscr{T}\}.$$

Checking the definitions in Sect. 37, we have:

Lemma 1. *If M is any set in $\mathscr{S} \times \mathscr{T}$, then $0 \le f_M \le \|\nu\|_\infty$ and $0 \le g^M \le \|\mu\|_\infty$.*

If M_n is a sequence of sets in $\mathscr{S} \times \mathscr{T}$, and if $M_n \downarrow M$, then

$$(M_n)_x \downarrow M_x$$

by 36.2, hence

$$\nu[(M_n)_x] \downarrow \nu(M_x)$$

by 4.2. Similarly for y-sections. Summarizing:

Lemma 2. *If $M_n \downarrow M$, then $f_{M_n} \downarrow f_M$ and $g^{M_n} \downarrow g^M$.*

The next result lies deeper:

Lemma 3. *If M is any set in $\mathscr{S} \times \mathscr{T}$, the functions f_M and g^M are measurable.*

Proof. The functions in question are real-valued by Lemma 1. Let $\mathscr{R}$ be the ring generated by the class of all measurable rectangles $E \times F$; thus,

$$\mathfrak{S}(\mathscr{R}) = \mathscr{S} \times \mathscr{T}.$$

Suppose first that M belongs to $\mathscr{R}$; by 34.1, M can be written as a finite union of disjoint measurable rectangles $E \times F$, hence the measurability of f_M and g^M follows at once from 37.1 and 37.2.

Let $\mathscr{M}$ be the class of all sets M such that both f_M and g^M are measurable. Since $\mathscr{M}$ contains $\mathscr{R}$, and is clearly a monotone class by 37.3 and Lemma 2 (and 14.3), it follows from the (LMC) that $\mathscr{M}$ contains $\mathfrak{S}(\mathscr{R}) = \mathscr{S} \times \mathscr{T}$. ▌

Lemma 4. *In a finite measure space, every essentially bounded measurable function is integrable.*

Proof. Suppose f is an essentially bounded measurable function defined on X; say

$$|f(x)| \leq K \text{ a.e.},$$

where K is a (finite) real number. If $E = N(f)$, then

$$|f| \leq K\chi_E \text{ a.e.};$$

since $K\chi_E$ is an integrable simple function, f is integrable by 25.7. ▌

Theorem 1. *If $(X, \mathscr{S}, \mu)$ and $(Y, \mathscr{T}, \nu)$ are finite measure spaces, then, for every M in $\mathscr{S} \times \mathscr{T}$, the functions f_M and g^M are integrable, and*

$$(*) \qquad \int f_M \, d\mu = \int g^M \, d\nu.$$

Proof. The functions f_M and g^M are bounded (Lemma 1), and measurable (Lemma 3), hence they are integrable by Lemma 4.

If $M = E \times F$, then M satisfies $(*)$ by 37.1. Suppose next that M belongs to the ring $\mathscr{R}$ generated by the $E \times F$; writing M as a finite disjoint union of measurable rectangles, it is clear from 37.2 (and the additivity of integrals) that M also satisfies $(*)$.

Let us define real-valued set functions φ and ψ on $\mathscr{S} \times \mathscr{T}$ by the formulas

$$\varphi(M) = \int f_M \, d\mu,$$

$$\psi(M) = \int g^M \, d\nu.$$

We know already that $\varphi = \psi$ on $\mathscr{R}$, and the problem is to show that $\varphi = \psi$ on $\mathscr{S} \times \mathscr{T}$.

We shall show that φ and ψ are (finite) measures on $\mathscr{S} \times \mathscr{T}$. The set function φ is additive by 37.2. If $M \subset N$, then $f_M \leq f_N$ by part (*iv*) of 36.2; hence $\varphi(M) \leq \varphi(N)$. Clearly $\varphi(\varnothing) = 0$. If $M_n \uparrow M$, then $f_{M_n} \uparrow f_M$ by 37.3, hence

$$\varphi(M_n) \uparrow \varphi(M)$$

by the (MCT). Summarizing, φ is a measure; by a similar argument, so is ψ. Since $\varphi = \psi$ on $\mathscr{R}$, it follows from the (UET) that $\varphi = \psi$ on

$$\mathfrak{S}(\mathscr{R}) = \mathscr{S} \times \mathscr{T}. \quad \blacksquare$$

Rearranging the results of Theorem 1, we arrive at "product measure":

Theorem 2. *If $(X, \mathscr{S}, \mu)$ and $(Y, \mathscr{T}, \nu)$ are finite measure spaces, there exists a unique measure τ on $\mathscr{S} \times \mathscr{T}$ such that*

$$\tau(E \times F) = \mu(E)\nu(F)$$

for every measurable rectangle $E \times F$. Moreover, τ is finite, and

$$\tau(M) = \int f_M \, d\mu = \int g^M \, d\nu$$

for every M in $\mathscr{S} \times \mathscr{T}$.

Proof. Define the set function τ on $\mathscr{S} \times \mathscr{T}$ by the formula

$$\tau(M) = \int f_M \, d\mu.$$

As shown in the proof of Theorem 1, τ is a finite measure on $\mathscr{S} \times \mathscr{T}$, and

$$\tau(M) = \int g^M \, d\nu.$$

By 37.1, $\tau(E \times F) = \mu(E)\nu(F)$.

Suppose now that ρ is any measure on $\mathscr{S} \times \mathscr{T}$ such that

$$\rho(E \times F) = \mu(E)\nu(F)$$

for every measurable rectangle. It follows from 34.1 that $\rho = \tau$ on the ring $\mathscr{R}$ generated by the measurable rectangles, hence $\rho = \tau$ on $\mathscr{S} \times \mathscr{T}$ by the (UET). $\blacksquare$

With notation as in Theorem 2, we shall write $\tau = \mu \times \nu$ for the unique (finite) measure on $\mathscr{S} \times \mathscr{T}$ such that

$$(\mu \times \nu)(E \times F) = \mu(E)\nu(F)$$

for every measurable rectangle.

It should be noted that the existence and uniqueness of $\mu \times \nu$ can be done entirely in terms of x-sections and the functions f_M. The point of a symmetrical development, especially the relation $(*)$ in Theorem 1, is to prepare the way for Fubini's theorem.

The following theorem will be needed in the next section:

Theorem 3. *If μ_1, μ_2 are finite measures on $\mathscr{S}$, and ν_1, ν_2 are finite measures on $\mathscr{T}$, such that $\mu_1 \leq \mu_2$ and $\nu_1 \leq \nu_2$, then*

$$\mu_1 \times \nu_1 \leq \mu_2 \times \nu_2.$$

Proof. We are assuming that $\mu_1(E) \leq \mu_2(E)$ for all E in $\mathcal{S}$, and $\nu_1(F) \leq \nu_2(F)$ for all F in $\mathcal{T}$, and hence

$$(\mu_1 \times \nu_1)(E \times F) \leq (\mu_2 \times \nu_2)(E \times F)$$

for every measurable rectangle. Let $\mathcal{M}$ be the class of all sets M in $\mathcal{S} \times \mathcal{T}$ such that

$$(\mu_1 \times \nu_1)(M) \leq (\mu_2 \times \nu_2)(M);$$

it is clear that $\mathcal{M}$ is closed under finite (even countable) disjoint unions, and contains every measurable rectangle, hence it contains the ring $\mathcal{R}$ they generate. Since

$$\mathfrak{S}(\mathcal{R}) = \mathcal{S} \times \mathcal{T},$$

our assertion that $\mathcal{S} \times \mathcal{T} \subset \mathcal{M}$ will follow from the (LMC) as soon as we show that $\mathcal{M}$ is a monotone class. Suppose M_n is a sequence of sets in $\mathcal{M}$. If $M_n \uparrow M$, then M belongs to $\mathcal{M}$ by a general property of measures; if $M_n \downarrow M$, then M belongs to $\mathcal{M}$ by a property of finite measures (4.2). ∎

39. The Product of Any Two Measure Spaces

In this section, we allow $(X, \mathcal{S}, \mu)$ and $(Y, \mathcal{T}, \nu)$ to be arbitrary measure spaces; the problem is to define a suitable measure on $\mathcal{S} \times \mathcal{T}$. Let $\mathcal{S}_\varphi$ be the class of all sets P in $\mathcal{S}$ such that $\mu(P) < \infty$; evidently $\mathcal{S}_\varphi$ is a ring, and is, moreover, an ideal in $\mathcal{S}$, in the sense that $P \cap E \in \mathcal{S}_\varphi$ whenever $P \in \mathcal{S}_\varphi$ and $E \in \mathcal{S}$. Similarly, $\mathcal{T}_\varphi$ is the class of all sets Q in $\mathcal{T}$ such that $\nu(Q) < \infty$. The use of the letters P and Q is meant to indicate the rings from which the sets are drawn. A set of the form $P \times Q$ will be called a **finite rectangle**.

Recall that for each set E in $\mathcal{S}$, the *contraction* of μ by E is the measure μ_E defined on $\mathcal{S}$ by the formula

$$\mu_E(E_1) = \mu(E \cap E_1);$$

in particular, if $P \in \mathcal{S}_\varphi$, then μ_P is a finite measure on $\mathcal{S}$ by 4.3.

Given $M \in \mathcal{S} \times \mathcal{T}$, our problem is to "measure" M; this will be done by "measuring" $(P \times Q) \cap M$, and taking the LUB over all finite rectangles.

Lemma 1. *For each finite rectangle $P \times Q$, there exists a unique (finite) measure $\tau^{P \times Q}$ on $\mathcal{S} \times \mathcal{T}$ such that*

$$\tau^{P \times Q}(E \times F) = \mu(P \cap E)\nu(Q \cap F)$$

for every measurable rectangle $E \times F$.

Proof. Apply 38.2 to the finite measure spaces $(X, \mathscr{S}, \mu_P)$ and $(Y, \mathscr{T}, \nu_Q)$; that is, take

$$\tau^{P \times Q} = \mu_P \times \nu_Q. \quad \blacksquare$$

The class of all finite rectangles, ordered by inclusion, is directed to the right; for, if $P_1 \times Q_1$ and $P_2 \times Q_2$ are finite rectangles, setting $P = P_1 \cup P_2$ and $Q = Q_1 \cup Q_2$ we have $P_i \times Q_i \subset P \times Q$. We propose to apply 10.1 to the family of measures $\tau^{P \times Q}$, hence we must first verify that the family is increasingly directed:

Lemma 2. *If* $P_1 \times Q_1 \subset P_2 \times Q_2$, *then* $\tau^{P_1 \times Q_1} \leq \tau^{P_2 \times Q_2}$.

Proof. If $P_1 \times Q_1 = \varnothing$, then either $P_1 = \varnothing$ or $Q_1 = \varnothing$, and in either case it is clear that $\tau^{P_1 \times Q_1} = 0 \leq \tau^{P_2 \times Q_2}$. Otherwise, $P_1 \subset P_2$ and $Q_1 \subset Q_2$. For any $E \in \mathscr{S}$, we have

$$\mu_{P_1}(E) = \mu(P_1 \cap E) \leq \mu(P_2 \cap E) = \mu_{P_2}(E),$$

thus $\mu_{P_1} \leq \mu_{P_2}$. Similarly $\nu_{Q_1} \leq \nu_{Q_2}$. It follows from 38.3 that

$$\tau^{P_1 \times Q_1} \leq \tau^{P_2 \times Q_2}. \quad \blacksquare$$

According to 10.1, the formula

$$\pi(M) = \text{LUB}\,\{\tau^{P \times Q}(M) : P \in \mathscr{S}_\varphi,\ Q \in \mathscr{T}_\varphi\}$$

defines a measure π on $\mathscr{S} \times \mathscr{T}$. Briefly, $\pi = \text{LUB}\ \tau^{P \times Q}$. It is now necessary to sharpen Lemma 2:

Lemma 3. *If* $P_1 \times Q_1 \subset P_2 \times Q_2$, *then the contraction*

$$(\tau^{P_2 \times Q_2})_{P_1 \times Q_1}$$

coincides with $\tau^{P_1 \times Q_1}$.

Proof. Let us write $\tau_i = \tau^{P_i \times Q_i}$ $(i = 1, 2)$. If $P_1 \times Q_1 = \varnothing$, the conclusion is clear; otherwise, $P_1 \subset P_2$ and $Q_1 \subset Q_2$. For any measurable rectangle $E \times F$, we have

$$\begin{aligned}(\tau_2)_{P_1 \times Q_1}(E \times F) &= \tau_2[(P_1 \times Q_1) \cap (E \times F)] \\ &= \tau_2[(P_1 \cap E) \times (Q_1 \cap F)] \\ &= \mu_{P_2}(P_1 \cap E)\nu_{Q_2}(Q_1 \cap F) \\ &= \mu(P_2 \cap P_1 \cap E)\nu(Q_2 \cap Q_1 \cap F) \\ &= \mu(P_1 \cap E)\nu(Q_1 \cap F) = \tau_1(E \times F).\end{aligned}$$

It follows from 34.1 that the finite measures $(\tau_2)_{P_1 \times Q_1}$ and τ_1 agree on the ring $\mathscr{R}$ generated by the measurable rectangles, hence they agree on $\mathfrak{S}(\mathscr{R}) = \mathscr{S} \times \mathscr{T}$ by the (UET). $\blacksquare$

Lemma 4. *Let* $\pi = \text{LUB}\ \tau^{P \times Q}$ *be the measure defined above. For each finite rectangle* $P \times Q$, *the contraction* $\pi_{P \times Q}$ *is identical with* $\tau^{P \times Q}$.

Proof. Let $M \in \mathscr{S} \times \mathscr{T}$. If $P_1 \times Q_1 \supset P \times Q$, then

$$\tau^{P_1 \times Q_1}[(P \times Q) \cap M] = (\tau^{P_1 \times Q_1})_{P \times Q}(M) = \tau^{P \times Q}(M)$$

by Lemma 3; taking LUB over all such $P_1 \times Q_1$, we have

$$\pi[(P \times Q) \cap M] = \tau^{P \times Q}(M)$$

by the definition of π. ▌

A general product measure is now at hand:

Theorem 1. *If* $(X, \mathscr{S}, \mu)$ *and* $(Y, \mathscr{T}, \nu)$ *are arbitrary measure spaces, there exists a unique measure* π *on* $\mathscr{S} \times \mathscr{T}$ *having the following two properties:*

(1) $\pi(P \times Q) = \mu(P)\nu(Q)$ *for every finite rectangle* $P \times Q$,

(2) $\pi(M) = \text{LUB}\,\{\pi[(P \times Q) \cap M] : P \in \mathscr{S}_\varphi,\ Q \in \mathscr{T}_\varphi\}$, *for each* M *in* $\mathscr{S} \times \mathscr{T}$.

Proof. Let $\pi = \text{LUB}\ \tau^{P \times Q}$ as defined above. If $P \times Q$ is any finite rectangle, quoting Lemma 4 at the appropriate step we have

$$\begin{aligned}\pi(P \times Q) &= \pi[(P \times Q) \cap (P \times Q)] = \pi_{P \times Q}(P \times Q) \\ &= \tau^{P \times Q}(P \times Q) = \mu(P \cap P)\nu(Q \cap Q) = \mu(P)\nu(Q);\end{aligned}$$

this establishes (1). If M is any set in $\mathscr{S} \times \mathscr{T}$, then by Lemma 4 and the definition of π, we have (all LUB's are taken over the family of all $P \times Q$'s):

$$\text{LUB}\ \pi[(P \times Q) \cap M] = \text{LUB}\ \pi_{P \times Q}(M) = \text{LUB}\ \tau^{P \times Q}(M) = \pi(M).$$

This establishes (2).

Suppose now that ρ is any measure on $\mathscr{S} \times \mathscr{T}$ having the properties (1) and (2). If $P \times Q$ is a finite rectangle, let us show first that the contractions $\rho_{P \times Q}$ and $\pi_{P \times Q}$ coincide on the ring $\mathscr{R}$ generated by the measurable rectangles. Indeed, if $E \times F$ is any measurable rectangle, then

$$\begin{aligned}\rho_{P \times Q}(E \times F) &= \rho[(P \times Q) \cap (E \times F)] = \rho[(P \cap E) \times (Q \cap F)] \\ &= \mu(P \cap E)\nu(Q \cap F) = \pi[(P \cap E) \times (Q \cap F)] \\ &= \pi_{P \times Q}(E \times F).\end{aligned}$$

It follows from 34.1 that $\rho_{P \times Q}$ and $\pi_{P \times Q}$ agree on $\mathscr{R}$; since they are finite-valued on $\mathscr{R}$, it follows from the (UET) that

$$\rho_{P \times Q} = \pi_{P \times Q}$$

on $\mathfrak{S}(\mathscr{R}) = \mathscr{S} \times \mathscr{T}$. Then $\rho = \pi$ results at once from the respective conditions (2). ▮

With notation as in Theorem 1, we shall write $\pi = \mu \times \nu$; thus, $\mu \times \nu$ is the unique measure on $\mathscr{S} \times \mathscr{T}$ such that:

(1) $(\mu \times \nu)(P \times Q) = \mu(P)\nu(Q)$ for every finite rectangle $P \times Q$,

(2) $(\mu \times \nu)(M) = \text{LUB}\{(\mu \times \nu)[(P \times Q) \cap M] : P \in \mathscr{S}_\varphi, Q \in \mathscr{T}_\varphi\}$, for each M in $\mathscr{S} \times \mathscr{T}$.

Incidentally, if μ and ν happen to be finite measures, then every measurable rectangle is a finite rectangle, hence condition (1) ensures that the measure $\mu \times \nu$ constructed in this section is identical with the measure $\mu \times \nu$ constructed in the preceding section. In the general case, the result of Lemma 4 may be written thus:

$$(\mu \times \nu)_{P \times Q} = \mu_P \times \nu_Q.$$

The results are simpler for the σ-finite case, as we shall see in the next theorem.

Lemma 5. *If $E \times F$ is a measurable rectangle with σ-finite sides, then*

$$(\mu \times \nu)(E \times F) = \mu(E)\nu(F).$$

Proof. We mean, of course, that each of the sets E and F is expressible as the union of a sequence of measurable sets of finite measure. Let P_n be a sequence of sets in $\mathscr{S}_\varphi$ such that $P_n \uparrow E$, and similarly let $Q_n \uparrow F$. Evidently

$$P_n \times Q_n \uparrow E \times F,$$

hence

$$\begin{aligned}(\mu \times \nu)(E \times F) &= \text{LUB}\,(\mu \times \nu)(P_n \times Q_n)\\ &= \text{LUB}\,\mu(P_n)\nu(Q_n) = \mu(E)\nu(F)\end{aligned}$$

by Lemma 2 in Sect. 10 (this is the first use made of that lemma). ▮

Theorem 2. *If $(X, \mathscr{S}, \mu)$ and $(Y, \mathscr{T}, \nu)$ are σ-finite measure spaces, there exists a unique measure $\mu \times \nu$ on $\mathscr{S} \times \mathscr{T}$ such that*

(1) $$(\mu \times \nu)(P \times Q) = \mu(P)\nu(Q)$$

for every finite rectangle $P \times Q$. Moreover, $\mu \times \nu$ is σ-finite, and

(1′) $$(\mu \times \nu)(E \times F) = \mu(E)\nu(F)$$

for every measurable rectangle $E \times F$.

Proof. Let us write $\mathscr{R}_\varphi$ for the ring generated by the finite rectangles $P \times Q$. Since $\mathfrak{S}(\mathscr{S}_\varphi) = \mathscr{S}$ and $\mathfrak{S}(\mathscr{T}_\varphi) = \mathscr{T}$ by the assumed σ-finiteness, it follows from 35.2 that $\mathfrak{S}(\mathscr{R}_\varphi) = \mathscr{S} \times \mathscr{T}$.

The measure $\mu \times \nu$ defined earlier in the section satisfies (1′) by Lemma 5. Since each set in $\mathscr{S} \times \mathscr{T}$ can be covered by a countable union of finite rectangles (1.4), it is clear that $\mu \times \nu$ is σ-finite.

If ρ is any measure on $\mathscr{S} \times \mathscr{T}$ which satisfies (1), it follows from 34.1 that the restriction of ρ to $\mathscr{R}_\varphi$ is a finite measure coinciding with $\mu \times \nu$, hence $\rho = \mu \times \nu$ on $\mathfrak{S}(\mathscr{R}_\varphi) = \mathscr{S} \times \mathscr{T}$ by the (UET). ▌

Our particular definition of product measure might well be called the "semifinite product" (see Exercise 15). It works quite smoothly for semifinite measures (see Exercise 18), and best of all for σ-finite measures (see the next three sections). However, the price one pays for uniqueness, namely condition (2) in Theorem 1, can lead to pathology (see Exercise 17).

EXERCISES

1. Suppose $\mu_1, \ldots, \mu_n$ are σ-finite measures defined on the σ-rings $\mathscr{S}_1, \ldots, \mathscr{S}_n$, respectively, and let $\mathscr{S}$ be the σ-ring generated by the class of all sets of the form

$$E = E_1 \times \cdots \times E_n$$

with $E_i \in \mathscr{S}_i$. Then there exists a unique measure μ on $\mathscr{S}$ such that

$$\mu(E) = \mu_1(E_1) \cdots \mu_n(E_n)$$

for all such sets E.

2. If $f \in \mathscr{L}^1(\mu)$ and $g \in \mathscr{L}^1(\nu)$, and one defines

$$h(x, y) = f(x)g(y),$$

then $h \in \mathscr{L}^1(\mu \times \nu)$, and $\int h = \left(\int f\right)\left(\int g\right)$.

3. If μ and ν are σ-finite, then

$$(\mu \times \nu)(E \times F) = \mu(E)\nu(F)$$

for every measurable rectangle $E \times F$ (Theorem 2). What is the situation for arbitrary measures?

***4.** If μ_1, μ_2 are finite measures on $\mathscr{S}$, and ν_1, ν_2 are finite measures on $\mathscr{T}$, and if $\mu_1 \leq \mu_2$ and $\nu_1 \leq \nu_2$, then

$$\mu_1 \times \nu_1 \leq \mu_2 \times \nu_2$$

by 38.3. On the basis of Exercise 6.1, it is easy to see that the same is true if the measures are merely assumed to be σ-finite. What if the measures are completely arbitrary?

***5.** If μ and ν are σ-finite, then

$$(\mu \times \nu)_{E \times F} = \mu_E \times \nu_F$$

for every measurable rectangle $E \times F$. What if μ and ν are arbitrary measures?

***6.** If μ is a σ-finite measure on $\mathscr{S}$, and ν_1, ν_2 are σ-finite measures on $\mathscr{T}$, then

$$\mu \times (\nu_1 + \nu_2) = \mu \times \nu_1 + \mu \times \nu_2.$$

Can the assumption of σ-finiteness be dropped?

***7.** Let $(\mu_i)_{i \in I}$ and $(\nu_j)_{j \in J}$ be increasingly directed families of measures on $\mathscr{S}$ and $\mathscr{T}$, respectively, and let $\mu = \text{LUB}\, \mu_i$, $\nu = \text{LUB}\, \nu_j$ (as in Sect. 10). Assuming the μ_i and ν_j are σ-finite, it follows from Exercise 4 that the family of measures

$$(\mu_i \times \nu_j)_{(i,j) \in I \times J}$$

is increasingly directed. If, moreover, μ and ν are σ-finite, it is easy to see that

$$\mu \times \nu = \text{LUB}\, \mu_i \times \nu_j.$$

Can the assumption of σ-finiteness be dropped? (It is understood that $(i,j) \leq (i',j')$ if and only if both $i \leq i'$ and $j \leq j'$.)

8. If $f \in \mathscr{L}^1(\mu)$ and $g \in \mathscr{L}^1(\nu)$, $f \geq 0$, $g \geq 0$, and if one defines

$$h(x,y) = f(x)g(y),$$

then $h \in \mathscr{L}^1(\mu \times \nu)$, and $(\mu \times \nu)_h = \mu_f \times \nu_g$ (see Sect. 26 for the notations).

***9.** If μ, ν, and ρ are σ-finite measures, then

$$(\mu \times \nu) \times \rho = \mu \times (\nu \times \rho)$$

in the obvious sense. Is σ-finiteness necessary?

***10.** Recall that if μ is a measure on a σ-ring $\mathscr{S}$, then μ_λ is the extension of μ to $\mathscr{S}_\lambda$ described in Exercise 17.1. A comparison of $\mu_\lambda \times \nu_\lambda$ with $(\mu \times \nu)_\lambda$ entails a comparison of $\mathscr{S}_\lambda \times \mathscr{T}_\lambda$ with $(\mathscr{S} \times \mathscr{T})_\lambda$. In any case, we have

$$\mathscr{S}_\lambda \times \mathscr{T}_\lambda \subset (\mathscr{S} \times \mathscr{T})_\lambda$$

by Exercise 35.1, and we may ask: is the restriction of $(\mu \times \nu)_\lambda$ to $\mathscr{S}_\lambda \times \mathscr{T}_\lambda$ equal to $\mu_\lambda \times \nu_\lambda$?

***11.** What is the relation between $\hat{\mu} \times \hat{\nu}$ and $(\mu \times \nu)^\wedge$, where $\hat{\mu}$ is the completion of μ?

***12.** If $\pi = \mu \times \nu$, and h is a function in $\mathscr{L}^1(\pi)$ such that

$$\int_{E \times F} h \, d\pi \geq 0$$

for all measurable rectangles $E \times F$, then $h \geq 0$ a.e. Is it sufficient to assume that

$$\int_{P \times Q} h \, d\pi \geq 0$$

for all finite rectangles $P \times Q$?

13. Let $(X, \mathscr{S}, \mu)$ be any σ-finite measure space, and consider the measure space $(R, \mathscr{B}, m)$, where m is the Borel restriction of Lebesgue measure. Form the product measure space

$$(X \times R, \mathscr{S} \times \mathscr{B}, \mu \times m).$$

If f is any real-valued function defined on X, such that $f \geq 0$, the *ordinate set* of f is the subset O_f of $X \times R$ consisting of all pairs (x, c) such that $0 \leq c < f(x)$; thus,

$$O_f = \bigcup_{x \in X} \{x\} \times [0, f(x)).$$

(*i*) If $f = \chi_E$, then $O_f = E \times [0, 1)$.

(*ii*) If $0 \leq f_n \uparrow f$, then $O_{f_n} \uparrow O_f$.

(*iii*) If f is measurable, then $O_f \in \mathscr{S} \times \mathscr{B}$.

(*iv*) If f is integrable, then $\int f \, d\mu = (\mu \times m)(O_f)$.

14. Let μ_1, μ_2 be measures on $\mathscr{S}$, and ν_1, ν_2 measures on $\mathscr{T}$, such that $\mu_1 \leq \mu_2$ and $\nu_1 \leq \nu_2$. Assume, moreover, that μ_1 and μ_2 admit the same sets of finite measure (that is, $\mu_1(E) < \infty$ if and only if $\mu_2(E) < \infty$), and similarly for ν_1 and ν_2. Then $\mu_1 \times \nu_1 \leq \mu_2 \times \nu_2$.

15. The measure $\mu \times \nu$ is always semifinite in the sense of Exercise 25.9.

16. Let $(X, \mathscr{S}, \mu)$ and $(Y, \mathscr{T}, \nu)$ be arbitrary measure spaces, and suppose h is a real-valued function on $X \times Y$ which is measurable with respect to $\mathscr{S} \times \mathscr{T}$. In order that h be $(\mu \times \nu)$-integrable, it is necessary and sufficient that it be $(\mu_P \times \nu_Q)$-integrable for all finite rectangles $P \times Q$, and that

$$\text{LUB}\left\{\int |h| \, d(\mu_P \times \nu_Q) : P \in \mathscr{S}_\varphi, Q \in \mathscr{T}_\varphi\right\} < \infty.$$

In this case, if $h \geq 0$, then

$$\int h \, d(\mu \times \nu) = \text{LUB}\left\{\int h \, d(\mu_P \times \nu_Q) : P \in \mathscr{S}_\varphi, Q \in \mathscr{T}_\varphi\right\}.$$

17. (Johnson) Condition (2) of Theorem 1 assures the uniqueness of π, but opens the door to pathology of the following sort. Let X be an uncountable set, $\mathscr{S}$ the class of all subsets of X, and define $\mu(A)$ to be 0 or ∞ according as A is countable or uncountable; then $\mu \times \nu = 0$ for every measure ν. (The trouble is that μ is highly nonsemifinite.)

18. (Johnson) Let us say that $\mu \times \nu$ is *multiplicative* in case

$$(\mu \times \nu)(E \times F) = \mu(E)\nu(F)$$

for every measurable rectangle $E \times F$. If μ and ν are semifinite, then $\mu \times \nu$ is multiplicative. If $\mu \times \nu$ is multiplicative, and if each of μ and ν possesses at least one nonzero finite value, then μ and ν are semifinite.

19. If $E \times F$ is a measurable rectangle with σ-finite sides, then

$$(\mu \times \nu)_{E \times F} = \mu_E \times \nu_F.$$

40. Product of Two σ-Finite Measure Spaces; Iterated Integrals

Throughout this section, $(X, \mathscr{S}, \mu)$ and $(Y, \mathscr{T}, \nu)$ are *σ-finite* measure spaces, and $\pi = \mu \times \nu$ is the unique measure on $\mathscr{S} \times \mathscr{T}$ such that $\pi(P \times Q) = \mu(P)\nu(Q)$ for every finite rectangle (39.2). We know, moreover, that π is σ-finite, and that $\pi(E \times F) = \mu(E)\nu(F)$ for every measurable rectangle. The role of σ-finiteness will be apparent in the proofs of Theorems 1 and 2; the need for σ-finiteness assumptions rests essentially in the fact that the (MCT) is formulated for *sequences* of integrable functions.

The general nature of the problem considered in this section is as follows. Suppose we are given a function $h: X \times Y \to R$ which is integrable with respect to π, that is, $h \in \mathscr{L}^1(\pi)$. We wish to evaluate $\int h\, d\pi$ by performing an "iterated integration," say by first integrating with respect to y (that is, with respect to ν), and then with respect to x (that is, with respect to μ). To simplify the problem, one looks first at the case that h is the characteristic function of a measurable set of finite measure (see Lemma 3 below). To make the problem even easier, we consider first of all the case that h is the characteristic function of a measurable set M such that $M \subset P \times Q$ for some finite rectangle $P \times Q$ (see Lemma 2 below); this case is treated essentially by the finite results of Sect. 38, and Lemma 1 is simply a technical device for applying them:

Lemma 1. *Let $G \in \mathscr{S}$, and consider the measure space $(X, \mathscr{S}, \mu_G)$, where μ_G is the contraction of μ by G. Suppose that $f: X \to R$ is integrable with respect to μ_G, and that $f = 0$ on $X - G$. Then f is also integrable with respect to μ, and*

$$\int f\, d\mu = \int f\, d\mu_G.$$

Proof. It is implicit in our assumptions that f is measurable with respect to $\mathscr{S}$. (Incidentally, this is an essential advantage of contraction—the concept of measurability remains the same.) By obvious linearity, we may assume that $f \geq 0$ (write $f = f^+ - f^-$, and note that f^+ and f^- also vanish on $X - G$).

Suppose first that $f = \chi_E$ for some measurable set E. Since f vanishes on $X - G$, we have $E \subset G$. By assumption, $f \in \mathscr{L}^1(\mu_G)$, hence $\mu_G(E) < \infty$, and

$$\int f\, d\mu_G = \mu_G(E) = \mu(G \cap E) = \mu(E);$$

in particular $\mu(E) < \infty$, hence $f \in \mathscr{L}^1(\mu)$ and

$$\int f\,d\mu = \mu(E) = \int f\,d\mu_G.$$

The case that f is simple follows at once from the case just considered. In general, there exists a sequence f_n of functions which are ISF with respect to μ_G, such that $0 \le f_n \uparrow f$ and $\int f_n\,d\mu_G$ is bounded. By the preceding case, we have $f_n \in \mathscr{L}^1(\mu)$, and

$$\int f_n\,d\mu = \int f_n\,d\mu_G$$

is bounded, hence $f \in \mathscr{L}^1(\mu)$ by the definition of integrability. By the definition of integral (or by the (MCT)), we have

$$\int f\,d\mu_G = \text{LUB} \int f_n\,d\mu_G = \text{LUB} \int f_n\,d\mu = \int f\,d\mu. \quad ▮$$

(Incidentally, σ-finiteness plays no role in Lemma 1.) Recall the notations introduced in Sect. 37: if $M \in \mathscr{S} \times \mathscr{T}$, the functions $f_M\colon X \to R_e$ and $g^M\colon Y \to R_e$ are defined by the formulas $f_M(x) = \nu(M_x)$ and $g^M(y) = \mu(M^y)$.

Lemma 2. *Suppose $M \subset P \times Q$, where M is in $\mathscr{S} \times \mathscr{T}$ and $P \times Q$ is a finite rectangle. Let $h = \chi_M$. Then:*

(*i*) *For each $x \in X$, one has $h_x \in \mathscr{L}^1(\nu)$ and*

$$\int h_x\,d\nu = f_M(x).$$

(*ii*) *For each $y \in Y$, one has $h^y \in \mathscr{L}^1(\mu)$ and*

$$\int h^y\,d\mu = g^M(y).$$

(*iii*) *Moreover, the functions h, f_M, and g^M are integrable, and*

$$\pi(M) = \int h\,d\pi = \int f_M\,d\mu = \int g^M\,d\nu.$$

Proof. Since $\pi(M) \le \pi(P \times Q) = \mu(P)\nu(Q) < \infty$, we have $h \in \mathscr{L}^1(\pi)$ and $\int h\,d\pi = \pi(M)$.

The plan of the proof is to apply 38.1 to the finite measure spaces

$$(X, \mathscr{S}, \mu_P) \qquad \text{and} \qquad (Y, \mathscr{T}, \nu_Q).$$

Since (see Sect. 39)

$$(\mu_P \times \nu_Q)(M) = (\mu \times \nu)_{P \times Q}(M) = \pi[(P \times Q) \cap M] = \pi(M) < \infty,$$

we have $h \in \mathscr{L}^1(\mu_P \times \nu_Q)$ and

$$\int h\,d(\mu_P \times \nu_Q) = (\mu_P \times \nu_Q)(M) = \pi(M).$$

Let us (temporarily) denote by $\hat{f}_M$ and $\hat{g}^M$ the constructs of Sect. 37 formed relative to the measure spaces

$$(X, \mathscr{S}, \mu_P) \qquad \text{and} \qquad (Y, \mathscr{T}, \nu_Q).$$

Suppose $x \in X$; since $M_x \subset (P \times Q)_x = Q$ or $\varnothing$ (according as $x \in P$ or $x \notin P$), we have $M_x \subset Q$ in any case, hence

$$\hat{f}_M(x) = \nu_Q(M_x) = \nu(Q \cap M_x) = \nu(M_x) = f_M(x).$$

Thus, $\hat{f}_M = f_M$, and similarly $\hat{g}^M = g^M$. Observe that $f_M = 0$ on $X - P$, and $g^M = 0$ on $Y - Q$. Now, by 38.1 we have

$$\hat{f}_M \in \mathscr{L}^1(\mu_P), \qquad \hat{g}^M \in \mathscr{L}^1(\nu_Q),$$

and

$$\int \hat{f}_M \, d\mu_P = \int \hat{g}^M \, d\nu_Q = (\mu_P \times \nu_Q)(M).$$

In other words,

$$f_M \in \mathscr{L}^1(\mu_P), \qquad g^M \in \mathscr{L}^1(\nu_Q),$$

and

$$\int f_M \, d\mu_P = \int g^M \, d\nu_Q = \pi(M).$$

Citing Lemma 1, we have

$$f_M \in \mathscr{L}^1(\mu), \qquad g^M \in \mathscr{L}^1(\nu),$$

and

$$\int f_M \, d\mu = \int g^M \, d\nu = \pi(M).$$

This proves *(iii)*.

If $x \in X$, then $h_x = \chi_{M_x}$; since $M_x \subset Q$, clearly $h_x \in \mathscr{L}^1(\nu)$ and

$$\int h_x \, d\nu = \nu(M_x) = f_M(x).$$

Similarly $y \in Y$ implies $h^y = \chi_{M^y}$, and $M^y \subset P$, hence $h^y \in \mathscr{L}^1(\mu)$ and

$$\int h^y \, d\mu = \mu(M^y) = g^M(y). \quad \blacksquare$$

Theorem 1. *Suppose M is in $\mathscr{S} \times \mathscr{T}$. The following conditions on M are equivalent:*

(a) $\pi(M) < \infty$.

(b) *There exists an $f \in \mathscr{L}^1(\mu)$ such that $f_M = f$ a.e. $[\mu]$.*

(c) *There exists a $g \in \mathscr{L}^1(\nu)$ such that $g^M = g$ a.e. $[\nu]$.*

In this case,

$$\int f \, d\mu = \int g \, d\nu = \pi(M).$$

Proof. Let $E \times F$ be a measurable rectangle such that $M \subset E \times F$ (35.1). By the assumed σ-finiteness, we may choose sequences $P_n \in \mathscr{S}_\varphi$ and $Q_n \in \mathscr{T}_\varphi$ such that $P_n \uparrow E$ and $Q_n \uparrow F$; then

$$P_n \times Q_n \uparrow E \times F.$$

Defining

$$M_n = M \cap (P_n \times Q_n),$$

we have $M_n \uparrow M$, where $M_n \subset P_n \times Q_n$. By this device, we are in a position to exploit Lemma 2 (catalyst: the (MCT)). By Lemma 2, the functions f_{M_n} and g^{M_n} are integrable, and

$$\int f_{M_n}\, d\mu = \int g^{M_n}\, d\nu = \pi(M_n).$$

By 37.3, we have $f_{M_n} \uparrow f_M$ and $g^{M_n} \uparrow g^M$.

(a) *implies* (b): Suppose $\pi(M) < \infty$. Since $f_{M_n} \uparrow f_M$ and

$$\int f_{M_n}\, d\mu = \pi(M_n) \leq \pi(M) < \infty$$

for all n, it follows from the (MCT) that there exists an $f \in \mathscr{L}^1(\mu)$ such that $f_{M_n} \uparrow f$ a.e. $[\mu]$. Evidently $f_M = f$ a.e. $[\mu]$.

(a) *implies* (c): Similarly.

(b) *implies* (a): Suppose $f \in \mathscr{L}^1(\mu)$ and $f_M = f$ a.e. $[\mu]$. Then $f_{M_n} \uparrow f$ a.e. $[\mu]$, hence by the (MCT) we have

$$\int f_{M_n}\, d\mu \uparrow \int f\, d\mu,$$

that is, $\pi(M_n) \uparrow \int f\, d\mu$. But $\pi(M_n) \uparrow \pi(M)$; hence $\pi(M) = \int f\, d\mu < \infty$.

(c) *implies* (a): Similarly, if $g \in \mathscr{L}^1(\nu)$ and $g^M = g$ a.e. $[\nu]$, we deduce that $\pi(M) = \int g\, d\nu < \infty$.

Combining (b) and (c), we have

$$\int f\, d\mu = \pi(M) = \int g\, d\nu. \quad \blacksquare$$

[The condition (b) can be expressed by saying that f_M is "a.e. integrable" with respect to μ. We shall not develop this idea, but simply mention that it is an easy way to discuss "measurability" and "integrability" for extended real valued functions.]

It will be convenient to have a notation for the class of null sets with respect to a measure. Let us write

$$\mathscr{S}_0 = \{E \in \mathscr{S} : \mu(E) = 0\};$$

thus, $\mathscr{S}_0$ is the class of all μ-null subsets of X. Similarly,

$$\mathscr{T}_0 = \{F \in \mathscr{T} : \nu(F) = 0\}$$

is the class of all ν-null subsets of Y. Generic elements of $\mathscr{S}_0$ and $\mathscr{T}_0$ will be denoted by E_0 and F_0, respectively.

Suppose h is an extended real valued function defined on $X \times Y$, that is,

$$h: X \times Y \to R_e.$$

We shall say that h possesses an **iterated integral** $\iint h \, d\nu \, d\mu$ in case there exists an $E_0 \in \mathscr{S}_0$ and an $f \in \mathscr{L}^1(\mu)$ such that

$$x \notin E_0 \quad \text{implies} \quad h_x \in \mathscr{L}^1(\nu) \quad \text{and} \quad \int h_x \, d\nu = f(x).$$

Suppose also $E_0^* \in \mathscr{S}_0$ and $f^* \in \mathscr{L}^1(\mu)$ have the property that

$$x \notin E_0^* \quad \text{implies} \quad h_x \in \mathscr{L}^1(\nu) \quad \text{and} \quad \int h_x \, d\nu = f^*(x).$$

Then $E_0 \cup E_0^*$ is a null set, and

$$x \notin E_0 \cup E_0^* \quad \text{implies} \quad f(x) = \int h_x \, d\nu = f^*(x);$$

thus, $f = f^*$ a.e. $[\mu]$. In view of 25.5, we may unambiguously define the *value* of $\iint h \, d\nu \, d\mu$ to be $\int f \, d\mu$; thus,

$$\iint h \, d\nu \, d\mu = \int f \, d\mu.$$

Similarly, h is said to possess an iterated integral $\iint h \, d\mu \, d\nu$ if there exist $F_0 \in \mathscr{T}_0$ and $g \in \mathscr{L}^1(\nu)$ such that

$$y \notin F_0 \quad \text{implies} \quad h^y \in \mathscr{L}^1(\mu) \quad \text{and} \quad \int h^y \, d\mu = g(y);$$

one then unambiguously defines

$$\iint h \, d\mu \, d\nu = \int g \, d\nu.$$

The final assertion in Theorem 1 can be stated in terms of iterated integrals:

Lemma 3. *Suppose $M \in \mathscr{S} \times \mathscr{T}$, and $\pi(M) < \infty$. Let $h = \chi_M$. Then, both iterated integrals of h exist, and*

$$\iint h \, d\nu \, d\mu = \iint h \, d\mu \, d\nu = \int h \, d\pi = \pi(M).$$

Proof. By Theorem 1, there exists an $f \in \mathscr{L}^1(\mu)$ such that $f_M = f$ a.e. $[\mu]$. Let E_0 be a μ-null set on whose complement $f_M(x) = f(x)$. Recall that

$$h_x = (\chi_M)_x = \chi_{M_x};$$

if $x \notin E_0$, then

$$\nu(M_x) = f_M(x) = f(x) < \infty,$$

hence $h_x \in \mathscr{L}^1(\nu)$ and

$$\int h_x \, d\nu = \nu(M_x) = f(x).$$

This shows that $\iint h \, d\nu \, d\mu$ exists, and is equal to $\int f \, d\mu$. Similarly, there exists a $g \in \mathscr{L}^1(\nu)$ such that $g^M = g$ a.e. $[\nu]$, and one deduces that $\iint h \, d\mu \, d\nu$ exists, and is equal to $\int g \, d\nu$. Finally,

$$\int f \, d\mu = \int g \, d\nu = \int h \, d\pi$$

by Theorem 1. ∎

The next two lemmas lead up to a generalization of Lemma 3. The first asserts that iterated integration is a linear process:

Lemma 4. *Let h and k be real-valued functions defined on $X \times Y$, such that the iterated integrals $\iint h \, d\nu \, d\mu$ and $\iint k \, d\nu \, d\mu$ exist, and let a and b be real numbers. Then the iterated integral*

$$\iint (ah + bk) \, d\nu \, d\mu$$

exists, and is equal to

$$a \iint h \, d\nu \, d\mu + b \iint k \, d\nu \, d\mu.$$

Proof. Choose $E_0, E_0^* \in \mathscr{S}_0$, and $f, f^* \in \mathscr{L}^1(\mu)$, such that

$$x \notin E_0 \quad \text{implies} \quad h_x \in \mathscr{L}^1(\nu) \quad \text{and} \quad \int h_x \, d\nu = f(x),$$

while

$$x \notin E_0^* \quad \text{implies} \quad k_x \in \mathscr{L}^1(\nu) \quad \text{and} \quad \int k_x \, d\nu = f^*(x).$$

Then $E_0 \cup E_0^*$ is a null set, $af + bf^*$ is integrable, and

$$x \notin E_0 \cup E_0^* \quad \text{implies} \quad (ah + bk)_x = ah_x + bk_x \in \mathscr{L}^1(\nu)$$

and

$$\begin{aligned} \int (ah + bk)_x \, d\nu &= \int (ah_x + bk_x) \, d\nu \\ &= a \int h_x \, d\nu + b \int k_x \, d\nu \\ &= af(x) + bf^*(x) = (af + bf^*)(x). \end{aligned}$$

This shows that $\iint (ah + bk)\, d\nu\, d\mu$ exists, and is equal to

$$\int (af + bf^*)\, d\mu = a \int f\, d\mu + b \int f^*\, d\mu$$
$$= a \iint h\, d\nu\, d\mu + b \iint k\, d\nu\, d\mu. \quad \blacksquare$$

The next lemma is a slight generalization of Lemma 2:

Lemma 5. *Suppose h is simple with respect to $\mathscr{S} \times \mathscr{T}$, and $N(h) \subset P \times Q$ for some finite rectangle $P \times Q$. Then:*

(*i*) *h is an* ISF *with respect to π.*

(*ii*) *Both iterated integrals of h exist, and*

$$\iint h\, d\nu\, d\mu = \iint h\, d\mu\, d\nu = \int h\, d\pi.$$

More precisely:

(*iii*) *For each $x \in X$, we have $h_x \in \mathscr{L}^1(\nu)$, and defining $f(x) = \int h_x\, d\nu$, we have $f \in \mathscr{L}^1(\mu)$.*

(*iv*) *For each $y \in Y$, we have $h^y \in \mathscr{L}^1(\mu)$, and defining $g(y) = \int h^y\, d\mu$, we have $g \in \mathscr{L}^1(\nu)$.*

(*v*) *Thus, $\int f\, d\mu = \iint h\, d\nu\, d\mu = \iint h\, d\mu\, d\nu = \int g\, d\nu$.*

Proof. (*i*) $\pi(N(h)) \le \pi(P \times Q) = \mu(P)\nu(Q) < \infty$.

(*ii*) is clear from Lemmas 3 and 4.

(*iii*) and (*iv*) are clear from Lemma 2. $\blacksquare$

We are now ready to prove what is essentially *Fubini's theorem:*

Theorem 2. *Suppose h is integrable with respect to π, and $h \ge 0$. Then both iterated integrals of h exist, and*

$$\iint h\, d\nu\, d\mu = \iint h\, d\mu\, d\nu = \int h\, d\pi.$$

Proof. In the statement of Fubini's theorem in the next section, the hypothesis will be weakened to $h \in \mathscr{L}^1(\pi)$, but for the present we are assuming moreover that $h \ge 0$ (everywhere on $X \times Y$). Since h is measurable, $N(h)$ is in $\mathscr{S} \times \mathscr{T}$, hence by 35.1 there exists a measurable rectangle $E \times F$ such that $N(h) \subset E \times F$. Choose sequences $P_n \in \mathscr{S}_\varphi$ and $Q_n \in \mathscr{T}_\varphi$ such that $P_n \uparrow E$ and $Q_n \uparrow F$ (possible by σ-finiteness); then

$$P_n \times Q_n \uparrow E \times F,$$

hence

$$\chi_{P_n \times Q_n} \uparrow \chi_{E \times F}.$$

Let h_n be a sequence of simple functions such that $0 \le h_n \uparrow h$ (16.4). Since $N(h) \subset E \times F$, we have

$$\chi_{P_n \times Q_n} h_n \uparrow \chi_{E \times F} h = h;$$

replacing h_n by $\chi_{P_n \times Q_n} h_n$, we may assume that $N(h_n) \subset P_n \times Q_n$. In particular,

$$\pi(N(h_n)) \le \pi(P_n \times Q_n) = \mu(P_n)\nu(Q_n) < \infty$$

shows that the h_n are ISF with respect to π.

Summarizing, we have $h_n \uparrow h$, where the h_n and h are integrable with respect to π; also each h_n is simple, and $N(h_n)$ is contained in a finite rectangle. By the (MCT), we know

$$\int h_n \, d\pi \uparrow \int h \, d\pi.$$

Now, Lemma 5 may be applied to each h_n: all x-sections and all y-sections of h_n are integrable, and defining

$$f_n(x) = \int (h_n)_x \, d\nu \qquad \text{and} \qquad g_n(y) = \int (h_n)^y \, d\mu,$$

we have $f_n \in \mathscr{L}^1(\mu)$, $g_n \in \mathscr{L}^1(\nu)$, and

$$\int f_n \, d\mu = \int g_n \, d\nu = \int h_n \, d\pi.$$

For each $x \in X$, we have $(h_n)_x \in \mathscr{L}^1(\nu)$, and $(h_n)_x \uparrow h_x$ pointwise on Y as $n \to \infty$; in particular,

$$f_n(x) = \int (h_n)_x \, d\nu \uparrow$$

as $n \to \infty$. Thus, $f_n \uparrow$ pointwise on X. But $f_n \in \mathscr{L}^1(\mu)$, and

$$\int f_n \, d\mu = \int h_n \, d\pi \uparrow \int h \, d\pi < \infty;$$

by the (MCT) for μ, there exists an $f \in \mathscr{L}^1(\mu)$ such that $f_n \uparrow f$ a.e. $[\mu]$. Then,

$$\int f \, d\mu = \text{LUB} \int f_n \, d\mu = \text{LUB} \int h_n \, d\pi = \int h \, d\pi.$$

Similarly, there exists a $g \in \mathscr{L}^1(\nu)$ such that

$$g_n \uparrow g \text{ a.e. } [\nu], \qquad \int g \, d\nu = \int h \, d\pi.$$

Let E_0 be a null set such that $x \notin E_0$ implies $f_n(x) \uparrow f(x)$. For *all* $x \in X$ we have $(h_n)_x \in \mathscr{L}^1(\nu)$ and $(h_n)_x \uparrow h_x$. Fix $x \in X - E_0$; then $(h_n)_x \in \mathscr{L}^1(\nu)$, and

$$\int (h_n)_x \, d\nu = f_n(x) \uparrow f(x) < \infty;$$

hence by the (MCT) for ν we conclude that the measurable (36.8) function h_x is integrable with respect to ν, and

$$\int h_x \, d\nu = \text{LUB} \int (h_n)_x \, d\nu = \text{LUB} f_n(x) = f(x).$$

It follows that $\iint h \, d\nu \, d\mu$ exists, and is equal to $\int f \, d\mu$. Similarly $\iint h \, d\mu \, d\nu$ exists and is equal to $\int g \, d\nu$. Finally,

$$\int f \, d\mu = \int h \, d\pi = \int g \, d\nu. \quad \blacksquare$$

EXERCISES

1. Let f, f_0 be measurable functions on a σ-finite measure space $(X, \mathscr{S}, \mu)$, and g, g_0 measurable functions on a σ-finite measure space $(Y, \mathscr{T}, \nu)$, and define

$$h(x, y) = f(x)g(y), \qquad h_0(x, y) = f_0(x)g_0(y).$$

If $f = f_0$ a.e. $[\mu]$ and $g = g_0$ a.e. $[\nu]$, then $h = h_0$ a.e. $[\mu \times \nu]$.

2. If the iterated integral $\iint h \, d\nu \, d\mu$ exists, and $M \in \mathscr{S} \times \mathscr{T}$, does $\iint \chi_M h \, d\nu \, d\mu$ necessarily exist?

3. Suppose $(X, \mathscr{S}, \mu)$ and $(Y, \mathscr{T}, \nu)$ are arbitrary measure spaces, f and f_0 are μ-integrable functions, g and g_0 are ν-integrable functions, and suppose that

$$f = f_0 \text{ a.e. } [\mu], \qquad g = g_0 \text{ a.e. } [\nu].$$

Define

$$h(x, y) = f(x)g(y) \qquad \text{and} \qquad h_0(x, y) = f_0(x)g_0(y).$$

Then $h = h_0$ a.e. $[\mu \times \nu]$.

41. Fubini's Theorem

As in the preceding section, we assume that $(X, \mathscr{S}, \mu)$ and $(Y, \mathscr{T}, \nu)$ are σ-finite measures spaces, and $\pi = \mu \times \nu$. The following result is known as *Fubini's theorem:*

Theorem 1. *If h is integrable with respect to π, then both iterated integrals of h exist, and*

$$\iint h \, d\nu \, d\mu = \iint h \, d\mu \, d\nu = \int h \, d\pi.$$

Proof. If $h \geq 0$, this is 40.2. In general, write $h = h^+ - h^-$ and cite Lemma 4 in Sect. 40. $\blacksquare$

There is a partial converse to the Fubini theorem:

Theorem 2. *Suppose* $h: X \times Y \to R$ *is measurable with respect to* $\mathscr{S} \times \mathscr{T}$, *and* $h \geq 0$. *If one of the iterated integrals of h exists, then h is integrable with respect to* π, *hence by Fubini's theorem the other iterated integral also exists, and*

$$\iint h\,d\nu\,d\mu = \iint h\,d\mu\,d\nu = \int h\,d\pi.$$

Proof. In the next section, the hypothesis is weakened to $h \geq 0$ a.e. $[\pi]$, but for the present we are assuming that $h \geq 0$ (everywhere).

Let $0 \leq h_n \uparrow h$, with all notations as in the proof of 40.2. In particular, all of the sections of h_n are integrable, and defining

$$f_n(x) = \int (h_n)_x\,d\nu \qquad \text{and} \qquad g_n(y) = \int (h_n)^y\,d\mu,$$

we have $f_n \in \mathscr{L}^1(\mu)$, $g_n \in \mathscr{L}^1(\nu)$, and

$$\int f_n\,d\mu = \int g_n\,d\nu = \int h_n\,d\pi.$$

Moreover, $f_n\uparrow$ and $g_n\uparrow$. (All that is needed for this is that h be a non-negative measurable function.)

Suppose, for instance, that $\iint h\,d\nu\,d\mu$ exists. Let E_0 be a μ-null set, and $f \in \mathscr{L}^1(\mu)$, such that

$$x \notin E_0 \qquad \text{implies} \qquad h_x \in \mathscr{L}^1(\nu) \qquad \text{and} \qquad \int h_x\,d\nu = f(x).$$

Fix an $x \in X - E_0$; since $(h_n)_x \uparrow h_x$, and these functions are all integrable, we have

$$\int (h_n)_x\,d\nu \uparrow \int h_x\,d\nu$$

by the (MCT), thus $f_n(x) \uparrow f(x)$. We have shown that $f_n \uparrow f$ a.e. $[\mu]$; again by the (MCT) we have

$$\int f\,d\mu = \text{LUB} \int f_n\,d\mu = \text{LUB} \int h_n\,d\pi.$$

In particular, $\int h_n\,d\pi$ is bounded; since $h_n \uparrow h$ and h is measurable, it follows from the (MCT) that h is integrable. ∎

EXERCISES

1. If h is a measurable function on $X \times Y$, then h is π-integrable if and only if the iterated integral $\iint \chi_M h\,d\nu\,d\mu$ exists for every M in $\mathscr{S} \times \mathscr{T}$.

2. If h is a measurable function on $X \times Y$, then h is π-integrable if and only if the iterated integral $\iint |h|\,d\nu\,d\mu$ exists.

3. It is easy to construct an example (for suitable μ and ν) of a measurable function h on $X \times Y$ such that (*i*) every section of h is integrable, with integral zero, and (hence) (*ii*) both iterated integrals of h exist, and are equal to zero, but (*iii*) h is not π-integrable.

***4.** What part, if any, of the Fubini theory survives for the product of arbitrary (not necessarily σ-finite) measures? Does it help to assume that the measures are semifinite?

*42. Complements

Notations and σ-finiteness assumptions are the same as in the preceding section.

We now consider several refinements of the preceding material. Most of these results are based directly on 40.1, and do not make use of Fubini's theorem or its converse (40.2, 41.1, 41.2); those proofs which depend only on 40.1 will be so marked.

Theorem 1. *If $M \in \mathscr{S} \times \mathscr{T}$, the following conditions on M are equivalent:*

(a) $\pi(M) = 0$.

(b) $\nu(M_x) = 0$ *for almost all x.*

(c) $\mu(M^y) = 0$ *for almost all y.*

Proof (based on 40.1). (a) *implies* (b) *and* (c): By 40.1, there exist $f \in \mathscr{L}^1(\mu)$ and $g \in \mathscr{L}^1(\nu)$ such that $f = f_M$ a.e. $[\mu]$, $g = g^M$ a.e. $[\nu]$, and

$$\int f\, d\mu = \int g\, d\nu = \pi(M) = 0.$$

Since $f_M \geq 0$, we have $f \geq 0$ a.e. $[\mu]$, hence $f = 0$ a.e. $[\mu]$ by 25.9. Similarly $g = 0$ a.e. $[\nu]$. Then, $f_M = 0$ a.e. $[\mu]$ and $g^M = 0$ a.e. $[\nu]$, and these are the conditions (b) and (c), respectively.

(b) *implies* (a): We are assuming that $f_M = 0$ a.e. $[\mu]$. Since $0 \in \mathscr{L}^1(\mu)$, it follows from 40.1 that $\pi(M) < \infty$, and $\pi(M) = \int 0\, d\mu = 0$. ∎

Corollary 1. *Suppose h and k are extended real valued functions defined on $X \times Y$, such that $h = k$ a.e. $[\pi]$. Then,*

(*i*) $h_x = k_x$ *a.e.* $[\nu]$, *for almost all x.*

(*ii*) $h^y = k^y$ *a.e.* $[\mu]$, *for almost all y.*

Proof (based on 40.1). By assumption there exists a π-null set M such that $h(x, y) = k(x, y)$ on the complement of M. Then, by Theorem 1, there exists a μ-null set E_0 such that $x \notin E_0$ implies $\nu(M_x) = 0$. Fix $x \in X - E_0$; if $y \notin M_x$, then $(x, y) \notin M$, hence

$$h_x(y) = h(x, y) = k(x, y) = k_x(y),$$

and since M_x is a ν-null set, this means that $h_x = k_x$ a.e. $[\nu]$. This proves (i), and (ii) follows similarly. ∎

Corollary 2. *Let h be an extended real valued function such that $\iint h\,d\nu\,d\mu$ exists, and suppose k is a measurable function such that $h = k$ a.e. $[\pi]$. Then $\iint k\,d\nu\,d\mu$ also exists, and*

$$\iint k\,d\nu\,d\mu = \iint h\,d\nu\,d\mu.$$

Proof (*based on 40.1*). By Corollary 1, there exists a μ-null set E_0^* such that $x \notin E_0^*$ implies $h_x = k_x$ a.e. $[\nu]$. On the other hand, since $\iint h\,d\nu\,d\mu$ exists, there is a μ-null set E_0 and an $f \in \mathscr{L}^1(\mu)$ such that

$$x \notin E_0 \qquad \text{implies} \qquad h_x \in \mathscr{L}^1(\nu) \qquad \text{and} \qquad \int h_x\,d\nu = f(x).$$

Now, every section of k is measurable by 36.8. Hence if $x \notin E_0^* \cup E_0$, the relations $h_x = k_x$ a.e. $[\nu]$ and $h_x \in \mathscr{L}^1(\nu)$ imply (25.5) that k_x is integrable, and

$$\int k_x\,d\nu = \int h_x\,d\nu = f(x).$$

Since $E_0^* \cup E_0$ is a μ-null set and $f \in \mathscr{L}^1(\mu)$, we have shown that $\iint k\,d\nu\,d\mu$ exists, and is equal to

$$\int f\,d\mu = \iint h\,d\nu\,d\mu. \quad ∎$$

Strengthening 41.2, we have:

Corollary 3. *Suppose h is measurable with respect to $\mathscr{S} \times \mathscr{T}$, and $h \geq 0$ a.e. $[\pi]$. If one of the iterated integrals of h exists, then h is integrable with respect to π, hence by Fubini's theorem the other iterated integral also exists, and*

$$\iint h\,d\nu\,d\mu = \iint h\,d\mu\,d\nu = \int h\,d\pi.$$

Proof. Say $\iint h\,d\nu\,d\mu$ exists. If M is a π-null set on whose complement $h(x, y) \geq 0$, then the function $k = \chi_{X \times Y - M} h$ is measurable (15.1), $k = h$ a.e. $[\pi]$, and $k \geq 0$. Then $\iint k\,d\nu\,d\mu$ exists by Corollary 2; since k is measurable and $k \geq 0$, it follows from 41.2 that $k \in \mathscr{L}^1(\pi)$, hence $h \in \mathscr{L}^1(\pi)$ by 25.5. ∎

The next two corollaries show how to replace certain functions h by functions k whose sections are well behaved:

Corollary 4. *Suppose h is a measurable function such that $\iint h\,d\nu\,d\mu$ exists. Then there exists a measurable function k such that:*

(i) $h = k$ a.e. $[\pi]$.

(ii) For each $x \in X$, we have $k_x \in \mathscr{L}^1(\nu)$, and defining $f(x) = \int k_x\,d\nu$ we have $f \in \mathscr{L}^1(\mu)$.

(iii) There exists a μ-null set E_0 such that $x \notin E_0$ implies $h_x = k_x$ (everywhere on Y); briefly, $h_x = k_x$, for almost all x.

(iv) $\iint k\,d\nu\,d\mu$ exists, and is equal to $\iint h\,d\nu\,d\mu$.

Proof (based on 40.1). Let E_0 be a μ-null set and $f \in \mathscr{L}^1(\mu)$, such that

$$x \notin E_0 \quad \text{implies} \quad h_x \in \mathscr{L}^1(\nu) \quad \text{and} \quad \int h_x\,d\nu = f(x).$$

Replacing f by $\chi_{X - E_0} f$, we may assume without loss of generality that $f = 0$ on E_0 (25.2). By definition,

$$\iint h\,d\nu\,d\mu = \int f\,d\mu.$$

Since $N(h) \in \mathscr{S} \times \mathscr{T}$, we have $N(h) \subset E \times F$ for a suitable measurable rectangle $E \times F$ (35.1). Evidently $h\chi_{E \times F} = h$. Define

$$k(x, y) = \begin{cases} h(x, y) & \text{if } x \notin E_0, \\ 0 & \text{if } x \in E_0. \end{cases}$$

Thus,

$$k = \begin{cases} h & \text{on } (X - E_0) \times Y, \\ 0 & \text{on } E_0 \times Y, \end{cases}$$

in other words,

$$k = h\chi_{(X - E_0) \times Y} = h\chi_{E \times F}\chi_{(X - E_0) \times Y}$$

$$= h\chi_{(E \times F) \cap [(X - E_0) \times Y]} = h\chi_{(E - E_0) \times F};$$

in particular, since $(E - E_0) \times F \in \mathscr{S} \times \mathscr{T}$, it follows from 15.1 that k is measurable with respect to $\mathscr{S} \times \mathscr{T}$.

We assert that $k = h$ a.e. $[\pi]$. Now, $k = h$ on $(X - E_0) \times Y$. Also, if $h(x, y) = 0$, then $k(x, y) = 0$ (k is a "multiple" of h). Hence if $h(x, y) \neq k(x, y)$, necessarily $(x, y) \in E_0 \times Y$ and $h(x, y) \neq 0$, that is,

$$(x, y) \in (E_0 \times Y) \cap N(h) \subset (E_0 \times Y) \cap (E \times F) = (E_0 \cap E) \times F.$$

But

$$\pi[(E_0 \cap E) \times F] \leq \pi(E_0 \times F) = \mu(E_0)\nu(F) = 0\nu(F) = 0,$$

thus $(E_0 \cap E) \times F$ is a π-null set on whose complement $h(x, y) = k(x, y)$.

Then $\iint k\, d\nu\, d\mu$ exists by Corollary 2, and is equal to $\iint h\, d\nu\, d\mu$. Indeed, since $k = h$ on $(X - E_0) \times Y$ and $k = 0$ on $E_0 \times Y$, we have: $x \notin E_0$ implies

$$k_x = h_x \in \mathscr{L}^1(\nu) \qquad \text{and} \qquad \int k_x\, d\nu = \int h_x\, d\nu = f(x),$$

whereas $x \in E_0$ implies

$$k_x = 0 \in \mathscr{L}^1(\nu) \qquad \text{and} \qquad \int k_x\, d\nu = 0 = f(x),$$

thus the conditions (*ii*) and (*iii*) are fulfilled. ▮

Finally, we show how to "smooth out" any h in $\mathscr{L}^1(\pi)$ so that all of its sections are pleasant:

Corollary 5. *Suppose $h \in \mathscr{L}^1(\pi)$. Then, there exists $k \in \mathscr{L}^1(\pi)$ such that:*

(*i*) *$k = h$ a.e. $[\pi]$.*

(*ii*) *Every section of k is integrable.*

(*iii*) *Defining $f(x) = \int k_x\, d\nu$, we have $f \in \mathscr{L}^1(\mu)$.*

(*iv*) *Defining $g(y) = \int k^y\, d\mu$, we have $g \in \mathscr{L}^1(\nu)$.*

(*v*) *There exists a μ-null set E_0 and a ν-null set F_0, such that*

$$x \notin E_0 \quad \textit{implies} \quad h_x = k_x \text{ a.e. } [\nu] \quad (\textit{hence } h_x \in \mathscr{L}^1(\nu)),$$

$$y \notin F_0 \quad \textit{implies} \quad h^y = k^y \text{ a.e. } [\mu] \quad (\textit{hence } h^y \in \mathscr{L}^1(\mu)).$$

Proof. By Fubini's theorem, $\iint h\, d\nu\, d\mu$ exists. Let E_0 be a μ-null set and $f \in \mathscr{L}^1(\mu)$, such that

$$x \notin E_0 \quad \text{implies} \quad h_x \in \mathscr{L}^1(\nu) \quad \text{and} \quad \int h_x\, d\nu = f(x).$$

Replacing f by $\chi_{X - E_0} f$, we may assume $f = 0$ on E_0. Similarly, there exists a ν-null set F_0 and $g \in \mathscr{L}^1(\nu)$, such that

$$y \notin F_0 \quad \text{implies} \quad h^y \in \mathscr{L}^1(\mu) \quad \text{and} \quad \int h^y\, d\mu = g(y),$$

and moreover $g = 0$ on F_0.

Let $E \times F$ be a measurable rectangle such that $N(h) \subset E \times F$. Define

$$k(x, y) = \begin{cases} h(x, y) & \text{on } (X - E_0) \times (Y - F_0), \\ 0 & \text{on } E_0 \times Y, \\ 0 & \text{on } (X - E_0) \times F_0; \end{cases}$$

in other words, $k = h\chi_{(X - E_0) \times (Y - F_0)}$.

Since $h = h\chi_{E \times F}$, we see that $k = h\chi_{(E - E_0) \times (F - F_0)}$, and in particular k is measurable; indeed, $k \in \mathscr{L}^1(\pi)$ by 25.2.

Since k vanishes wherever h does, and since $k = h$ on $(X - E_0) \times (Y - F_0)$, k and h can differ only on

$$\begin{aligned} N(h) \cap [(E_0 \times Y) \cup ((X - E_0) \times F_0)] & \\ &\subset (E \times F) \cap [(E_0 \times Y) \cup ((X - E_0) \times F_0)] \\ &= [(E \cap E_0) \times F] \cup [(E - E_0) \times (F \cap F_0)] \\ &\subset (E_0 \times F) \cup (E \times F_0); \end{aligned}$$

since $(E_0 \times F) \cup (E \times F_0)$ is clearly a π-null set, we see that $k = h$ a.e. $[\pi]$.

(ii) through (v): Suppose $x \in X - E_0$; then

$$k_x = \begin{cases} h_x & \text{on } Y - F_0, \\ 0 & \text{on } F_0, \end{cases}$$

hence $k_x = h_x$ a.e. $[\nu]$, and it follows from the measurability (36.8) of k_x and the integrability of h_x that $k_x \in \mathscr{L}^1(\nu)$ (25.5), and moreover

$$\int k_x \, d\nu = \int h_x \, d\nu = f(x).$$

Now suppose $x \in E_0$; then $k_x = 0 \in \mathscr{L}^1(\nu)$, and

$$\int k_x \, d\nu = 0 = f(x).$$

Summarizing: for each $x \in X$, we have $k_x \in \mathscr{L}^1(\nu)$ and $\int k_x \, d\nu = f(x)$; and for each $x \in X - E_0$, we have $k_x = h_x$ a.e. $[\nu]$. Similarly for the remaining assertions. ∎

Additional "complements" on product measure may be found in Sect. 88.

CHAPTER 7

Finite Signed Measures

Throughout this chapter, we work in the context of a given measurable space $(X, \mathscr{S})$. All set functions to be discussed will have the σ-ring $\mathscr{S}$ for their domain of definition. The sets $E, F, G, \ldots$ in $\mathscr{S}$ are called measurable, and $\mathscr{S}_\lambda$ denotes the class of locally measurable sets $A, B, C, \ldots$.

43. Absolute Continuity

Suppose μ is a measure on $\mathscr{S}$. The central concept in this chapter is as follows: an extended real valued set function ν defined on $\mathscr{S}$ is said to be **absolutely continuous** with respect to the measure μ in case, given any $\varepsilon > 0$, there exists a $\delta > 0$ such that the relations $E \in \mathscr{S}$ and $\mu(E) \leq \delta$ imply $|\nu(E)| \leq \varepsilon$. Briefly, ν is AC with respect to μ. When ν is itself a finite measure, the concept of absolute continuity has a simple reformulation:

Theorem 1. *If $(X, \mathscr{S}, \mu)$ is a measure space, and ν is a finite measure on $\mathscr{S}$, the following conditions on ν are equivalent:*

(a) *ν is* AC *with respect to μ.*

(b) *$\mu(E) = 0$ implies $\nu(E) = 0$.*

Proof. (a) *implies* (b): For each positive integer n, let $\delta_n > 0$ be such that the relations $E \in \mathscr{S}$, $\mu(E) \leq \delta_n$ imply $\nu(E) \leq 1/n$. Suppose now that E is a μ-null set; then $\mu(E) < \delta_n$ for all n, hence $\nu(E) \leq 1/n$ for all n, thus E is also a ν-null set.

(b) *implies* (a): Assume to the contrary that for some $\varepsilon > 0$, every $\delta > 0$ "fails," and hence there exists a sequence E_n of measurable sets such that $\mu(E_n) \leq 1/2^n$ but $\nu(E_n) > \varepsilon$. Defining

$$E = \limsup E_n,$$

we have $\nu(E) \geq \varepsilon$ by the Arzela-Young theorem (17.2) applied to the finite measure ν. Now, if

$$F_n = \bigcup_{k \geq n} E_k,$$

we have $F_n \downarrow E$ by the definition of E. In particular,

$$\mu(E) \leq \mu(F_n) \leq \sum_{k \geq n} \mu(E_k) \leq \sum_{k \geq n} 2^{-k} = 2^{-n+1}$$

for all n, hence $\mu(E) = 0$; but $\nu(E) \geq \varepsilon > 0$, contrary to (b). ▮

The most important example of an absolutely continuous set function is the indefinite integral of an integrable function:

Theorem 2. *If f is integrable with respect to μ, then μ_f is* AC *with respect to μ.*

Proof. Since $|f|$ is integrable by 25.3, and

$$|\mu_f(E)| \leq \mu_{|f|}(E)$$

for every measurable set E, it is clearly sufficient to consider the case that $f \geq 0$. Then μ_f is a finite measure by 26.2. If $\mu(E) = 0$, then $\chi_E f = 0$ a.e. $[\mu]$, hence

$$\mu_f(E) = \int \chi_E f \, d\mu = 0$$

by 25.5; it follows from Theorem 1 that μ_f is AC with respect to μ. ▮

In particular, it is clear that if $f \in \mathscr{L}^1(\mu)$ and E_n is a sequence of measurable sets such that $\mu(E_n) \to 0$, then $\mu_f(E_n) \to 0$.

Set functions such as μ_f are rather special: they are real-valued, and *countably additive* (26.2). The Radon-Nikodym theorem, to be proved in Sec. 52, is a partial converse to Theorem 2; it states that if $(X, \mathscr{S}, \mu)$ is a *σ-finite* measure space, and ν is a countably additive real-valued set function on $\mathscr{S}$ such that ν is AC with respect to μ, then $\nu = \mu_f$ for a suitable μ-integrable function f.

EXERCISE

***1.** If ν_i is AC with respect to μ_i $(i = 1, 2)$, is $\nu_1 \times \nu_2$ AC with respect to $\mu_1 \times \mu_2$? The answer is yes if the μ_i are σ-finite and the ν_i are finite (see Exercise 52.1).

44. Finite Signed Measures

A real-valued set function ν defined on $\mathscr{S}$ is said to be a **signed measure** if it is countably additive; that is, if E_n is any sequence of mutually disjoint measurable sets, and

$$E = \bigcup_1^\infty E_n,$$

then

$$\nu(E) = \sum_1^\infty \nu(E_n),$$

in the sense that the sequence of partial sums $\sum_1^n \nu(E_k)$ converges to the real number $\nu(E)$. The triple $(X, \mathscr{S}, \nu)$ is sometimes called a "signed measure space." It is possible to consider countably additive set functions with extended real values, but we shall not do so; the real-valued case is technically simpler, equally instructive, and actually yields a slightly different theory (see 45.2). To emphasize that the values of ν are to be real, we shall always use the term **finite signed measure.** Let us note two important examples of finite signed measures:

(1) Suppose $\mu_1, \ldots, \mu_n$ are finite measures on $\mathscr{S}$, and $a_1, \ldots, a_n$ are real numbers. Then, the set function

$$\nu = \sum_1^n a_k \mu_k$$

is a finite signed measure, by elementary properties of limits. Briefly, a finite linear combination of finite measures is a finite signed measure. Conversely, it will be shown in Sect. 49 that if ν is any finite signed measure, there always exist finite measures μ_1, μ_2 such that $\nu = \mu_1 - \mu_2$ (together with certain conditions that make the decomposition unique).

(2) If μ is a measure on $\mathscr{S}$, and $f \in \mathscr{L}^1(\mu)$, then the formula

$$\mu_f(E) = \int_E f \, d\mu$$

defines a finite signed measure μ_f on $\mathscr{S}$, and μ_f is AC with respect to μ by 43.2. In Sect. 52, a partial converse will be proved (Radon-Nikodym theorem): if $(X, \mathscr{S}, \mu)$ is a σ-finite measure space, and ν is a finite signed measure on $\mathscr{S}$ which is AC with respect to μ, then $\nu = \mu_f$ for a suitable f in $\mathscr{L}^1(\mu)$.

In the following theorem, we list some of the elementary properties of a finite signed measure:

Theorem 1. *Suppose ν is a finite signed measure on $\mathscr{S}$. Let E, F, and E_n ($n = 1, 2, 3, \ldots$) be measurable sets. Then:*

(1) $\nu(\varnothing) = 0$.

(2) *ν is finitely additive.*

(3) *ν is subtractive: if $E \subset F$, then $\nu(F - E) = \nu(F) - \nu(E)$.*

(4) *If the E_n are mutually disjoint, the series $\sum_1^\infty \nu(E_n)$ converges absolutely.*

(5) *If $E_n \uparrow E$, then $\nu(E_n) \to \nu(E)$.*

(6) *If $E_n \downarrow E$, then $\nu(E_n) \to \nu(E)$.*

(7) *Let $(E_i)_{i \in I}$ be a family of measurable sets such that $E_i \cap E_j = \varnothing$ when $i \neq j$. If $\nu(E_i) > 0$ for all i, then I is countable. Same conclusion if $\nu(E_i) < 0$ for all i. Same conclusion if $\nu(E_i) \neq 0$ for all i.*

Proof. (1) Since $\varnothing = \varnothing \cup \varnothing \cup \varnothing \cup \cdots$, the series $\sum_1^\infty \nu(\varnothing)$ is convergent, hence $\nu(\varnothing) = 0$.

(2) This is immediate from (1) and countable additivity.

(3) If $E \subset F$, then $F = (F - E) \cup E$, hence $\nu(F) = \nu(F - E) + \nu(E)$.

(4) Let $E_{n_1}, E_{n_2}, E_{n_3}, \ldots$ be the (possibly finite) subsequence with $\nu(E_{n_k}) > 0$, and let $E_{m_1}, E_{m_2}, E_{m_3}, \ldots$ be the (possibly finite) subsequence with $\nu(E_{m_j}) \leq 0$. Since the E_{n_k} are mutually disjoint, the series

$$\sum_k |\nu(E_{n_k})| = \sum_k \nu(E_{n_k})$$

is convergent. Similarly,

$$\sum_j |\nu(E_{m_j})| = -\sum_j \nu(E_{m_j})$$

is convergent. It is then clear that the partial sums of the series $\sum_1^\infty |\nu(E_n)|$ are bounded.

(5) Suppose $E_n \uparrow E$. Define $E_0 = \varnothing$, and consider

$$E = E_1 \cup (E_2 - E_1) \cup (E_3 - E_2) \cup \cdots = \bigcup_1^\infty (E_n - E_{n-1});$$

by countable additivity and subtractivity, we have

$$\nu(E) = \sum_1^\infty \nu(E_n - E_{n-1}) = \lim_n \sum_1^n \nu(E_k - E_{k-1})$$

$$= \lim_n \sum_1^n [\nu(E_k) - \nu(E_{k-1})] = \lim_n \nu(E_n).$$

(6) If $E_n \downarrow E$, then $E_1 - E_n \uparrow E_1 - E$; quote (5) and subtractivity.

(7) Suppose $\nu(E_i) > 0$ for all i. For each positive integer m, let

$$I_m = \{i \in I: \nu(E_i) > 1/m\};$$

clearly $I_m \uparrow I$. We assert that each I_m is finite. If not, choose an infinite sequence $i_1, i_2, i_3, \ldots$ in I_m. Since the E_{i_k} are disjoint, $\sum_1^\infty \nu(E_{i_k})$ converges; but $\nu(E_{i_k}) > 1/m$ for all k, a contradiction. ∎

EXERCISES

1. If ν is a finite signed measure on $\mathscr{S}$, the class of all sets E in $\mathscr{S}$ such that $\nu(E) = 0$ need not be a ring. This is a significant difference between measures and signed measures.

2. Let $\mathscr{R}$ be a ring, and $\mathscr{S}$ the σ-ring generated by $\mathscr{R}$. If ν is a finite signed measure on $\mathscr{S}$ such that $\nu = 0$ on $\mathscr{R}$, then $\nu = 0$ on $\mathscr{S}$. If ν_1, ν_2 are finite signed measures on $\mathscr{S}$ such that $\nu_1 = \nu_2$ on $\mathscr{R}$, then $\nu_1 = \nu_2$ on $\mathscr{S}$.

3. If $\mathscr{R}$ is a ring, and ν is a finite signed measure on $\mathfrak{S}(\mathscr{R})$ such that $\nu(E) \geq 0$ for all E in $\mathscr{R}$, then ν is a measure.

45. Contractions of a Finite Signed Measure

If ν is a finite signed measure, and A is a locally measurable set, the **contraction** of ν by A is the set function ν_A defined on $\mathscr{S}$ by the formula

$$\nu_A(E) = \nu(A \cap E).$$

It is clear that ν_A is itself a finite signed measure. The elementary properties of contraction are as follows:

Theorem 1. *Suppose ν is a finite signed measure on $\mathscr{S}$, and A, B are locally measurable sets. Then:*

(1) $(\nu_A)_B = \nu_{A \cap B}$.

(2) *If $A \cap B = \varnothing$, then $\nu_{A \cup B} = \nu_A + \nu_B$.*

(3) *If $A \subset B$, then $\nu_{B-A} = \nu_B - \nu_A$.*

(4) *In any case, $\nu_{A \cup B} + \nu_{A \cap B} = \nu_A + \nu_B$.*

(5) *In order that $\nu_{A \cup B} = \nu_A + \nu_B$, it is necessary and sufficient that $\nu_{A \cap B} = 0$.*

Proof. (1) follows at once from the relation

$$A \cap (B \cap E) = (A \cap B) \cap E.$$

(2) is clear from the additivity of ν.

(3) results on applying (2) to the relation $B = (B - A) \cup A$.

(4) follows from (3) and the relation $A \cup B - B = A - A \cap B$.

(5) is immediate from (4). ∎

The next result lies deeper, and plays a crucial role in Sect. 52. *The finiteness of* ν *is indispensable here:*

Theorem 2. *If* ν *is a finite signed measure, there exists a measurable set* E *such that* $\nu_E = \nu$.

Proof. By Zorn's lemma, let $(E_i)_{i \in I}$ be a maximal family of mutually disjoint measurable sets such that $\nu(E_i) \neq 0$ for all i. By 44.1, the family is countable, hence $E = \bigcup_i E_i$ is a measurable set.

If F is any measurable set, then $\nu(F - E) = 0$ by maximality, hence

$$\nu(F) = \nu(F - E) + \nu(F \cap E) = \nu(F \cap E) = \nu_E(F);$$

that is, $\nu = \nu_E$. ∎

EXERCISES

1. If μ is discrete measure on the class $\mathscr{S}$ of all countable subsets of an uncountable set X, then μ is σ-finite but there exists no measurable set E such that $\mu = \mu_E$.

2. If μ and ν are finite signed measures on $\mathscr{S}$ and $\mathscr{T}$, respectively, it can be shown (Exercise 49.1) that there exists a unique finite signed measure $\mu \times \nu$ on $\mathscr{S} \times \mathscr{T}$ such that

$$(\mu \times \nu)(E \times F) = \mu(E)\nu(F)$$

for every measurable rectangle $E \times F$. Granted this, one has

$$(\mu \times \nu)_{A \times B} = \mu_A \times \nu_B$$

for all locally measurable sets A and B.

46. Purely Positive and Purely Negative Sets

Let ν be a finite signed measure on $\mathscr{S}$, fixed for the rest of the section. In Sect. 49, it will be shown that ν can be expressed as a difference of two finite measures, $\nu = \mu_1 - \mu_2$. The idea of the proof is to dissect the underlying set X into two parts, so that ν behaves like a measure on one of the parts, and like the negative of a measure on the other part. The purpose of this section is to make precise the concept of such behavior.

Let A be a locally measurable set. We shall say that A is **purely positive** (with respect to ν) in case $\nu_A \geq 0$, that is, in case ν_A is a measure; briefly, $A \geq 0$ (with respect to ν). Similarly, A is said to be **purely**

negative (with respect to ν) if $\nu_A \leq 0$, and this is denoted $A \leq 0$. If $\nu_A = 0$, we say that A is **equivalent to zero** (with respect to ν), and we write $A \equiv 0$.

Thus, $A \geq 0$ if and only if $\nu(A \cap E) \geq 0$ for every measurable set E; this is clearly equivalent to the condition that $\nu(E) \geq 0$ for every measurable set E such that $E \subset A$. Similarly, $A \leq 0$ if and only if $\nu(E) \leq 0$ for every measurable subset E of A. Finally, $A \equiv 0$ if and only if $\nu(E) = 0$ for every measurable subset E of A. The elementary properties of purity are as follows:

Theorem 1. *Suppose ν is a finite signed measure, and let A, B, C, A_n $(n = 1, 2, 3, \ldots)$ be locally measurable sets. Then:*

(1) *$A \equiv 0$ if and only if both $A \geq 0$ and $A \leq 0$.*

(2) *If $A \geq 0$ and $C \subset A$, then $C \geq 0$.*
If $A \leq 0$ and $C \subset A$, then $C \leq 0$.
If $A \equiv 0$ and $C \subset A$, then $C \equiv 0$.

(3) *If $A \geq 0$ and $B \leq 0$, then $A \cap B \equiv 0$.*

(4) *If $A_n \geq 0$ for all n, and $A = \bigcup_1^\infty A_n$, then $A \geq 0$.*
If $A_n \leq 0$ for all n, and $A = \bigcup_1^\infty A_n$, then $A \leq 0$.
If $A_n \equiv 0$ for all n, and $A = \bigcup_1^\infty A_n$, then $A \equiv 0$.

Proof. (1) $\nu_A = 0$ if and only if both $\nu_A \geq 0$ and $\nu_A \leq 0$.

(2) This is clear from the relation $\nu_C = \nu_{C \cap A} = (\nu_A)_C$.

(3) If $A \geq 0$, then $A \cap B \geq 0$ by (2); if moreover $B \leq 0$, then also $A \cap B \leq 0$, hence $A \cap B \equiv 0$ by (1).

(4) Suppose first that $A \geq 0$, $B \geq 0$, and $A \cap B = 0$; then $A \cup B \geq 0$ results from $\nu_{A \cup B} = \nu_A + \nu_B$. Next, if $A \geq 0$ and $B \geq 0$, then $A \cup B \geq 0$; this follows from the preceding remark and the relation

$$A \cup B = (A - B) \cup (B - A) \cup (A \cap B)$$

(all terms of the union are ≥ 0 by (2)). Suppose now that $A_n \geq 0$ $(n = 1, 2, 3, \ldots)$, and $A = \bigcup_1^\infty A_n$. Let $B_n = \bigcup_1^n A_k$; then $B_n \uparrow A$, and $B_n \geq 0$ by the preceding remarks. If E is any measurable set, then

$$B_n \cap E \uparrow A \cap E,$$

hence

$$\nu(B_n \cap E) \to \nu(A \cap E)$$

by 44.1; since $\nu(B_n \cap E) \geq 0$ for all n, we conclude that $\nu(A \cap E) \geq 0$. Thus, $\nu_A(E) \geq 0$ for every measurable set E. The second part of (4) is proved similarly, and the third part is a consequence of the first two.

Incidentally, the second part could be deduced from the first by considering the finite signed measure $-\nu$. ▮

The following "existence theorem" for pure sets is a fundamental tool for the whole chapter:

Theorem 2. *Let ν be a finite signed measure on $\mathcal{S}$, and let E be a measurable set. If $\nu(E) > 0$, there exists a measurable set E_0 such that $E_0 \subset E$, $E_0 \geq 0$, and $\nu(E_0) > 0$. Similarly, if $\nu(E) < 0$, there exists a measurable set E_0 such that $E_0 \subset E$, $E_0 \leq 0$, and $\nu(E_0) < 0$.*

Proof. Suppose $\nu(E) > 0$. If $\nu(F) \geq 0$ for every measurable subset F of E, simply take $E_0 = E$. Otherwise, by Zorn's lemma, let (F_i) be a maximal family of mutually disjoint measurable subsets of E such that $\nu(F_i) < 0$ for all i. The family is countable by 44.1, hence the set $F = \bigcup F_i$ is measurable. Then, $F \subset E$ and

$$\nu(F) = \sum \nu(F_i) < 0.$$

Define $E_0 = E - F$; since $\nu(E) > 0$, $\nu(F) < 0$, and

$$\nu(E_0) = \nu(E) - \nu(F),$$

we have $\nu(E_0) > 0$. We assert that $E_0 \geq 0$; for, if G is a measurable subset of E_0, then $G \cap F = \varnothing$, hence $G \cap F_i = \varnothing$ for all i, and it follows from maximality that $\nu(G) \geq 0$.

The second assertion of the theorem follows from applying the first to $-\nu$. ▮

47. Comparison of Finite Measures

A concept which is closely related to absolute continuity is the following: if ν and μ are measures defined on $\mathcal{S}$, let us write $\nu \ll \mu$ in case every μ-null set is ν-null, that is, $\mu(E) = 0$ implies $\nu(E) = 0$. Suggested verbalization of $\nu \ll \mu$: ν is **dominated** by μ. If in particular ν is a finite measure, we know from 43.1 that $\nu \ll \mu$ if and only if ν is AC with respect to μ. It is also clear from the proof of 43.1 that if ν is AC with respect to μ, then $\nu \ll \mu$ even if ν is not finite. For our purposes, the concept of absolute continuity would suffice, but the $\nu \ll \mu$ relation is simpler to work with, and so we shall formulate our results in terms of this notation. We summarize some elementary properties of the $\ll$ relation:

Theorem 1. *Let μ, ν, and ρ be measures defined on $\mathcal{S}$. Then:*

(1) $\mu \ll \mu$.

(2) *If $\nu \ll \mu$ and $\mu \ll \rho$, then $\nu \ll \rho$.*

(3) *If* $\mu \ll 0$, *then* $\mu = 0$.

(4) *If* $\nu \leq \mu$, *then* $\nu \ll \mu$.

(5) *If* $\nu \ll \mu$, *then* $\nu_A \ll \mu_A$ *for every locally measurable set* A.

(6) $\mu_A \ll \mu$, *for every locally measurable set* A.

(7) *If* ν *is* AC *with respect to* μ, *then* $\nu \ll \mu$.

(8) *If* ν *is a finite measure, then* $\nu \ll \mu$ *if and only if* ν *is* AC *with respect to* μ.

(9) *If* $f \in \mathscr{L}^1(\mu)$, *and* $f \geq 0$ *a.e.* $[\mu]$, *then* $\mu_f \ll \mu$.

Proof. (1) through (6) are obvious.

(7) See the proof of (a) implies (b) in 43.1.

(8) This is 43.1.

(9) μ_f is a (finite) measure by 26.2; quote 43.2 and (7). ∎

The next theorem indicates an extent to which a relation of the form $\nu \ll \mu$ implies a relation of the form $\nu \geq \mu$:

Theorem 2. *Suppose* ν *and* μ *are finite measures on* $\mathscr{S}$ *such that* $\nu \ll \mu$ *and* $\nu \neq 0$. *Then there exists an* $\varepsilon > 0$ *and a measurable set* E *such that:*

(*i*) $E \geq 0$ *with respect to the finite signed measure* $\nu - \varepsilon\mu$.

(*ii*) $(\nu - \varepsilon\mu)(E) > 0$, $\nu(E) > 0$, $\mu(E) > 0$.

Proof. Since $\nu \neq 0$, let F be a measurable set such that $\nu(F) > 0$. Choose a number $\varepsilon > 0$ so small that

$$\nu(F) - \varepsilon\mu(F) > 0.$$

For the finite signed measure $\rho = \nu - \varepsilon\mu$, we have $\rho(F) > 0$. By 46.2, there exists a measurable set E such that $E \subset F$, $E \geq 0$ with respect to ρ, and $\rho(E) > 0$. Thus,

$$\nu(E) - \varepsilon\mu(E) > 0.$$

Necessarily $\mu(E) > 0$; for if $\mu(E) = 0$, then $\nu \ll \mu$ would imply $\nu(E) = 0$, and hence

$$\rho(E) = \nu(E) - \varepsilon\mu(E) = 0,$$

contrary to the choice of E. Thus, $\mu(E) > 0$, and finally

$$\nu(E) > \varepsilon\mu(E) > 0. \quad ∎$$

With notation as in the above theorem, we have $\varepsilon\mu_E \leq \nu_E$ (this is simply another way of saying that $E \geq 0$ with respect to $\nu - \varepsilon\mu$). Obviously this implies $\mu_E \ll \nu_E$; on the other hand, $\nu_E \ll \mu_E$ results from $\nu \ll \mu$. Thus, we are led to the following definition: two measures μ and ν are said to be **equivalent** in case both $\nu \ll \mu$ and $\mu \ll \nu$; this relation will be denoted $\nu \equiv \mu$, and clearly holds if and only if ν and μ have the same null sets. If ν, μ, ρ are measures on $\mathscr{S}$, the following remarks are evident: (*i*) $\mu \equiv \mu$, (*ii*) $\nu \equiv \mu$ if and only if $\mu \equiv \nu$, (*iii*) if $\nu \equiv \mu$ and $\mu \equiv \rho$, then $\nu \equiv \rho$, and (*iv*) $\mu \equiv 0$ implies $\mu = 0$.

The remarks immediately following Theorem 2 may be expressed thus: if ν and μ are finite measures on $\mathscr{S}$ such that $\nu \ll \mu$ and $\nu \neq 0$, then there exists a measurable set E such that $\nu_E \equiv \mu_E$ and $\nu_E \neq 0$. We may continue by a process of "exhaustion":

Theorem 3. *If ν and μ are finite measures on $\mathscr{S}$ such that $\nu \ll \mu$, there exists a measurable set E such that $\nu \equiv \mu_E$ and $\nu = \nu_E$.*

Proof. If $\nu = 0$, take $E = \varnothing$. Otherwise, by Zorn's lemma, let (E_i) be a maximal family of mutually disjoint measurable sets such that $\nu(E_i) > 0$, and such that for each i there is an $\varepsilon_i > 0$ satisfying

$$\nu_{E_i} \geq \varepsilon_i \mu_{E_i}$$

(use Theorem 2 to get started). As noted in the above remarks, we have $\nu_{E_i} \equiv \mu_{E_i}$ for all i. The family (E_i) is countable by 44.1, hence the set $E = \bigcup E_i$ is measurable.

We assert that $\nu = \nu_E$. Since $\nu = \nu_E + \nu_{X-E}$, it will suffice to show that $\nu_{X-E} = 0$. Suppose F is any measurable subset of $X - E$; since $\nu_F \ll \mu_F$, it is clear from Theorem 2, and the maximality of the family (E_i), that $\nu_F = 0$, and in particular $\nu(F) = \nu_F(F) = 0$. Thus, $\nu_{X-E} = 0$.

Finally, let us show that $\nu \equiv \mu_E$. If G is any measurable set, then

$$E \cap G = \bigcup E_i \cap G,$$

hence

$$\nu(G) = \nu_E(G) = \nu(E \cap G) = \sum \nu(E_i \cap G) = \sum_i \nu_{E_i}(G).$$

The following statements are therefore equivalent: $\nu(G) = 0$, $\nu_{E_i}(G) = 0$ for all i, $\mu_{E_i}(G) = 0$ for all i (recall that $\nu_{E_i} \equiv \mu_{E_i}$), $\mu(E_i \cap G) = 0$ for all i, $\sum \mu(E_i \cap G) = 0$, $\mu(\bigcup E_i \cap G) = 0$, $\mu(E \cap G) = 0$, $\mu_E(G) = 0$. Thus, ν and μ_E have the same null sets. ▮

We now come to the *comparison theorem* for any two finite measures defined on the same σ-ring:

Theorem 4. *Let ν and μ be any two finite measures defined on $\mathscr{S}$. Then there exists a decomposition of X into mutually disjoint locally measurable sets E, A, B, such that*

$$\nu_E \equiv \mu_E,$$

$$\nu_A = 0,$$

$$\mu_B = 0,$$

and hence $\nu = \nu_E + \nu_B$, $\mu = \mu_E + \mu_A$. *We may suppose, moreover, that E and B (or E and A) are measurable.*

Proof. Since $\nu \le \nu + \mu$, obviously $\nu \ll \nu + \mu$, hence by Theorem 3 there exists a measurable set F such that $\nu \equiv (\nu + \mu)_F$ and $\nu = \nu_F$. Setting $A = X - F$, we have

$$\nu_A = \nu_X - \nu_F = \nu - \nu_F = 0.$$

Now

$$(\nu + \mu)_F = \nu_F + \mu_F = \nu + \mu_F,$$

hence $\nu \equiv \nu + \mu_F$. Then $\mu_F \le \nu + \mu_F \equiv \nu$, therefore it is clear that $\mu_F \ll \nu$; applying Theorem 3 again, there exists a measurable set G such that $\mu_F \equiv \nu_G$ and $\mu_F = (\mu_F)_G$. Setting $E = F \cap G$, we have

$$\mu_F = (\mu_F)_G = \mu_{F\cap G} = \mu_E,$$

and

$$\nu_E = \nu_{F\cap G} = (\nu_F)_G = \nu_G \equiv \mu_F = \mu_E;$$

thus, $\mu_{F-E} = 0$ and $\nu_E \equiv \mu_E$. Setting $B = F - E$, we have $\mu_B = 0$.

Summarizing, we have

$$X = (X - F) \cup F = A \cup F = A \cup (F - E) \cup E = A \cup B \cup E,$$

clearly a disjoint union, where A is locally measurable and B, E are measurable. Moreover, $\nu_E \equiv \mu_E$, $\nu_A = 0$, and $\mu_B = 0$. (If one wants E and A to be measurable, apply the theorem to the pair μ, ν instead of ν, μ.) ∎

The decomposition $\nu = \nu_E + \nu_B$ is called the **Lebesgue decomposition** of ν with respect to μ.

EXERCISES

1. If $\nu \ll \mu$ and $A \subset B$, then $\nu_A \ll \mu_B$.

2. Suppose $\nu = \sum_i \nu_i$ and $\mu = \sum_i \mu_i$ in the sense of Sect. 10. If $\nu_i \ll \mu_i$ for all i, then $\nu \ll \mu$. One has $\mu \ll \rho$ if and only if $\mu_i \ll \rho$ for all i.

***3.** If $\nu_i \ll \mu_i$ $(i = 1, 2)$, does it follow that

$$\nu_1 \times \nu_2 \ll \mu_1 \times \mu_2?$$

The answer is yes when the μ_i are σ-finite and the ν_i are finite (see Exercise 52.1).

48. A Preliminary Radon-Nikodym Theorem

The general Radon-Nikodym theorem to be proved in Sect. 52 will be based on the following special case:

Theorem 1. *If $(X, \mathscr{S}, \mu)$ is a finite measure space, and ν is a finite measure on $\mathscr{S}$ such that $\nu \ll \mu$, then there exists an $f \in \mathscr{L}^1(\mu)$ such that*

$$\nu = \mu_f$$

and $f \geq 0$. If also $g \in \mathscr{L}^1(\mu)$ and $\nu = \mu_g$, then $g = f$ a.e. $[\mu]$.

Proof. Let $\mathscr{K} = \{f \in \mathscr{L}^1(\mu) : f \geq 0 \text{ and } \mu_f \leq \nu\}$. For instance, $0 \in \mathscr{K}$. The proof of the theorem will be broken up into several steps.

(*i*) *If $f \in \mathscr{K}$ and $g \in \mathscr{K}$, then $f \cup g \in \mathscr{K}$.* In any case, $f \cup g \in \mathscr{L}^1(\mu)$ and $f \cup g \geq 0$. Let E be any measurable set; the problem is to show that $\mu_{f \cup g}(E) \leq \nu(E)$, that is,

$$\int_E (f \cup g)\, d\mu \leq \nu(E).$$

Let

$$F = \{x \in E: (f \cup g)(x) = f(x)\},$$

and

$$G = \{x \in E: (f \cup g)(x) = g(x)\};$$

obviously $E = F \cup G$. Since

$$F = E \cap (f \cup g - f)^{-1}(\{0\}) = E - N(f \cup g - f),$$

it is clear that F is measurable, and similarly G is measurable. Define $E_1 = F$ and $E_2 = E - F$; then $E = E_1 \cup E_2$ is a disjoint union. Since $E_1 = F$ and $E_2 \subset G$, we have $f \cup g = f$ on E_1, and $f \cup g = g$ on E_2. Then

$$\chi_E(f \cup g) = \chi_{E_1}(f \cup g) + \chi_{E_2}(f \cup g) = \chi_{E_1} f + \chi_{E_2} g;$$

integrating with respect to μ, we have

$$\mu_{f \cup g}(E) = \mu_f(E_1) + \mu_g(E_2) \leq \nu(E_1) + \nu(E_2) = \nu(E_1 \cup E_2) = \nu(E).$$

(*ii*) *$\int f\, d\mu$ is bounded, as f varies over $\mathscr{K}$.* Since ν is a finite measure, by 17.1 there exists a constant K such that $\nu(E) \leq K$ for all E in $\mathscr{S}$. For any $f \in \mathscr{K}$,

$$\int f\, d\mu = \int_{N(f)} f\, d\mu = \mu_f(N(f)) \leq \nu(N(f)) \leq K.$$

(*iii*) *If $M = \text{LUB}\{\int f\, d\mu : f \in \mathscr{K}\}$, there exists an $f \in \mathscr{K}$ such that $\int f\, d\mu = M$.* By (*ii*), $M < \infty$. Let g_n be a sequence in $\mathscr{K}$ such that

$$\int g_n\, d\mu \to M.$$

Defining $f_n = g_1 \cup \cdots \cup g_n$, we have $f_n \uparrow$ and $f_n \in \mathscr{K}$ by (*i*); since $g_n \leq f_n$, the relation

$$\int g_n\, d\mu \leq \int f_n\, d\mu \leq M$$

shows that $\int f_n\, d\mu \uparrow M$.

Since $f_n\uparrow$ and $\int f_n\,d\mu$ is bounded, it follows from the (MCT) that there exists an $f \in \mathscr{L}^1(\mu)$ such that $f_n \uparrow f$ a.e. $[\mu]$. Modifying f and the f_n on a suitable μ-null set, we may assume $f_n \uparrow f$. Then also $f \geq 0$. Of course

$$\int f_n\,d\mu \uparrow \int f\,d\mu,$$

hence $\int f\,d\mu = M$. In order to show that $f \in \mathscr{K}$, it remains to show that $\mu_f \leq \nu$. If E is any measurable set, then

$$\chi_E f_n \uparrow \chi_E f,$$

hence

$$\int \chi_E f_n\,d\mu \uparrow \int \chi_E f\,d\mu$$

by the (MCT), that is, $\mu_{f_n}(E) \uparrow \mu_f(E)$; since $\mu_{f_n}(E) \leq \nu(E)$ for all n, it follows that $\mu_f(E) \leq \nu(E)$.

(*iv*) *Summarizing, there exists an* $f \in \mathscr{K}$ *such that*

$$\int g\,d\mu \leq \int f\,d\mu$$

for all $g \in \mathscr{K}$.

(*v*) *With* f *as in* (*iv*), *we assert that* $\mu_f = \nu$. Define $\rho = \nu - \mu_f$; since $f \in \mathscr{K}$, we have $\rho \geq 0$, that is, ρ is a finite measure. Since $\nu \ll \mu$ (first use of this hypothesis) and $\mu_f \ll \mu$ (see 47.1), it is clear that $\rho \ll \mu$.

Assume to the contrary that $\rho \neq 0$. Then by 47.2 there exists an $\varepsilon > 0$ and a measurable set E such that $\varepsilon\mu_E \leq \rho_E$ and $\mu(E) > 0$. Define

$$g = f + \varepsilon\chi_E;$$

then

$$\int g\,d\mu = \int f\,d\mu + \varepsilon\mu(E) > \int f\,d\mu,$$

hence we shall obtain a contradiction by showing that $g \in \mathscr{K}$. Of course $g \in \mathscr{L}^1(\mu)$ and $g \geq 0$; given a measurable set F, the problem is to show that $\mu_g(F) \leq \nu(F)$. Indeed,

$$\begin{aligned}
\mu_g(F) &= \int_F (f + \varepsilon\chi_E)\,d\mu = \int \chi_F f\,d\mu + \varepsilon \int \chi_F\chi_E\,d\mu \\
&= \mu_f(F) + \varepsilon\mu(F \cap E) = \mu_f(F) + \varepsilon\mu_E(F) \leq \mu_f(F) + \rho_E(F) \\
&= \mu_f(F) + \rho(E \cap F) = \mu_f(F) + [\nu(E \cap F) - \mu_f(E \cap F)] \\
&= [\mu_f(F) - \mu_f(E \cap F)] + \nu(E \cap F) \\
&= \mu_f(F - E \cap F) + \nu(E \cap F) \leq \nu(F - E \cap F) + \nu(E \cap F) \\
&= \nu(F).
\end{aligned}$$

The last assertion of the theorem is immediate from 26.1. ▮

We may use Theorem 1 to obtain a second proof of 47.3: if ν and μ are finite measures on $\mathscr{S}$ such that $\nu \ll \mu$, there exists a measurable set E such that $\nu \equiv \mu_E$ and $\nu = \nu_E$. Indeed, by Theorem 1, there exists an $f \in \mathscr{L}^1(\mu)$ such that $f \geq 0$ and $\nu = \mu_f$, and our assertion follows at once from a general lemma:

Lemma. *If $(X, \mathscr{S}, \mu)$ is any measure space, $f \in \mathscr{L}^1(\mu)$, and $f \geq 0$ a.e. $[\mu]$, then $\mu_f \equiv \mu_E$, where $E = N(f)$.*

Proof. In view of the definition of equivalence of measures, it must be shown that μ_f and μ_E have the same null sets. Indeed, if F is a measurable set, the following statements are equivalent: $\mu_f(F) = 0$, $\int \chi_F f \, d\mu = 0$, $\chi_F f = 0$ a.e. $[\mu]$ (see 25.9), $\chi_F \chi_E = 0$ a.e. $[\mu]$, $\chi_{F \cap E} = 0$ a.e. $[\mu]$, $\mu(F \cap E) = 0$, $\mu_E(F) = 0$. ▮

49. Jordan-Hahn Decomposition of a Finite Signed Measure

If ν is a finite signed measure on $(X, \mathscr{S})$, we shall show in this section how to decompose X into two parts, so that ν behaves like a measure on one part, and like the negative of a measure on the other part. The following lemma will be used to prove the essential uniqueness of such a decomposition:

Lemma. *Suppose ν is a finite signed measure on $\mathscr{S}$, and suppose that A and B are locally measurable sets such that $A \geq 0$ and $B \geq 0$ with respect to ν. Then, the following conditions are equivalent:*

(a) $\nu_A = \nu_B$.

(b) $A \triangle B \equiv 0$ *with respect to* ν.

In this case, $\nu_A = \nu_B = \nu_{A \cup B} = \nu_{A \cap B}$.

Proof. (a) *implies* (b): (This part of the proof does not make use of the assumption $A \geq 0$, $B \geq 0$.) Suppose $\nu_A = \nu_B$. Since

$$A \triangle B = (A - A \cap B) \cup (B - A \cap B)$$

is a disjoint union, we have

$$\nu_{A \triangle B} = \nu_{A - A \cap B} + \nu_{B - A \cap B} = [\nu_A - \nu_{A \cap B}] + [\nu_B - \nu_{A \cap B}]$$

by 45.1. But

$$\nu_{A \cap B} = (\nu_A)_B = (\nu_B)_B = \nu_B,$$

hence

$$\nu_A = \nu_B = \nu_{A \cap B};$$

it follows that $\nu_{A \triangle B} = 0$, that is, $A \triangle B \equiv 0$ with respect to ν.

(b) *implies* (a): Suppose $A \geq 0$, $B \geq 0$, and $A \triangle B \equiv 0$. Thus, $\nu_{A \triangle B} = 0$, and ν_A, ν_B are measures. Now,

$$0 = \nu_{A \triangle B} = \nu_{A - A \cap B} + \nu_{B - A \cap B};$$

since $A - A \cap B \geq 0$ and $B - A \cap B \geq 0$ by 46.1, it follows that

$$\nu_{A - A \cap B} = \nu_{B - A \cap B} = 0,$$

that is,

$$\nu_A - \nu_{A \cap B} = \nu_B - \nu_{A \cap B} = 0.$$

Thus, $\nu_A = \nu_B = \nu_{A \cap B}$. Finally,

$$A \cup B = (A \triangle B) \cup (A \cap B),$$

hence

$$\nu_{A \cup B} = \nu_{A \triangle B} + \nu_{A \cap B} = 0 + \nu_{A \cap B}. \quad \blacksquare$$

Theorem 1. *If ν is a finite signed measure on $\mathscr{S}$, there exists a locally measurable set A such that $A \geq 0$ and $X - A \leq 0$ with respect to ν. If B is another locally measurable set such that $B \geq 0$ and $X - B \leq 0$, then $A \triangle B \equiv 0$, and hence $\nu_A = \nu_B$ and $\nu_{X-A} = \nu_{X-B}$. One can arrange to have A measurable.*

Proof. Define $\alpha = \text{LUB}\,\{\nu(E) : E \in \mathscr{S}, E \geq 0\}$ (for instance, $\varnothing \geq 0$). Choose any sequence of measurable sets E_n such that $E_n \geq 0$ and $\text{LUB}\,\nu(E_n) = \alpha$, and define

$$A_n = \bigcup_1^n E_k, \qquad A = \bigcup_1^\infty E_k.$$

Then the A_n and A are measurable, and $A \geq 0$ by 46.1. Since $E_n \subset A_n \subset A$, the monotonicity of the measure ν_A implies that $\nu(E_n) \leq \nu(A_n) \leq \nu(A)$ for all n; then $\alpha \leq \nu(A)$, and in particular α is finite. Since $A \geq 0$ implies $\nu(A) \leq \alpha$, we have $\nu(A) = \alpha$.

We assert that $X - A \leq 0$. If E is a measurable subset of $X - A$, the problem is to show that $\nu(E) \leq 0$. If on the contrary $\nu(E) > 0$, then by 46.2 there exists a measurable set F such that $F \subset E$, $F \geq 0$, and $\nu(F) > 0$. But $F \cap A = \varnothing$, hence

$$\nu(F \cup A) = \nu(F) + \nu(A) = \nu(F) + \alpha > \alpha;$$

since $F \cup A \geq 0$ by 46.1, this contradicts the definition of α.

Suppose B is another locally measurable set such that $B \geq 0$ and $X - B \leq 0$. Quoting 46.1, we have

$$A - B = A \cap (X - B) \equiv 0$$

and

$$B - A = B \cap (X - A) \equiv 0,$$

hence $A \triangle B = (A - B) \cup (B - A) \equiv 0$. Then $\nu_A = \nu_B$ by the lemma, and

$$\nu_{X-A} = \nu_X - \nu_A = \nu_X - \nu_B = \nu_{X-B}. \quad \blacksquare$$

If ν is a finite signed measure, a locally measurable set A is said to define a **Hahn decomposition** of X with respect to ν in case $A \geq 0$ and $X - A \leq 0$ with respect to ν. Define $\nu^+ = \nu_A$ and $\nu^- = -\nu_{X-A}$; then ν^+ and ν^- are finite measures on $\mathscr{S}$, independent of the particular Hahn decomposition. Evidently

$$\nu = \nu^+ - \nu^-;$$

this is called the **Jordan decomposition** of ν. The finite measures

$$\nu^+, \nu^-, \quad \text{and} \quad \nu^+ + \nu^-$$

are called the **upper variation, lower variation**, and **total variation** of ν. The total variation of ν is denoted $|\nu|$, thus

$$|\nu| = \nu^+ + \nu^-.$$

Corollary. *A finite signed measure ν is bounded, that is, there exists a real number $M > 0$ such that $|\nu(E)| \leq M$ for every measurable set E.*

Proof. Writing ν as the difference of two finite measures, our assertion is immediate from 17.1. $\blacksquare$

The Hahn decomposition may be further refined as follows. Let A be a locally measurable set such that $A \geq 0$ and $X - A \leq 0$ with respect to ν. By 45.2, there exists a measurable set E such that $\nu = \nu_E$. Define

$$E_1 = A \cap E, \qquad E_2 = (X - A) \cap E, \qquad \text{and} \qquad C = X - E.$$

Then we have a disjoint union

$$X = E_1 \cup E_2 \cup C,$$

where E_1 and E_2 are measurable, C is locally measurable, and moreover $E_1 \geq 0$ and $E_2 \leq 0$ by 46.1. Indeed,

$$\nu_{E_1} = \nu_{A \cap E} = (\nu_E)_A = \nu_A,$$

and similarly $\nu_{E_2} = \nu_{X-A}$; finally,

$$\nu_C = \nu_X - \nu_E = \nu - \nu = 0,$$

and so $C \equiv 0$. Summarizing: if ν is a finite signed measure, there exists a disjoint union

$$X = E_1 \cup E_2 \cup C,$$

where the E_i are measurable and C is locally measurable, and $E_1 \geq 0$, $E_2 \leq 0$, $C \equiv 0$.

EXERCISES

1. If μ and ν are finite signed measures on $\mathcal{S}$ and $\mathcal{T}$, respectively, there exists a unique finite signed measure $\mu \times \nu$ on $\mathcal{S} \times \mathcal{T}$ such that

$$(\mu \times \nu)(E \times F) = \mu(E)\nu(F)$$

for every measurabie rectangle $E \times F$.

2. Suppose μ and ν are measures on $\mathcal{S}$ and $\mathcal{T}$, respectively. If $f \in \mathcal{L}^1(\mu)$, $g \in \mathcal{L}^1(\nu)$, and one defines

$$h(x, y) = f(x)g(y),$$

then $h \in \mathcal{L}^1(\mu \times \nu)$, and

$$(\mu \times \nu)_h = \mu_f \times \nu_g.$$

3. If $(X, \mathcal{S}, \mu)$ is any measure space, and f is an integrable function, the Jordan decomposition of the indefinite integral μ_f may be obtained by considering the set $A = \{x : f(x) \geq 0\}$.

4. If μ and ν are finite signed measures on $\mathcal{S}$ and $\mathcal{T}$, respectively, such that

$$(\mu \times \nu)(E \times F) \geq 0$$

for every measurable rectangle, then $\mu \times \nu$ is a measure. If, moreover, neither μ nor ν is identically zero, then either both μ and ν are measures, or else both $-\mu$ and $-\nu$ are measures.

5. If μ and ν are finite signed measures on $\mathcal{S}$ and $\mathcal{T}$, respectively, then

$$|\mu \times \nu| = |\mu| \times |\nu|.$$

50. Domination of Finite Signed Measures

We begin with a summary of elementary properties of the "variations" associated with a finite signed measure, as defined in the preceding section:

Theorem 1. *Let ν be a finite signed measure on $\mathcal{S}$. Then:*

(1) $|\nu(E)| \leq |\nu|(E)$ *for every measurable set E.*

(2) $\nu = 0$ *if and only if* $|\nu| = 0$.

(3) $\nu^+ \ll |\nu|$ *and* $\nu^- \ll |\nu|$.

(4) *If μ is a measure on $\mathcal{S}$, then $|\nu| \ll \mu$ if and only if both $\nu^+ \ll \mu$ and $\nu^- \ll \mu$.*

Proof. (1) If E is any measurable set, then

$$\nu(E) = \nu^+(E) - \nu^-(E),$$

while

$$|\nu|(E) = \nu^+(E) + \nu^-(E).$$

(2) If $\nu = 0$, then every contraction ν_A of ν is 0, and in particular $\nu^+ = 0$, $\nu^- = 0$; then $|\nu| = \nu^+ + \nu^- = 0$. If conversely $|\nu| = 0$, that is, if $\nu^+ + \nu^- = 0$, then $\nu^+ = \nu^- = 0$, hence $\nu = \nu^+ - \nu^- = 0$. (Alternatively, $\nu = 0$ results from (1).)

(4) If $|\nu| \ll \mu$, then it is clear from $\nu^+ \leq |\nu|$ and $\nu^- \leq |\nu|$ that $\nu^+ \ll \mu$ and $\nu^- \ll \mu$.

If conversely $\nu^+ \ll \mu$ and $\nu^- \ll \mu$, then $\mu(E) = 0$ implies $\nu^+(E) = 0$ and $\nu^-(E) = 0$, hence

$$|\nu|(E) = \nu^+(E) + \nu^-(E) = 0;$$

that is, $|\nu| \ll \mu$. ∎

Recall that the relation $\ll$ was defined in Sect. 47 for two *measures*. If ν is a finite signed measure, we have available the finite measure $|\nu|$ by Sect. 49. If ν is a finite signed measure and μ is a measure, we shall write

$$\nu \ll \mu$$

in case the relation

$$|\nu| \ll \mu$$

holds in the sense defined for measures. Here are several reformulations of this relation of "domination" of a finite signed measure by a measure:

Theorem 2. *If $(X, \mathscr{S}, \mu)$ is a measure space, and ν is a finite signed measure on $\mathscr{S}$, the following conditions are equivalent:*

(a) $\nu \ll \mu$.

(b) $|\nu| \ll \mu$.

(c) $\mu(E) = 0$ *implies* $|\nu|(E) = 0$.

(d) $\nu^+ \ll \mu$ *and* $\nu^- \ll \mu$.

(e) $|\nu|$ *is* AC *with respect to* μ*; that is, given any* $\varepsilon > 0$, *there exists a* $\delta > 0$ *such that* $\mu(E) \leq \delta$ *implies* $|\nu|(E) \leq \varepsilon$.

(f) ν *is* AC *with respect to* μ*; that is, given any* $\varepsilon > 0$, *there exists a* $\delta > 0$ *such that* $\mu(E) \leq \delta$ *implies* $|\nu(E)| \leq \varepsilon$.

(g) $\mu(E) = 0$ *implies* $\nu(E) = 0$.

Proof. (a), (b), and (c) are equivalent by the definitions, while (b) and (d) are equivalent by Theorem 1. Since $|\nu|$ is a finite measure, (c) and (e) are equivalent by 43.1. Summarizing, (a) through (e) are equivalent.

(e) *implies* (f): This is immediate from $|\nu(E)| \leq |\nu|(E)$.

(f) *implies* (g): See the proof of 43.1.

(g) *implies* (c): Let A be a locally measurable set which defines a Hahn-Jordan decomposition of ν, that is, $\nu^+ = \nu_A$ and $\nu^- = -\nu_{X-A}$. Suppose $\mu(E) = 0$. Define

$$E_1 = A \cap E \quad \text{and} \quad E_2 = (X - A) \cap E;$$

since $E_i \subset E$, we have $\mu(E_i) = 0$ by the monotonicity of μ. By the assumption (g), we have $\nu(E_i) = 0$. Thus,

$$0 = \nu(E_1) = \nu(A \cap E) = \nu_A(E) = \nu^+(E),$$

and

$$0 = \nu(E_2) = \nu((X - A) \cap E) = \nu_{X-A}(E) = -\nu^-(E),$$

hence $|\nu|(E) = \nu^+(E) + \nu^-(E) = 0$. ▌

51. The Radon-Nikodym Theorem for a Finite Measure Space

Combining the Jordan-Hahn decomposition (Sect. 49) with the "preliminary" Radon-Nikodym theorem (Sect. 48), we obtain an "intermediate" Radon-Nikodym theorem:

Theorem 1. *If $(X, \mathscr{S}, \mu)$ is a finite measure space, and ν is a finite signed measure on $\mathscr{S}$ such that $\nu \ll \mu$, then there exists an $f \in \mathscr{L}^1(\mu)$ such that $\nu = \mu_f$, that is,*

$$\nu(E) = \int_E f \, d\mu$$

for every measurable set E. This f is unique a.e. $[\mu]$.

Proof. In view of 50.2, our assumption is that $\nu^+ \ll \mu$ and $\nu^- \ll \mu$. By 48.1, there exist functions $g, h \in \mathscr{L}^1(\mu)$ such that $\nu^+ = \mu_g$ and $\nu^- = \mu_h$. Defining $f = g - h$, we have $f \in \mathscr{L}^1(\mu)$ and

$$\mu_f = \mu_g - \mu_h = \nu^+ - \nu^- = \nu.$$

The a.e. uniqueness of f results from 26.1. ▌

The proof of the general Radon-Nikodym theorem in the next section will be based directly on the "preliminary" 48.1, rather than the above "intermediate" result.

52. The Radon-Nikodym Theorem for a σ-Finite Measure Space

Theorem 1. *Suppose $(X, \mathscr{S}, \mu)$ is a σ-finite measure space, and ν is a finite signed measure on $\mathscr{S}$. Then, the following conditions are equivalent:*

(a) $\nu \ll \mu$

(b) $|\nu| \ll \mu$

(c) $\nu^+ \ll \mu$ *and* $\nu^- \ll \mu$

(d) ν *is* AC *with respect to* μ

(e) $|\nu|$ *is* AC *with respect to* μ

(f) $\mu(E) = 0$ *implies* $\nu(E) = 0$

(g) *There exists an* $f \in \mathscr{L}^1(\mu)$ *such that* $\nu(E) = \int_E f\,d\mu$ *for every measurable set* E.

In this case, f *is unique a.e.* $[\mu]$.

Proof. The equivalence of (a) through (f) was noted in 50.2; (g) implies (d) by 43.2, and such an f is unique a.e. by 26.1.

(a) *implies* (g): Suppose $\nu \ll \mu$. Since $\nu^+ \ll \mu$ and $\nu^- \ll \mu$, we may assume without loss of generality that ν is a finite *measure.* Let F be a measurable set such that $\nu_F = \nu$ (45.2). Since μ is σ-finite, there exists a sequence E_n of mutually disjoint measurable sets such that $F = \bigcup_1^\infty E_n$ and $\mu(E_n) < \infty$ for all n.

For each n, define $\mu_n = \mu_{E_n}$ and $\nu_n = \nu_{E_n}$; then $(X, \mathscr{S}, \mu_n)$ is a finite measure space, ν_n is a finite measure on $\mathscr{S}$, and it is clear that $\nu_n \ll \mu_n$ (47.1). By 48.1, there exists a $g_n \in \mathscr{L}^1(\mu_n)$ such that $g_n \geq 0$ and $(\mu_n)_{g_n} = \nu_n$, that is,

$$\int_E g_n\,d\mu_n = \nu_n(E)$$

for every measurable set E. Now,

$$\nu_n(E) = \nu(E_n \cap E) = \nu_n(E_n \cap E) = \int_{E_n \cap E} g_n\,d\mu_n = \int_E \chi_{E_n} g_n\,d\mu_n;$$

defining $f_n = \chi_{E_n} g_n$, we have $f_n \in \mathscr{L}^1(\mu_n)$ by 25.2, and

$$\nu_n(E) = \int_E f_n\,d\mu_n$$

for every measurable set E. Moreover, $f_n = 0$ on $X - E_n$. If E is any measurable set, then

$$\chi_E f_n \in \mathscr{L}^1(\mu_n) \qquad \text{and} \qquad \chi_E f_n = 0 \text{ on } X - E_n,$$

hence by Lemma 1 to 40.1 we have

$$\chi_E f_n \in \mathscr{L}^1(\mu) \qquad \text{and} \qquad \int \chi_E f_n\,d\mu = \int \chi_E f_n\,d\mu_n;$$

that is,

$$\mu_{f_n}(E) = \nu_n(E).$$

Summarizing, we have

$$f_n \in \mathscr{L}^1(\mu), \qquad \mu_{f_n} = \nu_n, \qquad f_n \geq 0, \qquad \text{and} \qquad \chi_{E_n} f_n = f_n.$$

In particular,

$$\nu(E_n) = \nu(E_n \cap E_n) = \nu_n(E_n) = \mu_{f_n}(E_n) = \int \chi_{E_n} f_n \, d\mu = \int f_n \, d\mu;$$

that is, $\int f_n \, d\mu = \nu(E_n)$ for all n. Defining $h_n = \sum_1^n f_k$, we have $h_n \uparrow$ and

$$\int h_n \, d\mu = \sum_1^n \int f_k \, d\mu = \sum_1^n \nu(E_k) = \nu\left(\bigcup_1^n E_k\right) \leq \nu(F) < \infty$$

for all n, hence by the (MCT) there exists an $f \in \mathscr{L}^1(\mu)$ such that $h_n \uparrow f$ a.e. $[\mu]$.

If E is any measurable set, then $\chi_E h_n \uparrow \chi_E f$ a.e. $[\mu]$, hence

$$\int_E h_n \, d\mu \uparrow \int_E f \, d\mu$$

by the (MCT), that is, $\mu_{h_n}(E) \uparrow \mu_f(E)$. Since

$$\int_E h_n \, d\mu = \sum_1^n \int_E f_k \, d\mu = \sum_1^n \nu_k(E) = \sum_1^n \nu(E_k \cap E)$$

$$= \sum_1^n \nu_E(E_k) = \nu_E\left(\bigcup_1^n E_k\right) \uparrow \nu_E(F),$$

we conclude that

$$\mu_f(E) = \nu_E(F) = \nu_F(E) = \nu(E);$$

that is, $\mu_f = \nu$. ▮

EXERCISE

***1.** If ν_i is AC with respect to μ_i $(i = 1, 2)$, where the μ_i are σ-finite measures, and the ν_i are finite measures (or signed measures), then $\nu_1 \times \nu_2$ is AC with respect to $\mu_1 \times \mu_2$. How can the finiteness conditions be relaxed?

*53. Riesz Representation Theorem

Let us consider an important application of the Radon-Nikodym theorem. For simplicity, let us suppose that $(X, \mathscr{S}, \mu)$ is a finite measure space. Let g be a fixed essentially bounded measurable function such that $g \geq 0$ a.e. For each $f \in \mathscr{L}^1(\mu)$, we have $gf \in \mathscr{L}^1(\mu)$; for, if $M \geq 0$ is a real number such that $|g| \leq M$ a.e., then $|gf| \leq M|f|$ a.e., hence the measurable (13.6) function gf is integrable by 25.7. If moreover $f \geq 0$ a.e., then $gf \geq 0$ a.e., hence $\int gf\, d\mu \geq 0$ by 25.6. Defining

$$\varphi_g \colon \mathscr{L}^1(\mu) \to R$$

by the formula

$$\varphi_g(f) = \int gf\, d\mu,$$

we obtain a "positive linear form" on $\mathscr{L}^1(\mu)$:

$$\varphi_g(f_1 + f_2) = \varphi_g(f_1) + \varphi_g(f_2),$$

$$\varphi_g(cf) = c\varphi_g(f),$$

and

$$\varphi_g(f) \geq 0 \text{ whenever } f \geq 0 \text{ a.e.}$$

Moreover, φ_g is "bounded" in the sense that

$$|\varphi_g(f)| \leq M \int |f|\, d\mu$$

for all $f \in \mathscr{L}^1(\mu)$.

We are concerned in this section with the converse:

Theorem 1. *Let $(X, \mathscr{S}, \mu)$ be a finite measure space, and suppose φ is a positive linear form on $\mathscr{L}^1(\mu)$, that is, φ is a real-valued function defined on $\mathscr{L}^1(\mu)$ such that*

(1) $$\varphi(f_1 + f_2) = \varphi(f_1) + \varphi(f_2),$$

(2) $$\varphi(cf) = c\varphi(f),$$

(3) $$\varphi(f) \geq 0 \quad \textit{whenever} \quad f \geq 0.$$

Assume, moreover, that φ is bounded; that is, assume there exists a real number $M \geq 0$ such that

(4) $$|\varphi(f)| \leq M \int |f|\, d\mu$$

for all f in $\mathscr{L}^1(\mu)$. Then there exists a bounded measurable function g, $g \geq 0$, such that

$$\varphi(f) = \int gf\, d\mu$$

for all f in $\mathscr{L}^1(\mu)$.

In the following proof, which will be broken up into a series of lemmas, the use of the letter f is meant to imply that $f \in \mathscr{L}^1(\mu)$.

Lemma 1. *If $f = 0$ a.e., then $\varphi(f) = 0$. If $f_1 \leq f_2$ a.e., then $\varphi(f_1) \leq \varphi(f_2)$.*

Proof. The first assertion is clear from (4). If $f_1 \le f_2$ a.e., then $f_2 - f_1 \ge 0$ a.e. Say $f_2 - f_1 = f$ a.e., where $f \ge 0$. Then $\varphi(f) \ge 0$ by (3). Since $f_2 - f_1 - f = 0$ a.e., we have $\varphi(f_2 - f_1 - f) = 0$, thus

$$\varphi(f_2) - \varphi(f_1) = \varphi(f_2 - f_1) = \varphi(f) \ge 0. \quad \blacksquare$$

Lemma 2. *If $f_n \uparrow f$ a.e., then $\varphi(f_n) \uparrow \varphi(f)$.*

Proof. We know that $\varphi(f_n) \uparrow$ by Lemma 1. Since

$$\int f_n \, d\mu \uparrow \int f \, d\mu$$

by the (MCT), and since $f - f_n = |f - f_n|$ a.e., we have

$$|\varphi(f) - \varphi(f_n)| = |\varphi(f - f_n)| \le M \int |f - f_n| \, d\mu$$
$$= M \left(\int f \, d\mu - \int f_n \, d\mu \right) \to 0,$$

that is, $\varphi(f_n) \to \varphi(f)$. $\blacksquare$

Lemma 3. *Define a set function ν on $\mathscr{S}$ by the formula*

$$\nu(E) = \varphi(\chi_E).$$

Then ν is a finite measure on $\mathscr{S}$, and $\nu \ll \mu$.

Proof. Of course every χ_E is integrable with respect to μ. Since $\chi_E \ge 0$ we have $0 \le \nu(E) < \infty$. If E and F are disjoint measurable sets, then

$$\nu(E \cup F) = \varphi(\chi_{E \cup F}) = \varphi(\chi_E + \chi_F) = \varphi(\chi_E) + \varphi(\chi_F) = \nu(E) + \nu(F),$$

thus ν is additive. It follows that ν is monotone: $E \subset F$ implies $\nu(E) \le \nu(F)$. If $E_n \uparrow E$, then $\chi_{E_n} \uparrow \chi_E$, hence $\nu(E_n) \uparrow \nu(E)$ by Lemma 2. Thus, ν is a finite measure. If $\mu(E) = 0$, then $\chi_E = 0$ a.e. $[\mu]$, hence $\nu(E) = \varphi(\chi_E) = 0$ by Lemma 1; thus, $\nu \ll \mu$. $\blacksquare$

Since $\nu \ll \mu$, there exists, by the Radon-Nikodym theorem (48.1), a function $g \in \mathscr{L}^1(\mu)$ such that $g \ge 0$ and

$$\nu(E) = \int_E g \, d\mu$$

for every measurable set E.

Lemma 4. *If g is the function described above, then g is essentially bounded; indeed, $|g| \le M$ a.e. $[\mu]$.*

Proof. Recall that $g \ge 0$. Defining $E = \{x\colon g(x) > M\}$, we have $\chi_E g \ge M\chi_E$. However, $M\mu(E) = \int M\chi_E \, d\mu \le \int \chi_E g \, d\mu = \int_E g \, d\mu = \nu(E) =$

$\varphi(\chi_E) \le M \int \chi_E \, d\mu$ (by the assumption (4)) $= M\mu(E)$, hence $\int \chi_E g \, d\mu = M\mu(E)$. Thus,

$$\int (\chi_E g - M\chi_E) \, d\mu = 0;$$

since $\chi_E g - M\chi_E \ge 0$, it follows from 25.9 that

$$\chi_E g - M\chi_E = 0 \text{ a.e.},$$

and since $g(x) - M > 0$ on E, necessarily $\mu(E) = 0$. In other words, $g \le M$ a.e. ▮

Replacing g by $\chi_{X-E}g$, where E is the μ-null set described in Lemma 4, we may assume without loss of generality that $0 \le g \le M$.

Lemma 5. *For every simple function f, $\varphi(f) = \int gf \, d\mu$.*

Proof. Sinces μ is finite, every such f is an ISF. By linearity, it suffices to consider $f = \chi_F$, where F is any measurable set. Then

$$\varphi(\chi_F) = \nu(F) = \int_F g \, d\mu = \int \chi_F g \, d\mu. \quad ▮$$

Finally,

Lemma 6. *For every f in $\mathscr{L}^1(\mu)$, $\varphi(f) = \int gf \, d\mu$.*

Proof. Writing $f = f^+ - f^-$, we may assume $f \ge 0$. Let f_n be a sequence of simple functions such that $0 \le f_n \uparrow f$. Then

$$0 \le gf_n \uparrow gf.$$

Quoting Lemma 5, we have

$$\int gf_n \, d\mu = \varphi(f_n) \le M \int f_n \, d\mu \le M \int f \, d\mu < \infty$$

for all n; it follows from the (MCT) that gf is integrable, and

$$\int gf_n \, d\mu \uparrow \int gf \, d\mu,$$

that is, $\varphi(f_n) \uparrow \int gf \, d\mu$. Since $\varphi(f_n) \uparrow \varphi(f)$ by Lemma 2 we conclude that $\int gf \, d\mu = \varphi(f)$. ▮

EXERCISE

1. Let $(X, \mathscr{S}, \mu)$ be a finite measure space. Let $p > 1$, with notation as in Exercise 33.3. If φ is a positive linear form on $\mathscr{L}^p$, and M is a real number ≥ 0 such that $|\varphi(f)| \le M\|f\|_p$ for all f in $\mathscr{L}^p$, then there exists a function g in $\mathscr{L}^q$ (sic) such that $g \ge 0$ and

$$\varphi(f) = \int fg \, d\mu$$

for all f in $\mathscr{L}^p$.

CHAPTER 8

Integration over Locally Compact Spaces

The fundamental issue in the theory of measure over locally compact spaces is the relationship between measurability and the topological structure (or between measurable sets and open sets, or between measurable functions and continuous functions). Thus the first problem is to specify a topologically natural σ-ring of sets to be called measurable. Of the several natural choices, we shall consider two: the class of Baire sets (Sect. 56) and the class of Borel sets (Sect. 57). Much of the chapter is devoted to a study of the relationship between the two resulting classes of measures, namely, Baire measures and Borel measures. The class of Baire measures is by far the better behaved of the two. The strongest bond between the two classes is established through the concept of regularity. The Riesz-Markoff theorem shows that the theory of Baire measures (or, of regular Borel measures) can be founded entirely on the most elementary formal properties of integration of continuous functions with compact support.

Throughout the chapter, X denotes a **locally compact** topological space; it is assumed that X is a separated space, that is, a Hausdorff space. It follows from the theory of uniformizable spaces that X is **completely regular**; thus, if x is a point of X and U is a neighborhood (not necessarily open) of x, there exists a real-valued continuous function f on X such that $0 \leq f \leq 1$, $f(x) = 1$, and $f = 0$ on $X - U$.

54. Continuous Functions with Compact Support

A real-valued function f defined on X is said to have **compact support** if there exists a compact set C in X such that $f = 0$ on $X - C$; briefly, f vanishes outside some compact set.

We shall write $\mathscr{L}(X)$, or briefly $\mathscr{L}$, for the class of all continuous real-valued functions on X with compact support.

Theorem 1. *If f and g belong to $\mathcal{L}$, and c is a real number, then the following functions all belong to $\mathcal{L}$: $f + g$, cf, fg, $|f|$, $f \cup g$, $f \cap g$, f^+, and f^-.*

Proof. It is clear that the functions in question are continuous. If C and D are compact sets outside of which f and g vanish, respectively, then all of the functions vanish outside the compact set $C \cup D$. ∎

Theorem 2. *If $x \in X$, and V is any neighborhood of x, there exists an f in $\mathcal{L}$ such that $f(x) = 1$, $f = 0$ on $X - V$, and $0 \leq f \leq 1$.*

Proof. Let U be a neighborhood of x such that the closure of U is compact and is contained in V, and let f be a real-valued continuous function on X such that $f(x) = 1$, $f = 0$ on $X - U$, and $0 \leq f \leq 1$; since f also vanishes outside the closure of U, we have $f \in \mathcal{L}$. ∎

Theorem 3. *If $C \subset U$, where C is compact and U is open, there exists an f in $\mathcal{L}$ such that $0 \leq f \leq 1$, $f = 1$ on C, and $f = 0$ on $X - U$.*

Proof. For each $x \in C$, there exists an f_x in $\mathcal{L}$ such that $f_x(x) = 2$, $f_x = 0$ on $X - U$, and $0 \leq f_x \leq 2$. Then the set

$$U_x = \{y : f_x(y) > 1\}$$

is an open neighborhood of x, and $U_x \subset U$. By the compactness of C, we have

$$C \subset U_{x_1} \cup \cdots \cup U_{x_n}$$

for suitable points $x_1, \ldots, x_n$ of C. Defining

$$g = f_{x_1} + \cdots + f_{x_n},$$

we have $g \in \mathcal{L}$, $g \geq 0$, $g > 1$ on C, and $g = 0$ on $X - U$; the function $f = \min\{g, 1\}$ meets all requirements. ∎

55. G_δ's and F_σ's

A subset A of X is called a G_δ in case there exists a sequence U_n of open sets such that $A = \bigcap_1^\infty U_n$. Replacing U_n by $U_1 \cap \cdots \cap U_n$, one can further suppose (as is sometimes convenient) that $U_n \downarrow A$. Every open set is a G_δ. The intersection of any sequence of G_δ's is itself a G_δ. The union of finitely many G_δ's is a G_δ (if $U_n \downarrow A$ and $V_n \downarrow B$, then $U_n \cup V_n \downarrow A \cup B$).

Theorem 1. *If f is a continuous real-valued function on X, and c is a real number, then each of the following sets is a closed G_δ:*

$$A = \{x: f(x) \geq c\}$$

$$B = \{x: f(x) \leq c\}$$

$$C = \{x: f(x) = c\}.$$

If moreover f has compact support, and $c > 0$, then the sets A and C are compact G_δ's.

Proof. Since

$$A = \bigcap_1^\infty \{x: f(x) > c - 1/n\},$$

it is clear from the continuity of f that A is a closed G_δ. Similarly for B (or consider $-f$); also, $C = A \cap B$. Finally, if f vanishes outside some compact set D, and $c > 0$, then $A \subset D$ and $C \subset D$. ▌

The next theorem is perhaps the most important single result for the theory of integration in locally compact spaces:

Theorem 2. *If C is a compact G_δ, there exists a sequence f_n in $\mathscr{L}$ such that $f_n \downarrow \chi_C$.*

Proof. Let U_n be a sequence of open sets such that $C = \bigcap_1^\infty U_n$. According to 54.3, there exists for each n a function g_n in $\mathscr{L}$ such that $0 \leq g_n \leq 1$, $g_n = 1$ on C, and $g_n = 0$ on $X - U_n$. Defining

$$f_n = g_1 \cap \cdots \cap g_n,$$

we have $f_n \in \mathscr{L}$, $f_n \downarrow$, $g_n \geq f_n \geq \chi_C$, $f_n = 1$ on C, and $f_n = 0$ on $X - U_n$.

We assert that $f_n \downarrow \chi_C$. If $x \in C$, then $f_n(x) = 1$ for all n, hence $f_n(x) \downarrow 1 = \chi_C(x)$. If $x \notin C$, there exists an index m such that $x \notin U_m$, and so $f_m(x) = 0$; it follows from monotonicity that $f_n(x) = 0$ for all $n \geq m$, thus $f_n(x) \downarrow 0 = \chi_C(x)$. ▌

A subset A of X is called an F_σ in case there exists a sequence F_n of closed sets such that $A = \bigcup_1^\infty F_n$; equivalently, $X - A$ is a G_δ. Replacing F_n by $F_1 \cup \cdots \cup F_n$, one can further suppose that $F_n \uparrow A$. Every closed set is an F_σ. The union of any sequence of F_σ's is itself an F_σ. The intersection of finitely many F_σ's is an F_σ. If f is a continuous real-valued function on X, and c is a real number, the set

$$\{x: f(x) > c\}$$

is an open F_σ; in particular, the set

$$N(f) = \{x: |f|(x) > 0\}$$

is an open F_σ.

EXERCISES

1. In a first countable Hausdorff space, every singleton $\{x\}$ is a G_δ. If every singleton in the locally compact space X is a G_δ, then X is first countable.

2. In a metric space, every closed set is a G_δ.

3. If A is a compact G_δ in X, there exists an f in $\mathscr{L}$ such that $0 \leq f \leq 1$ and $A = \{x: f(x) = 1\}$.

4. If $A \subset X$, and f_n is a sequence of continuous real-valued functions such that $f_n \downarrow \chi_A$, then A is a closed G_δ.

5. If $f: X \to Y$ is a continuous mapping, then the inverse image of every G_δ in Y is a G_δ in X. Similarly for F_σ's.

6. Every open set in X is the union of compact G_δ's.

7. The open sets in X which are expressible as a union of a sequence of compact G_δ's are basic for the topology of X.

8. If X is compact, then X is metrizable if and only if the diagonal of $X \times X$ is a G_δ.

56. Baire Sets

The σ-ring generated by the class of all compact G_δ's in X is called the class of **Baire sets** in X. A real-valued function on X which is measurable with respect to this σ-ring is called a **Baire function.**

Our first theorem concerning Baire sets is central for the theory of regularity to be given later in the chapter. It asserts, roughly speaking, that between a compact set and an open set containing it, one may interpolate Baire sets with special topological properties:

Theorem 1 (“**Baire-sandwich theorem**”). *If $C \subset U$, where C is compact and U is open, there exist Baire sets V and D such that:*

(1) $C \subset V \subset D \subset U$,

(2) *V is open, and is the union of a sequence of compact G_δ's,*

(3) *D is a compact G_δ.*

Proof. According to 54.3, there exists an f in $\mathscr{L}$ such that $f = 1$ on C and $f = 0$ on $X - U$. Define

$$V = \{x: f(x) > 1/2\} \quad \text{and} \quad D = \{x: f(x) \geq 1/2\}.$$

Evidently $C \subset V \subset D \subset U$, V is open, and D is a compact G_δ by 55.1. The formula

$$V = \bigcup_{n=1}^{\infty} \{x : f(x) \geq 1/2 + 1/n\}$$

shows that V is the union of a sequence of compact G_δ's, and is therefore a Baire set. ∎

It follows that the open Baire sets are basic for the topology of X:

Corollary. *Every open set is a union of open Baire sets.*

Proof. If U is an open set, and $x \in U$, the problem is to find an open Baire set V such that $x \in V \subset U$. Indeed, taking $C = \{x\}$ in Theorem 1, we may suppose further that V is the union of a sequence of compact G_δ's. ∎

Theorem 2. *Every f in $\mathscr{L}$ is a Baire function.*

Proof. Defining $A_n = \{x : |f(x)| \geq 1/n\}$, we have

$$N(f) = \bigcup_{1}^{\infty} A_n,$$

where the A_n are compact G_δ's by 55.1. Also, for any real number c, the set $\{x : f(x) \leq c\}$ is a closed G_δ; since

$$N(f) \cap \{x : f(x) \leq c\} = \bigcup_{1}^{\infty} A_n \cap \{x : f(x) \leq c\},$$

and since each term of the union is evidently a compact G_δ, it follows at once (Sect. 12) that f is a Baire function. ∎

The precise role of Baire sets in the theory of measure in locally compact spaces is indicated by the following corollaries:

Corollary 1. *If $\mathscr{S}$ is a σ-ring of subsets of X, the following conditions are equivalent:*

(a) *Every f in $\mathscr{L}$ is measurable with respect to $\mathscr{S}$.*

(b) *$\mathscr{S}$ contains every Baire set.*

(c) *$\mathscr{S}$ contains every compact G_δ.*

Proof. The equivalence of (b) and (c) is immediate from the definition of the Baire sets.

(a) *implies* (c): If C is any compact G_δ, choose a sequence f_n in $\mathscr{L}$ such that $f_n \downarrow \chi_C$ (55.2). Since the f_n are assumed to be measurable with respect to $\mathscr{S}$, so is their pointwise limit χ_C (14.3), and this means $C \in \mathscr{S}$.

(b) *implies* (a): This is immediate from Theorem 2. ∎

Corollary 2. *The class of all Baire functions is the smallest class of real-valued functions on X which contains $\mathscr{L}$ and is closed under sequential pointwise limits.*

Proof. Let $\mathscr{G}$ be the class of all Baire functions, and $\mathscr{F}$ the other class mentioned in the statement of the corollary. Since $\mathscr{G} \supset \mathscr{L}$ by Theorem 2, and since $\mathscr{G}$ is closed under sequential pointwise limits (14.3), we have $\mathscr{G} \supset \mathscr{F}$ by the definition of $\mathscr{F}$.

Observe that if $f, g \in \mathscr{F}$ and c is a real number, then the functions $f + g$, fg, and cf also belong to $\mathscr{F}$. As a sample of the type of argument that is needed, we show that if $f \in \mathscr{L}$, then $f + g \in \mathscr{F}$ for all $g \in \mathscr{F}$; indeed, if $\mathscr{F}_0$ is the class of all real-valued functions g on X such that $f + g \in \mathscr{F}$, clearly $\mathscr{F}_0$ contains $\mathscr{L}$ and is closed under sequential pointwise limits, and so $\mathscr{F} \subset \mathscr{F}_0$ as asserted.

Let $\mathscr{M}$ be the class of all sets A such that $\chi_A \in \mathscr{F}$. In view of the relations

$$\chi_{A \cap B} = \chi_A \chi_B$$

$$\chi_{A - B} = \chi_A - \chi_{A \cap B}$$

$$\chi_{A \cup B} = \chi_A + \chi_B - \chi_{A \cap B},$$

it is clear from the preceding paragraph that $\mathscr{M}$ is a ring of sets. Since $\mathscr{F}$ contains sequential pointwise limits, it follows that $\mathscr{M}$ is a monotone class, and so $\mathscr{M}$ is a σ-ring.

Since $\mathscr{M}$ contains every compact G_δ (55.2), it follows that $\mathscr{M}$ contains every Baire set. Then $\mathscr{F}$ contains every simple Baire function, and hence every sequential pointwise limit of simple Baire functions; thus $\mathscr{F}$ contains every Baire function (16.4). That is, $\mathscr{F} \supset \mathscr{G}$. ∎

A **Baire measure** on X is a measure μ defined on the class of *all Baire sets*, such that

$$\mu(C) < \infty$$

for every compact G_δ C. It follows from 1.4 that every Baire measure is σ-finite.

The rest of the section is for application in Chapter 9, and may be omitted until that time. To be specific, Theorem 3 will be needed just once, in the proof of 78.5; nevertheless, its role in Chapter 9 is decisive.

Lemma 1. *If $\mathscr{E}$ is any class of open sets which generates the topology of X, then the σ-ring generated by $\mathscr{E}$ contains every Baire set.*

Proof. Let $\mathscr{A}$ be the class of all finite intersections $U_1 \cap \cdots \cap U_n$ of sets U_i in $\mathscr{E}$. Our assumption is that each open set in X is the union of a suitable subclass of $\mathscr{A}$. The problem is to show that $\mathfrak{S}(\mathscr{E})$ contains every compact G_δ. Of course $\mathscr{A} \subset \mathfrak{S}(\mathscr{E})$.

Suppose $C \subset U$, where C is compact and U is open. Since $\mathscr{A}$ is a base for the open sets, there exists, by an easy open covering argument, a finite sequence $V_1, \ldots, V_n$ in $\mathscr{A}$ such that

$$C \subset V_1 \cup \cdots \cup V_n \subset U$$

(compare with the proof of 54.3). Writing $S = V_1 \cup \cdots \cup V_n$, we have $C \subset S \subset U$, and clearly $S \in \mathfrak{S}(\mathscr{E})$.

Suppose now that C is any compact G_δ, and let U_n be a sequence of open sets such that $C = \bigcap_1^\infty U_n$. Since $C \subset U_n$, there exists, by the preceding argument, a set S_n in $\mathfrak{S}(\mathscr{E})$ such that $C \subset S_n \subset U_n$. Evidently $C = \bigcap_1^\infty S_n$, and so $C \in \mathfrak{S}(\mathscr{E})$. ∎

Lemma 2. *If X and Y are locally compact spaces, then the σ-ring of Baire sets of the product topological space $X \times Y$ is equal to the Cartesian product of the σ-ring of Baire sets of X with the σ-ring of Baire sets of Y.*

Proof. Let us write $\mathscr{S}(X)$, $\mathscr{S}(Y)$, and $\mathscr{S}(X \times Y)$ for the σ-rings of Baire sets of X, Y, and $X \times Y$, respectively. The problem is to show that

$$\mathscr{S}(X \times Y) = \mathscr{S}(X) \times \mathscr{S}(Y).$$

If A is a compact G_δ in X, and B is a compact G_δ in Y, then the compact set $A \times B$ is a G_δ in $X \times Y$; indeed, if U_n and V_n are sequences of open sets such that $U_n \downarrow A$ and $V_n \downarrow B$, then

$$U_n \times V_n \downarrow A \times B$$

(Lemma 2 in Sect. 34). In particular, $A \times B$ is a Baire set, and it now follows from 35.2 that

$$\mathscr{S}(X) \times \mathscr{S}(Y) \subset \mathscr{S}(X \times Y).$$

From the corollary of Theorem 1, it is clear that the sets $U \times V$, where U is an open Baire set in X and V is an open Baire set in Y, are basic for the topology of $X \times Y$. Since every such $U \times V$ belongs to $\mathscr{S}(X) \times \mathscr{S}(Y)$, it follows from Lemma 1 that

$$\mathscr{S}(X \times Y) \subset \mathscr{S}(X) \times \mathscr{S}(Y). \quad ∎$$

Theorem 3. *If μ_i is a Baire measure on the locally compact space X_i ($i = 1, 2$), then $\mu_1 \times \mu_2$ is a Baire measure on the product topological space $X_1 \times X_2$.*

Proof. Let $\pi = \mu_1 \times \mu_2$; by Lemma 2, the domain of definition of π is precisely the class of all Baire sets in the (locally compact) space $X_1 \times X_2$. If C is any compact G_δ in $X_1 \times X_2$, it remains to show that $\pi(C) < \infty$. Since each point of $X_1 \times X_2$ is contained in a set $U \times V$, where U and V are open sets with compact closure, it follows that

$$C \subset (U_1 \times V_1) \cup \cdots \cup (U_n \times V_n)$$

for a suitable finite sequence of such rectangles. By Theorem 1, there exist compact G_δ's C_k and D_k such that $U_k \subset C_k$ and $V_k \subset D_k$; then

$$C \subset (C_1 \times D_1) \cup \cdots \cup (C_n \times D_n),$$

and so

$$\pi(C) \leq \sum_1^n \pi(C_k \times D_k) = \sum_1^n \mu_1(C_k)\mu_2(D_k) < \infty. \quad \blacksquare$$

EXERCISES

1. Every point of X belongs to at least one Baire set (indeed, to a compact G_δ).

2. Each Baire set is contained in the union of a sequence of compact G_δ's.

3. If E is any Baire set, there exists a sequence A_n of compact G_δ's such that E belongs to the σ-ring generated by the A_n.

4. Every compact G_δ is the intersection of a decreasing sequence of open Baire sets, with each of the latter the union of an increasing sequence of compact G_δ's.

5. If U is an open F_σ in X, and is moreover contained in the union of a sequence of compact sets, then U is a Baire set; indeed, U is the union of a sequence of compact G_δ's.

6. Every compact Baire set in X is a G_δ.

7. If f is a continuous function on X such that $N(f)$ is contained in the union of a sequence of compact sets, then f is a Baire function.

8. A real-valued function f on X is said to *vanish at infinity* in case for each $\varepsilon > 0$, the set

$$\{x: |f(x)| \geq \varepsilon\}$$

has compact closure. Show that every continuous function which vanishes at infinity is a Baire function.

9. The bounded Baire functions are the smallest class of bounded functions on X which contains $\mathscr{L}$ and is closed under uniformly bounded sequential pointwise limits.

10. If X and Y are locally compact spaces, then every function in $\mathscr{L}(X \times Y)$ is the uniform limit of functions of the form

$$f_1(x)g_1(y) + \cdots + f_n(x)g_n(y),$$

where $f_i \in \mathscr{L}(X)$ and $g_i \in \mathscr{L}(Y)$. An alternate proof of Lemma 2 can be based on this observation.

11. The class of Baire sets in X is the monotone class generated by the class of all compact G_δ's.

57. Borel Sets

The σ-ring generated by the class of all compact sets in X is called the class of **Borel sets** in X. A real-valued function on X which is measurable with respect to this σ-ring is called a **Borel function.** Evidently every Baire set is a Borel set, and therefore every Baire function is a Borel function. In particular, every f in $\mathscr{L}$ is a Borel function (56.2).

A subset A of X is said to be **bounded** if there exists a compact set C such that $A \subset C$ (equivalently, the closure of A is compact); A is said to be σ-**bounded** if there exists a sequence C_n of compact sets such that

$$A \subset \bigcup_1^\infty C_n.$$

In view of 1.4, we have:

Theorem 1. *Every Borel set is σ-bounded. In particular, if f is a Borel function, then $N(f)$ is σ-bounded.*

The σ-ring generated by the class of all *closed sets* in X will be called the class of **weakly Borel sets.** This is evidently a σ-algebra, and is also the σ-ring generated by the class of all open sets. A real-valued function which is measurable with respect to this σ-algebra will be called a **weakly Borel function.** The value of these concepts is didactic; our motive for introducing them is to throw light on the concept of Borel set:

Theorem 2. *The class of Borel sets coincides with the class of all σ-bounded weakly Borel sets.*

Proof. Since every compact set (in a separated space) is closed, it follows that every Borel set is weakly Borel; moreover, every Borel set is σ-bounded by Theorem 1.

If C is compact and F is closed, then $C \cap F$ is compact, and is therefore a Borel set; it follows from the corollary of 1.6 that $C \cap A$ is a Borel set when A is any weakly Borel set. By another application of the same reasoning, $E \cap A$ is a Borel set when E is any Borel set, and A is a weakly Borel set.

Suppose now that A is a σ-bounded weakly Borel set, and let C_n be a sequence of compact sets such that $A \subset \bigcup_1^\infty C_n$. Then $E = \bigcup_1^\infty C_n$ is a

Borel set, and $A = E \cap A$, hence A is a Borel set by the preceding remarks. ∎

A **Borel measure** on X is a measure μ defined on the class of *all Borel sets*, such that

$$\mu(C) < \infty$$

for every compact set C. It follows from 1.4 that every Borel measure is σ-finite. If μ is a Borel measure on X, and ν is the restriction of μ to the class of all Baire sets, evidently ν is a Baire measure; we shall call ν the **Baire restriction** of μ.

EXERCISES

1. In a metric space, the concepts of Baire set and Borel set coincide.

2. If U is an open F_σ in X, and is moreover a Borel set, then U is a Baire set.

3. Every σ-bounded set in X is contained in the union of a sequence of compact G_δ's.

4. Every continuous Borel function on X is a Baire function.

5. Show directly (avoiding Theorem 2) that every σ-bounded open set in X is a Borel set.

6. If X is the union of a sequence of compact sets, then the concepts of Borel set and weakly Borel set coincide.

7. Every continuous real-valued function on X is a weakly Borel function.

8. The ring of all Borel sets is an ideal in the ring of all weakly Borel sets.

9. Every weakly Borel set is locally measurable with respect to the σ-ring of Borel sets.

10. If f is a weakly Borel function such that $N(f)$ is σ-bounded, then f is a Borel function.

11. If f is a weakly Borel function, and g is a Borel function, then fg is a Borel function.

12. If ν is a Baire measure on X, and μ is any measure which is defined on the class of Borel sets and extends ν, then μ is a Borel measure.

13. Let us define the *weakly Baire sets* in X to be the σ-ring generated by the class of all closed G_δ's. Then:

(*i*) The class of Baire sets coincides with the class of all σ-bounded weakly Baire sets.

(*ii*) The ring of Baire sets is an ideal in the ring of weakly Baire sets.

(*iii*) Every weakly Baire set is locally measurable with respect to the σ-ring of Baire sets.

(*iv*) Every continuous real-valued function on X is a *weakly Baire function*, that is, it is measurable with respect to the class of weakly Baire sets.

(*v*) If f is a weakly Baire function, and g is a Baire function, then fg is a Baire function.

***14.** Let $\mathscr{S}$ and $\mathscr{T}$ be the σ-algebras of weakly Baire sets in the locally compact spaces X and Y, respectively. It is easy to see that the σ-algebra of weakly Baire sets in $X \times Y$ contains $\mathscr{S} \times \mathscr{T}$; is it equal to $\mathscr{S} \times \mathscr{T}$?

15. Let us write $\mathscr{B}(X)$ for the class of Borel sets in X. It is easy to see that

$$\mathscr{B}(X \times Y) \supset \mathscr{B}(X) \times \mathscr{B}(Y)$$

for every pair of locally compact spaces X and Y.

16. (Johnson) With notation as in Exercise 15, the inclusion

$$\mathscr{B}(X \times Y) \supset \mathscr{B}(X) \times \mathscr{B}(Y)$$

can be proper. For example, suppose $X = Y$ is a nonmetrizable compact space in which all closed sets are G_δ's. Then (*i*) $\mathscr{B}(X)$ coincides with the class of Baire sets in X, (*ii*) $\mathscr{B}(X) \times \mathscr{B}(X)$ coincides with the class of Baire sets in $X \times X$, and (*iii*) the diagonal of $X \times X$ is compact but is not a G_δ (thus it is a Borel set, but not a Baire set).

17. If A is a closed subset of the locally compact space X, then the Borel sets of the (locally compact) space A are precisely the sets $E \cap A$, where E is an arbitrary Borel set in X. These are also precisely the Borel sets of X which are contained in A.

18. If A is a closed G_δ in the locally compact space X, then the Baire sets of the (locally compact) space A are precisely the sets $F \cap A$, where F is an arbitrary Baire set in X. These are also precisely the Baire sets of X which are contained in A.

***19.** Every compact weakly Baire set is necessarily a G_δ. Can "compact" be replaced by "closed"?

20. If $X = R$, then the class of Borel sets in X, as defined in this section, coincides with the class of Borel sets defined in Sect. 1. Moreover, the concepts "Baire set," "Borel set," "weakly Baire set," and "weakly Borel set" all coincide (as they do in any σ-bounded metric space).

21. Any Borel measure μ on X can be extended to a weakly Borel measure μ_w as follows: let μ_λ be the extension of μ described in Exercise 17.1, and let μ_w be the restriction of μ_λ to the class of weakly Borel sets. Thus,

$$\mu_w(A) = \text{LUB}\,\{\mu(E) : E \subset A,\ E \text{ Borel}\}$$

for each weakly Borel set A. (A *weakly Borel measure* is a measure, defined on the class of all weakly Borel sets, which is finite-valued for compact sets.)

58. Preliminaries on Rings

In arriving at the class of all Baire sets (respectively Borel sets), one can first form the ring $\mathscr{R}$ generated by the compact G_δ's (respectively compact sets), and then take the σ-ring generated by $\mathscr{R}$. For technical reasons, it is of interest to have a more explicit description of the sets in $\mathscr{R}$. Now, the class of compact G_δ's (respectively compact sets) contains the union and intersection of any two of its members, and the description we are after is given by the following set theoretic theorem:

Theorem 1. *Let $\mathscr{A}$ be a class of subsets of X such that the empty set belongs to $\mathscr{A}$, and such that $A \cup B$ and $A \cap B$ belong to $\mathscr{A}$ whenever A and B do. If $\mathscr{R}$ is the ring generated by $\mathscr{A}$, then every element of $\mathscr{R}$ can be written as a finite union*

$$\bigcup_1^n (A_i - B_i),$$

where $A_i, B_i \in \mathscr{A}$, $A_i \supset B_i$, and the $A_i - B_i$ are mutually disjoint.

Proof. If $\mathscr{R}_0$ is the class of all sets which have such a representation, evidently

$$\mathscr{A} \subset \mathscr{R}_0 \subset \mathscr{R}.$$

To establish $\mathscr{R} \subset \mathscr{R}_0$, it will suffice to show that $\mathscr{R}_0$ is a ring.

It is obvious that $\mathscr{R}_0$ is closed under finite disjoint unions. Moreover, $\mathscr{R}_0$ contains the difference of any two sets A, B in $\mathscr{A}$, as we see from the relation $A - B = A - A \cap B$. The rest of the proof will be carried out in three steps.

(1) *$\mathscr{R}_0$ contains finite intersections.* Suppose E and F belong to $\mathscr{R}_0$, and let

$$E = \bigcup_1^n (A_i - B_i) \qquad \text{and} \qquad F = \bigcup_1^m (C_j - D_j)$$

be representations of the indicated type. From the identity

$$(A - B) \cap (C - D) = (A \cap C) - (B \cup D),$$

it follows that

$$E \cap F = \bigcup_{i,j} [(A_i \cap C_j) - (B_i \cup D_j)];$$

clearly the mn terms on the right are mutually disjoint, and each of them belongs to $\mathscr{R}_0$, hence so does their union. (Incidentally, our use of the term "finite intersection" must exclude the case of the intersection of an empty family of sets, for we do not assume that X is in $\mathscr{A}$. See Sect. 0.)

(2) *$\mathscr{R}_0$ contains differences.* Assuming E and F are notated as above, our problem is to show that $E - F$ belongs to $\mathscr{R}_0$. It is easy to see that

$$E - F = \bigcup_1^n G_i,$$

where

$$G_i = \bigcap_{j=1}^m [(A_i - B_i) - (C_j - D_j)].$$

Since $G_i \subset A_i - B_i$, the G_i are mutually disjoint, and so it will suffice to show that each G_i belongs to $\mathscr{R}_0$. By (1), it is even sufficient to show that each of the *mn* sets

$$(A_i - B_i) - (C_j - D_j)$$

belongs to $\mathscr{R}_0$. In general,

$$(A - B) - (C - D) = [A - (B \cup C)] \cup [(A \cap D) - B];$$

moreover, the two terms on the right are disjoint if one assumes that $C \supset D$, for in this case

$$A - (B \cup C) \subset A - D \subset X - D,$$

whereas

$$(A \cap D) - B \subset D.$$

The assertion (2) is now clear.

(3) *$\mathscr{R}_0$ contains finite unions.* Indeed, $E \cup F$ is the disjoint union of $E \cap F$, $E - F$, and $F - E$. ∎

The following consequence of Theorem 1 is useful in the study of Baire and Borel measures:

Theorem 2. *Suppose $\mathscr{A}$ is a nonempty class of subsets of X such that $A \cup B$ and $A \cap B$ belong to $\mathscr{A}$ whenever A and B do, and let $\mathscr{S}$ be the σ-ring generated by $\mathscr{A}$. If μ_1 and μ_2 are measures on $\mathscr{S}$ such that*

(i) $\mu_1(A) = \mu_2(A)$ for all A in $\mathscr{A}$,

and

(ii) the μ_i are finite valued on $\mathscr{A}$,

then $\mu_1 = \mu_2$, that is, $\mu_1(E) = \mu_2(E)$ for every E in $\mathscr{S}$.

Proof. If $\mathscr{R}$ is the ring generated by $\mathscr{A}$, it is clear from Theorem 1 (and the additivity and subtractivity of the μ_i) that

$$\mu_1(E) = \mu_2(E)$$

for all E in $\mathscr{R}$, and so $\mu_1 = \mu_2$ on $\mathfrak{S}(\mathscr{R}) = \mathscr{S}$ by the (UET). ∎

The application of Theorem 2 to Baire and Borel measures is immediate:

Theorem 3. *If μ_1 and μ_2 are Baire (respectively Borel) measures on the locally compact space X, such that*

$$\mu_1(C) = \mu_2(C)$$

for every compact G_δ *(respectively compact set)* C, *then* $\mu_1 = \mu_2$, *that is,*

$$\mu_1(E) = \mu_2(E)$$

for every Baire set (respectively Borel set) E.

In other words, a Baire measure is uniquely determined by its values for compact G_δ's; a Borel measure is uniquely determined by its values for compact sets. It is *not* asserted (nor is it true) that a Borel measure is determined by its values for Baire sets (cf. 59.2).

EXERCISES

1. If μ is a Borel measure such that $\mu(D) = 0$ for all compact G_δ's D, then $\mu(E) = 0$ for all Borel sets E. Hence: if the Baire restriction of μ is identically zero, then so is μ.

2. If μ_1, μ_2 are finite Borel measures on X such that $\mu_1 \leq \mu_2$ (i.e., $\mu_1(E) \leq \mu_2(E)$ for all Borel sets E), and if $\mu_1(D) = \mu_2(D)$ for all compact G_δ's D, then $\mu_1 = \mu_2$.

3. If μ is a Borel measure on X, and f is an integrable Borel function such that

$$\int_{C-D} f\,d\mu \geq 0$$

for all compact sets C and D, then $f \geq 0$ a.e.

4. (Johnson) In Theorem 2, it is sufficient to assume that the class $\mathscr{A}$ is closed under finite intersections. Finite unions then take care of themselves, in view of the identity $E \cup F - E = F - E \cap F$.

59. Regularity

The theory of regularity to be launched in this section will ultimately be applied to measures defined in locally compact spaces; however, it is essentially combinatorial in nature, and easy to treat axiomatically. We therefore introduce the following axioms (others will be added to the list as needed, in Sect. 61):

(I) *μ is a measure defined on a σ-ring $\mathscr{S}$.*

(II) *$\mathscr{C}$ is a subclass of $\mathscr{S}$ which is closed under countable intersections and finite unions. We assume that the empty set belongs to $\mathscr{C}$, and that $\mu(C) < \infty$ for all C in $\mathscr{C}$.*

(III) *$\mathscr{U}$ is a subclass of $\mathscr{S}$ which is closed under countable unions and finite intersections. We assume that for each E in $\mathscr{S}$, there is a set U in $\mathscr{U}$ such that $E \subset U$.*

To save notation, we shall not specify the underlying set of which the sets in $\mathcal{S}$ are subsets. All of our applications will be to the following two examples:

Example 1. Let $\mathcal{C}$ be the class of all compact sets in the locally compact space X, $\mathcal{S}$ the class of all Borel sets, $\mathcal{U}$ the class of all open Borel sets, and μ any Borel measure on X. Incidentally, it follows from 57.2 that $\mathcal{U}$ is precisely the class of all σ-bounded open sets.

Example 2. Let $\mathcal{C}$ be the class of all compact G_δ's in the locally compact space X, $\mathcal{S}$ the class of all Baire sets, $\mathcal{U}$ the class of all open Baire sets, and μ any Baire measure on X.

The validity of Axiom III for these two examples is verified simultaneously by the following argument. If E is any Borel set, there exists a sequence C_n of compact sets such that $E \subset \bigcup_1^\infty C_n$ (57.1). Let V_n be an open Baire set such that $C_n \subset V_n$ (56.1). Then $V = \bigcup_1^\infty V_n$ is an open Baire set such that $E \subset V$.

Returning to the axiomatic situation, it is appropriate to note that we are not assuming that $\mathcal{C}$ generates the σ-ring $\mathcal{S}$ (this assumption will eventually be made in Sect. 61).

A set E in $\mathcal{S}$ is said to be **outer regular** in case

$$\mu(E) = \text{GLB}\,\{\mu(U) : E \subset U,\ U \in \mathcal{U}\},$$

inner regular in case

$$\mu(E) = \text{LUB}\,\{\mu(C) : C \subset E,\ C \in \mathcal{C}\},$$

and **regular** if it is both outer and inner regular. We say that μ is a **regular** measure if every E in $\mathcal{S}$ is regular. All of these definitions are relative to the given system

$$(\mathcal{S}, \mu, \mathcal{C}, \mathcal{U}).$$

It will be shown in the next section that regularity is automatically verified in Example 2 above.

It should be noted that for any E in $\mathcal{S}$, the indicated GLB and LUB are defined (by Axioms III and II, respectively). The following remarks are often useful. If E is outer regular, then given any $\varepsilon > 0$ there exists a U in $\mathcal{U}$ such that $E \subset U$ and

$$\mu(U) \leq \mu(E) + \varepsilon$$

(even if $\mu(E) = \infty$). If E is inner regular, and $\mu(E) < \infty$, then given any $\varepsilon > 0$ there exists a C in $\mathcal{C}$ such that $C \subset E$ and

$$\mu(E) \leq \mu(C) + \varepsilon.$$

The following is a useful verbal guide: outer regularity of E means that we may approximate the measure of E from the outside by means of sets U; inner regularity means that we may approximate the measure of E from the inside by means of sets C.

We interrupt the abstract theory of regularity to present two highly useful and instructive results on regular Borel measures. The first of these is at the heart of the relation between Baire measures and regular Borel measures:

Theorem 1. *If μ is a regular Borel measure on the locally compact space X, and C is any compact set, there exists a compact G_δ D such that $C \subset D$ and $\mu(D - C) = 0$.*

Proof. By regularity we may choose a sequence U_n of open Borel sets such that $C \subset U_n$ and $\mu(C) = \text{GLB}\, \mu(U_n)$. For each n, there exists (by 56.1) a compact G_δ D_n such that $C \subset D_n \subset U_n$. Then $D = \bigcap_1^\infty D_n$ is a compact G_δ, $C \subset D$, and

$$\mu(C) \leq \mu(D) \leq \mu(D_n) \leq \mu(U_n)$$

for all n, and hence $\mu(C) = \mu(D)$; by finiteness, $\mu(D - C) = 0$. ∎

Theorem 2. *If μ_1 and μ_2 are regular Borel measures on the locally compact space X, such that*

$$\mu_1(D) = \mu_2(D)$$

for every compact G_δ D, then $\mu_1 = \mu_2$, that is,

$$\mu_1(E) = \mu_2(E)$$

for every Borel set E.

Proof. Given any compact set C, it will suffice by 58.3 (or regularity) to show that $\mu_1(C) = \mu_2(C)$. By Theorem 1 there exists a compact G_δ D_i such that $C \subset D_i$ and $\mu_i(D_i - C) = 0$. Then $D = D_1 \cap D_2$ is a compact G_δ,

$$\mu_i(D - C) \leq \mu_i(D_i - C) = 0,$$

and so

$$\mu_1(C) = \mu_1(D) = \mu_2(D) = \mu_2(C). \quad ∎$$

The rest of the section is devoted to the abstract discussion of regularity with respect to the system $(\mathscr{S}, \mu, \mathscr{C}, \mathscr{U})$ satisfying Axioms I–III.

Theorem 3.

(1) *If $\mu(E) = \infty$, then E is outer regular.*

(2) *If $\mu(E) = 0$, then E is inner regular.*

(3) *Every set in $\mathcal{U}$ is outer regular.*

(4) *Every set in $\mathcal{C}$ is inner regular.*

(5) *If C_n is a sequence of sets in $\mathcal{C}$, then $\bigcup_1^\infty C_n$ is inner regular.*

(6) *If U_n is a sequence of sets in $\mathcal{U}$, and $\mu(U_1) < \infty$, then $\bigcap_1^\infty U_n$ is outer regular.*

Proof. The statements (1) through (4) are obvious from the definitions and the monotonicity of μ.

(5) Replacing C_n by $C_1 \cup \cdots \cup C_n$, we may suppose (by Axiom II) that $C_n \uparrow E$, where $E = \bigcup_1^\infty C_n$. Then $\mu(C_n) \uparrow \mu(E)$, and it is clear that E is inner regular.

(6) Replacing U_n by $U_1 \cap \cdots \cap U_n$, we may suppose (by Axiom III) that $U_n \downarrow E$, where $E = \bigcap_1^\infty U_n$. Then $\mu(U_n) \downarrow \mu(E)$ by finiteness, and so E is outer regular. ∎

Theorem 4. *The union of a sequence of outer regular sets is outer regular.*

Proof. Suppose $E = \bigcup_1^\infty E_n$, where each E_n is outer regular. If $\mu(E) = \infty$, then E is outer regular by Theorem 3. Assuming $\mu(E) < \infty$, we have $\mu(E_n) < \infty$ for all n. Given any $\varepsilon > 0$, we may choose, for each n, a set U_n in $\mathcal{U}$ such that $E_n \subset U_n$ and

$$\mu(U_n) \leq \mu(E_n) + \varepsilon/2^n,$$

that is,

$$\mu(U_n - E_n) \leq \varepsilon/2^n.$$

Defining $U = \bigcup_1^\infty U_n$, we have $E \subset U$, and $U \in \mathcal{U}$ by Axiom III. Since

$$U - E \subset \bigcup_1^\infty U_n - E_n,$$

it follows that

$$\mu(U - E) \leq \sum_1^\infty \mu(U_n - E_n) \leq \varepsilon;$$

thus

$$\mu(U) \leq \mu(E) + \varepsilon,$$

and it is clear that E is outer regular. ∎

Lemma. *The union of finitely many inner regular sets is inner regular.*

Proof. Assuming E and F are inner regular, it will suffice to show that $E \cup F$ is inner regular.

If $\mu(E) = \infty$, we may choose (by the inner regularity of E) a sequence C_n in $\mathscr{C}$ such that $C_n \subset E$ and $\mu(C_n) \geq n$. Since $\mu(E \cup F) = \infty$ and $C_n \subset E \cup F$, it is clear that $E \cup F$ is inner regular. Similarly if $\mu(F) = \infty$.

Assuming $\mu(E) < \infty$ and $\mu(F) < \infty$, we have also $\mu(E \cup F) < \infty$. Given any $\varepsilon > 0$, choose C, D in $\mathscr{C}$ so that

$$C \subset E, D \subset F, \qquad \text{and} \qquad \mu(E - C) \leq \varepsilon/2, \mu(F - D) \leq \varepsilon/2.$$

Then $C \cup D \in \mathscr{C}$, $C \cup D \subset E \cup F$, and

$$(E \cup F) - (C \cup D) \subset (E - C) \cup (F - D),$$

and so

$$\mu[(E \cup F) - (C \cup D)] \leq \mu(E - C) + \mu(F - D) \leq \varepsilon/2 + \varepsilon/2.$$

Thus

$$\mu(E \cup F) \leq \mu(C \cup D) + \varepsilon,$$

and it is clear that $E \cup F$ is inner regular. ▮

Theorem 5. *The union of a sequence of inner regular sets is inner regular.*

Proof. Suppose $E = \bigcup_1^\infty E_n$, where each E_n is inner regular. Since $E_1 \cup \cdots \cup E_n$ is inner regular by the lemma, we may further suppose that $E_n \uparrow E$. Then $\mu(E_n) \uparrow \mu(E)$.

If $\mu(E) = \infty$, then given any real number $\eta > 0$ there is an index m such that $\mu(E_m) > \eta$. Since E_m is inner regular, we may choose C in $\mathscr{C}$ so that $C \subset E_m$ and $\mu(C) > \eta$. Then also $C \subset E$; since η is arbitrary and $\mu(C) > \eta$, we conclude that E is inner regular.

If $\mu(E) < \infty$, then $\mu(E_n) < \infty$ for all n. Given $\varepsilon > 0$, choose an index m so that

$$\mu(E - E_m) \leq \varepsilon/2.$$

By the inner regularity of E_m, there exists a C in $\mathscr{C}$ such that $C \subset E_m$ and

$$\mu(E_m - C) \leq \varepsilon/2.$$

Then also $C \subset E$, and

$$\mu(E - C) = \mu(E - E_m) + \mu(E_m - C) \leq \varepsilon/2 + \varepsilon/2,$$

thus $\mu(E) \leq \mu(C) + \varepsilon$. ▮

Theorem 6. *The intersection of a sequence of inner regular sets of finite measure is inner regular.*

Proof. Suppose $E = \bigcap_1^\infty E_n$, where the E_n are inner regular, and $\mu(E_n) < \infty$ for all n.

Given any $\varepsilon > 0$, we may choose, for each n, a set C_n in $\mathscr{C}$ such that $C_n \subset E_n$ and

$$\mu(E_n - C_n) \leq \varepsilon/2^n.$$

Defining $C = \bigcap_1^\infty C_n$, we have $C \in \mathscr{C}$ by Axiom II, and $C \subset E$; since

$$E - C \subset \bigcup_1^\infty E_n - C_n,$$

it follows that

$$\mu(E - C) \leq \sum_1^\infty \mu(E_n - C_n) \leq \varepsilon,$$

and so E is inner regular. ▌

Lemma. *The intersection of finitely many outer regular sets of finite measure is outer regular.*

Proof. Assuming E and F are outer regular sets of finite measure, it will suffice to show that $E \cap F$ is outer regular.

Given any $\varepsilon > 0$, choose U and V in $\mathscr{U}$ so that

$$E \subset U, F \subset V, \quad \text{and} \quad \mu(U - E) \leq \varepsilon/2, \mu(V - F) \leq \varepsilon/2.$$

Then $U \cap V \in \mathscr{U}$, $E \cap F \subset U \cap V$, and

$$U \cap V - E \cap F \subset (U - E) \cup (V - F),$$

hence

$$\mu(U \cap V - E \cap F) \leq \mu(U - E) + \mu(V - F) \leq \varepsilon/2 + \varepsilon/2. \ ▌$$

Theorem 7. *The intersection of a sequence of outer regular sets of finite measure is outer regular.*

Proof. Suppose $E = \bigcap_1^\infty E_n$, where the E_n are outer regular, and $\mu(E_n) < \infty$ for all n. Since $E_1 \cap \cdots \cap E_n$ is outer regular by the lemma, we may further suppose that $E_n \downarrow E$. Then $\mu(E_n) \downarrow \mu(E)$ by finiteness.

Given any $\varepsilon > 0$, let m be an index such that

$$\mu(E_m - E) \leq \varepsilon/2.$$

Since E_m is outer regular, we may choose U in $\mathscr{U}$ so that $E_m \subset U$ and

$$\mu(U - E_m) \leq \varepsilon/2.$$

Then also $E \subset U$, and

$$\mu(U - E) = \mu(U - E_m) + \mu(E_m - E) \leq \varepsilon/2 + \varepsilon/2. \ ▌$$

We may summarize Theorems 3 through 7 as follows. Let E_n be a sequence of sets in $\mathscr{S}$:

(1) If E_n is outer regular for all n, then $\bigcup_1^\infty E_n$ is also outer regular.

(2) If E_n is outer regular and $\mu(E_n) < \infty$ for all n, then $\bigcap_1^\infty E_n$ is also outer regular.

(3) If E_n is inner regular for all n, then $\bigcup_1^\infty E_n$ is also inner regular.

(4) If E_n is inner regular and $\mu(E_n) < \infty$ for all n, then $\bigcap_1^\infty E_n$ is also inner regular.

Moreover:

(5) If $\mu(E) = \infty$, then E is outer regular.

(6) If $\mu(E) = 0$, then E is inner regular.

(7) Every set in $\mathscr{U}$ is outer regular.

(8) Every set in $\mathscr{C}$ is inner regular.

We note in particular the following consequence of (1) through (4):

Theorem 8. *If E_n is a sequence of regular sets, then $\bigcup_1^\infty E_n$ is regular. If moreover $\mu(E_n) < \infty$ for all n, then $\bigcap_1^\infty E_n$ is also regular.*

EXERCISES

1. Let μ be a regular Borel measure on X, and T a homeomorphism of X. Suppose that for compact G_δ's D, the relation $\mu(D) = 0$ implies $\mu(T(D)) = 0$. Then for a Borel set E, the relation $\mu(E) = 0$ implies $\mu(T(E)) = 0$.

2. Suppose μ_1 and μ_2 are regular Borel measures on X such that for compact G_δ's D, the relation $\mu_2(D) = 0$ implies $\mu_1(D) = 0$. Then $\mu_1 \ll \mu_2$.

3. If μ_1 and μ_2 are regular Borel measures on X such that $\mu_1(D) \leq \mu_2(D)$ for every compact G_δ D, then $\mu_1(E) \leq \mu_2(E)$ for all Borel sets E.

60. Regularity of Baire Measures

Baire measures are automatically regular, as we shall see in this section. Let us suppose that we are given a Baire measure μ on the locally compact space X. Thus, the domain of definition of μ is the class of all Baire sets, and $\mu(C)$ is finite for every compact G_δ C. Our assertion is that for each Baire set E, we have

$$\mu(E) = \text{LUB}\,\{\mu(C): C \subset E,\ C \text{ compact } G_\delta\},$$

$$\mu(E) = \text{GLB}\,\{\mu(U): E \subset U,\ U \text{ open Baire set}\}.$$

In the notation of the preceding section, we are considering the system

$$(\mathscr{S}, \mu, \mathscr{C}, \mathscr{U}),$$

where $\mathscr{S}$ is the class of all Baire sets, $\mathscr{C}$ is the class of all compact G_δ's, and $\mathscr{U}$ is the class of all open Baire sets. Our problem is to show that each E in $\mathscr{S}$ is regular with respect to this system.

Lemma. *If C and D are compact G_δ's, then $C - D$ is regular.*

Proof. Since $C - D = C - C \cap D$, we may suppose further that $C \supset D$.

Outer regularity. Let U_n be a sequence of open sets such that $C = \bigcap_1^\infty U_n$. For each n, there exists (56.1) an open Baire set V_n and a compact G_δ D_n such that

$$C \subset V_n \subset D_n \subset U_n.$$

Since $C = \bigcap_1^\infty V_n$, and since $\mu(V_n) \leq \mu(D_n) < \infty$, we conclude from 59.3 that C is outer regular.

Accordingly, given any $\varepsilon > 0$, there exists an open Baire set U such that $C \subset U$ and $\mu(U - C) \leq \varepsilon$. Then $C - D \subset U - D$, where $U - D$ is an open Baire set. Since

$$\mu[(U - D) - (C - D)] = \mu(U - C) \leq \varepsilon,$$

we conclude that $C - D$ is outer regular.

Inner regularity. Let U_n be a sequence of open sets such that $D = \bigcap_1^\infty U_n$. For each n, let V_n be an open Baire set such that $D \subset V_n \subset U_n$; as noted in 56.1, we may suppose further that each V_n is the union of a sequence of compact G_δ's, and hence is an F_σ. Now

$$D = \bigcap_1^\infty V_n,$$

and so

$$C - D = \bigcup_1^\infty C - V_n;$$

to show that $C - D$ is inner regular, it will suffice by 59.3 to show that each $C - V_n$ is a compact G_δ. Indeed, since C is a compact G_δ, and $X - V_n$ is a closed G_δ, it follows that the set

$$C - V_n = C \cap (X - V_n)$$

is a compact G_δ. ∎

Theorem 1. *Every Baire measure is regular.*

Proof. We continue with the notations established above. Let $\mathscr{R}$ be the ring generated by the compact G_δ's; it is clear from 58.1 that $\mathscr{R}$ is the class of all finite unions

$$E = \bigcup_i (C_i - D_i),$$

where C_i, D_i are compact G_δ's. It follows from 59.4, 59.5, and the lemma that every E in $\mathscr{R}$ is regular.

Given any compact G_δ C, let us show first that $C \cap E$ is regular for every Baire set E. If $\mathscr{M}$ is the class of all Baire sets E such that $C \cap E$ is regular, it is clear from the preceding paragraph that $\mathscr{R} \subset \mathscr{M}$. To prove that $\mathscr{M}$ contains every Baire set, it will suffice by the (LMC) to show that $\mathscr{M}$ is a monotone class. Indeed, since $\mu(C \cap E) < \infty$ for every Baire set E, and in particular for E in $\mathscr{M}$, our assertion is clear from 59.8.

Suppose now that E is an arbitrary Baire set, and let C_n be a sequence of compact G_δ's such that $E \subset \bigcup_1^\infty C_n$ (1.4). Since

$$E = \bigcup_1^\infty C_n \cap E,$$

and each $C_n \cap E$ is regular by the foregoing discussion, we conclude from 59.8 that E is regular. ▮

EXERCISES

1. If ν is a Baire measure on X, and f is an integrable Baire function such that $\int_D f\,d\nu \geq 0$ for every compact G_δ D, then $f \geq 0$ a.e.

2. If ν_1 and ν_2 are Baire measures on X such that $\nu_1(D) \leq \nu_2(D)$ for every compact G_δ D, then $\nu_1(F) \leq \nu_2(F)$ for every Baire set F.

3. Any Baire measure ν on X can be extended to a weakly Baire measure ν_w as follows: let ν_λ be the extension of ν described in Exercise 17.1, and let ν_w be the restriction of ν_λ to the class of weakly Baire sets (cf. Exercise 57.21). Thus,

$$\nu_w(A) = \text{LUB}\,\{\nu(F) : F \subset A,\ F \text{ Baire}\}$$

for each weakly Baire set A. Better yet,

$$\nu_w(A) = \text{LUB}\,\{\nu(D) : D \subset A,\ D \text{ compact } G_\delta\}$$

for each weakly Baire set A.

61. Regularity (*Continued*)

We resume the abstract discussion of regularity relative to a system

$$(\mathscr{S}, \mu, \mathscr{C}, \mathscr{U})$$

satisfying Axioms I–III of Sect. 59. Our goal in this section is to obtain simple criteria for the regularity of μ. To proceed further, we introduce another axiom:

(IV) *If $U \in \mathscr{U}$ and $C \in \mathscr{C}$, then $U - C \in \mathscr{U}$.*

Example 1. In the case of a Borel measure, if U is an open Borel set and C is compact, then $U - C$ is an open Borel set.

Example 2. In the case of a Baire measure, if U is an open Baire set and C is a compact G_δ, then $U - C$ is an open Baire set.

The eventual object of the abstract discussion is to obtain simple criteria for regularity. Since every Baire measure is already known to be regular (60.1), the results of the present section are not needed for the Baire case. It is nevertheless true that all axioms to be introduced are verified by every Baire measure and every Borel measure, and so it is in principle possible to develop the theory of regularity simultaneously for Baire and Borel measures. This is, however, not convenient (see the discussion following Axiom VI below), and *the results of the present section will be applied only to Borel measures.*

Theorem 1 (Assuming Axioms I-IV). *If every set in $\mathscr{C}$ is outer regular (and hence regular), then $C - D$ is outer regular for all C, D in $\mathscr{C}$.*

Proof. Given C, D in $\mathscr{C}$, it is to be shown that $C - D$ is outer regular. Since $C - D = C - C \cap D$, we may assume $C \supset D$. Given any $\varepsilon > 0$, the assumed outer regularity of C implies that $U \in \mathscr{U}$ may be chosen so that $C \subset U$ and $\mu(U) \leq \mu(C) + \varepsilon$, that is, $\mu(U - C) \leq \varepsilon$. We have $C - D \subset U - D$, where $U - D \in \mathscr{U}$ by Axiom IV. Moreover,

$$\mu[(U - D) - (C - D)] = \mu(U - C) \leq \varepsilon,$$

and so $C - D$ is outer regular. ∎

We shall say that a set E in $\mathscr{S}$ is "bounded" (relative to the given system) if there exists a C in $\mathscr{C}$ such that $E \subset C$. Evidently every "bounded" set has finite measure.

In view of 56.1, it is easy to see that in both the Borel measure and Baire measure examples, the above notion of "boundedness" is consistent with the topological notion of boundedness defined in Sect. 57.

We now introduce another axiom:

(V) *If $C \in \mathscr{C}$, there exists a "bounded" U in $\mathscr{U}$ such that $C \subset U$. In other words, for each $C \in \mathscr{C}$, there exist sets $U \in \mathscr{U}$ and $D \in \mathscr{C}$ such that $C \subset U \subset D$.*

If C is a compact set in the locally compact space X, there exists (56.1) an open Baire set U and a compact G_δ D such that $C \subset U \subset D$. This shows simultaneously that both the Borel and the Baire examples satisfy Axiom V.

Theorem 2 (Assuming Axioms I-V). *If every "bounded" set in $\mathscr{U}$ is inner regular (and hence regular), then $C - D$ is inner regular for all C, D in $\mathscr{C}$.*

Proof. Let $C, D \in \mathscr{C}$. Since $C - D = C - C \cap D$, we may suppose $C \supset D$. By Axiom V, we may choose a "bounded" U in $\mathscr{U}$ such that $C \subset U$.

The set $U - D$ is clearly "bounded," and belongs to $\mathscr{U}$ by Axiom IV, and so by assumption it is inner regular. Given any $\varepsilon > 0$, we may therefore choose $E \in \mathscr{C}$ so that

$$E \subset U - D \qquad \text{and} \qquad \mu[(U - D) - E] \leq \varepsilon.$$

We have

$$C - D = (C \cap U) - D = C \cap (U - D) \supset C \cap E;$$

thus $C - D \supset C \cap E$, where $C \cap E \in \mathscr{C}$. Moreover,

$$(C - D) - (C \cap E) = (C - D) - E \subset (U - D) - E,$$

and so

$$\mu[(C - D) - (C \cap E)] \leq \mu[(U - D) - E] \leq \varepsilon.$$

It follows that $C - D$ is inner regular. ∎

Let us for a moment introduce the notations $\mathscr{O}$, $\mathscr{I}$, and bdd $\mathscr{U}$ for the classes of all outer regular sets, inner regular sets, and "bounded" sets in $\mathscr{U}$, respectively. We may summarize Theorems 1 and 2 concisely as follows: the hypothesis $\mathscr{C} \subset \mathscr{O}$ leads to the conclusion $\mathscr{C} - \mathscr{C} \subset \mathscr{O}$ and the hypothesis bdd $\mathscr{U} \subset \mathscr{I}$ leads to the conclusion $\mathscr{C} - \mathscr{C} \subset \mathscr{I}$. We are interested in the simultaneous validity of these conclusions, that is to say, the regularity of every set in $\mathscr{C} - \mathscr{C}$. The next axiom will guarantee that the two hypotheses in question are equivalent (hence either hypothesis will imply that every set in $\mathscr{C} - \mathscr{C}$ is regular):

(VI) *If $C \in \mathscr{C}$ and $U \in \mathscr{U}$, then $C - U \in \mathscr{C}$.*

Example 1. In the case of a Borel measure, if C is compact and U is an open Borel set (or any open set), then $C - U$ is compact. Thus, Axiom VI is verified.

**Example* 2. In the case of a Baire measure, if C is a compact G_δ, and U is an open Baire set, then $C - U$ is a compact Baire set. It is possible to show that every compact Baire set is a G_δ, and hence that Axiom VI is verified. However, the object of the abstract discussion is to obtain criteria for regularity; since it has already been established in 60.1 that every Baire measure is regular, for our purposes the Borel example is the only pertinent one. By applying the results of this section only to the Borel case, we avoid using the rather complicated theorem that compact Baire sets are G_δ's (Exercise 56.6).

Theorem 3 (Assuming Axioms I-VI). *The following conditions are equivalent:*

(a) *Every set in $\mathscr{C}$ is outer regular.*

(b) *Every "bounded" set in $\mathscr{U}$ is inner regular.*

Proof. (b) *implies* (a): Let $C \in \mathscr{C}$. By Axiom V there is a "bounded" U in $\mathscr{U}$ such that $C \subset U$. The set $U - C$ belongs to $\mathscr{U}$ by Axiom IV, and is clearly "bounded," and so by assumption it is inner regular. Accordingly, given $\varepsilon > 0$, we may choose D in $\mathscr{C}$ so that

$$D \subset U - C \qquad \text{and} \qquad \mu[(U - C) - D] \leq \varepsilon.$$

We have $C = U - (U - C) \subset U - D$, and $U - D$ belongs to $\mathscr{U}$ by Axiom IV; moreover, $(U - D) - C = (U - C) - D$, and so

$$\mu[(U - D) - C] = \mu[(U - C) - D] \leq \varepsilon.$$

It follows that C is outer regular. (Incidentally, this part of the argument does not use Axiom VI.)

(a) *implies* (b): Let U be a "bounded" set in $\mathscr{U}$. Choose C in $\mathscr{C}$ so that $U \subset C$. The set $C - U$ belongs to $\mathscr{C}$ by Axiom VI, and so by assumption it is outer regular. Accordingly, given $\varepsilon > 0$, we may choose V in $\mathscr{U}$ so that

$$C - U \subset V \qquad \text{and} \qquad \mu[V - (C - U)] \leq \varepsilon.$$

We have $C - V \subset C - (C - U) = U$, and $C - V \in \mathscr{C}$ by Axiom VI. Also,

$$U - (C - V) = (U - C) \cup (U \cap V) = \varnothing \cup (U \cap V) = U \cap V,$$

and obviously $U \cap V \subset V - (C - U)$. Summarizing, we have

$$U - (C - V) = U \cap V \subset V - (C - U),$$

and so

$$\mu[U - (C - V)] \leq \mu[V - (C - U)] \leq \varepsilon;$$

it follows that U is inner regular. ∎

The next axiom will ensure that regularity properties of $\mathscr{C}$, or of the "bounded" sets in $\mathscr{U}$, carry over to regularity properties of $\mathscr{S}$:

(VII) *The σ-ring generated by $\mathscr{C}$ is $\mathscr{S}$. That is, $\mathscr{S} = \mathfrak{S}(\mathscr{C})$.*

Example 1. If $\mathscr{C}$ is the class of all compact sets in the locally compact space X, then $\mathfrak{S}(\mathscr{C})$ is the class of all Borel sets. Thus, Axiom VII is verified in the case of a Borel measure.

Example 2. If $\mathscr{C}$ is the class of all compact G_δ's in the locally compact space X, then $\mathfrak{S}(\mathscr{C})$ is the class of all Baire sets. Thus, Axiom VII is verified in the case of a Baire measure.

Theorem 4 (Assuming Axioms I-VII) *The following conditions on the system $(\mathscr{S}, \mu, \mathscr{C}, \mathscr{U})$ are equivalent:*

(a) *μ is regular.*

(b) *Every set in $\mathscr{C}$ is outer regular.*

(c) *Every "bounded" set in $\mathscr{U}$ is inner regular.*

Proof. It is trivial that (a) implies (b) and (c); (b) and (c) are equivalent by Theorem 3.

(c) *implies* (a): Suppose (c) holds. Then (b) also holds by Theorem 3. It follows from Theorems 1 and 2 that $C - D$ is regular, for all C, D in $\mathscr{C}$. Let $\mathscr{R}$ be the ring generated by $\mathscr{C}$. Since (58.1) $\mathscr{R}$ is the class of all finite unions

$$E = \bigcup_i C_i - D_i,$$

where $C_i, D_i \in \mathscr{C}$, it follows from 59.8 that every E in $\mathscr{R}$ is regular.

If $C \in \mathscr{C}$, then $C \cap E$ is regular for every E in $\mathscr{S}$. For, let $\mathscr{M}$ be the class of all E in $\mathscr{S}$ such that $C \cap E$ is regular; since $\mu(C \cap E)$ is finite for all E in $\mathscr{S}$, and in particular for all E in $\mathscr{M}$, it follows from 59.8 that $\mathscr{M}$ is a monotone class. Since it is clear from the preceding paragraph that $\mathscr{R} \subset \mathscr{M}$, our assertion $\mathscr{S} \subset \mathscr{M}$ follows from the (LMC).

Finally, if E is any set in $\mathscr{S}$, there exists (1.4) a sequence C_n in $\mathscr{C}$ such that $E \subset \bigcup_1^\infty C_n$. Since

$$E = \bigcup_1^\infty C_n \cap E,$$

and since each $C_n \cap E$ is regular by the preceding paragraph, it follows from 59.8 that E is regular.

Summarizing, every E in $\mathscr{S}$ is regular; in other words, μ is regular. ∎

For our purposes, the following result is the important application of Theorem 4:

Theorem 5. *If μ is a Borel measure on the locally compact space X, the following conditions are equivalent:*

(a) *μ is a regular Borel measure.*

(b) *Every compact set C is outer regular, that is,*

$$\mu(C) = \text{GLB}\,\{\mu(U): C \subset U,\ U \textit{ open Borel set}\}.$$

(c) *Every bounded open Borel set U is inner regular, that is,*

$$\mu(U) = \text{LUB}\,\{\mu(C): C \subset U,\ C \textit{ compact}\}.$$

EXERCISES

1. If μ_1 and μ_2 are Borel measures on X such that $\mu_1 \ll \mu_2$, and μ_2 is regular, then μ_1 is also regular. (*Hint:* 59.1.)

2. Assuming Axioms I, II, and VII, μ is necessarily σ-finite.

3. Assuming Axioms I–V and VII, the σ-ring $\mathscr{S}$ is generated by the class of all "bounded" sets in $\mathscr{U}$.

4. In part (c) of Theorem 5: (*i*) the word "Borel" is redundant, and (*ii*) one can replace "compact" by "compact G_δ."

5. Every open Baire set is the union of a sequence of compact G_δ's.

6. Every bounded open Baire set is the difference of two compact G_δ's.

7. Given two systems

$$(\mathscr{S}_1, \mu_1, \mathscr{C}_1, \mathscr{U}_1) \qquad \text{and} \qquad (\mathscr{S}_2, \mu_2, \mathscr{C}_2, \mathscr{U}_2),$$

each satisfying Axioms I–III of Sect. 59, we may define a *product system*

$$(\mathscr{S}_1 \times \mathscr{S}_2, \mu_1 \times \mu_2, \mathscr{C}, \mathscr{U})$$

as follows. Let $\mathscr{C}$ be the smallest subclass of $\mathscr{S}_1 \times \mathscr{S}_2$ which contains every rectangle $C_1 \times C_2$ with $C_i \in \mathscr{C}_i$, and is closed under finite unions and countable intersections. Let $\mathscr{U}$ be the smallest subclass of $\mathscr{S}_1 \times \mathscr{S}_2$ which contains every rectangle $U_1 \times U_2$ with $U_i \in \mathscr{U}_i$, and is closed under finite intersections and countable unions. Then:

(*i*) Every set in $\mathscr{C}$ is contained in some rectangle $C_1 \times C_2$, for suitable $C_i \in \mathscr{C}_i$.

(*ii*) Every set in $\mathscr{S}_1 \times \mathscr{S}_2$ is contained in some rectangle $U_1 \times U_2$, for suitable $U_i \in \mathscr{U}_i$.

(*iii*) The system $(\mathscr{S}_1 \times \mathscr{S}_2, \mu_1 \times \mu_2, \mathscr{C}, \mathscr{U})$ also satisfies Axioms I–III.

(*iv*) If the μ_i are regular, and are moreover σ-finite, then $\mu_1 \times \mu_2$ is also regular.

(*v*) If $\mathfrak{S}(\mathscr{C}_i) = \mathscr{S}_i$ for $i = 1, 2$, then also $\mathfrak{S}(\mathscr{C}) = \mathscr{S}_1 \times \mathscr{S}_2$.

(*vi*) If $\mathfrak{S}(\mathscr{C}_i) = \mathscr{S}_i$ for $i = 1, 2$, and if the μ_i are regular, then $\mu_1 \times \mu_2$ is also regular.

62. Regular Borel Measures

We have a variety of criteria for the equality of two regular Borel measures on the same locally compact space:

Theorem 1. *If μ_1 and μ_2 are regular Borel measures on the locally compact space X, the following conditions are equivalent:*

(a) $\mu_1 = \mu_2$, *that is,* $\mu_1(E) = \mu_2(E)$ *for every Borel set* E.

(a′) $\mu_1(C) = \mu_2(C)$ *for every compact set* C.

(a″) $\mu_1(U) = \mu_2(U)$ *for every bounded open set* U.

(b) $\mu_1(F) = \mu_2(F)$ *for every Baire set* F.

(b′) $\mu_1(D) = \mu_2(D)$ *for every compact* G_δ D.

(b″) $\mu_1(V) = \mu_2(V)$ *for every bounded open Baire set* V.

Proof. The following assertions are trivial: condition (a) implies all the rest, (b) implies (b′) and (b″), (a′) implies (b′), and (a″) implies (b″). By 59.2, (b′) implies (a). We thus have the following diagram of implications:

$$\begin{array}{ccccc} (a'') & \Leftarrow & (a) & \Rightarrow & (a') \\ \Downarrow & & \Downarrow \quad \nwarrow & & \Downarrow \\ (b'') & \Leftarrow & (b) & \Rightarrow & (b') \end{array}$$

To complete the proof, it will clearly suffice to show that (b″) implies (a).

Assuming $\mu_1(V) = \mu_2(V)$ for all bounded open Baire sets V, it is to be shown that $\mu_1(E) = \mu_2(E)$ for every Borel set E. Given any compact G_δ D, it will suffice by 59.2 to show that $\mu_1(D) = \mu_2(D)$.

Let U_n be a sequence of open sets such that $D = \bigcap_1^\infty U_n$. By 56.1 there exists, for each n, an open Baire set V_n and a compact G_δ D_n such that

$$D \subset V_n \subset D_n \subset U_n.$$

Summarizing, we have a sequence V_n of bounded open Baire sets such that $D = \bigcap_1^\infty V_n$; replacing V_n by $V_1 \cap \cdots \cap V_n$, we may suppose further that $V_n \downarrow D$. Since

$$\mu_1(V_n) = \mu_2(V_n)$$

for all n by our assumption, and since

$$\mu_i(V_n) \downarrow \mu_i(D)$$

for each i (by finiteness), we conclude that $\mu_1(D) = \mu_2(D)$. ∎

In particular, two regular Borel measures on X are equal if and only if their Baire restrictions are equal. It follows that a Baire measure ν can be extended to *at most one* regular Borel measure μ. It will be shown in Sect. 65 that such an extension is in fact possible.

The remaining theorems in this section are for application in Chapter 9, and may be omitted until that time. If T is a homeomorphism of the locally compact space X, then a subset A of X will be a compact (respectively open, closed, G_δ, F_σ, Baire, Borel, weakly Borel, bounded, σ-bounded) set if and only if $T(A)$ has the same property. Since T^{-1} is also a homeomorphism, the same remarks apply to the sets A and $T^{-1}(A)$.

Theorem 2. *If T is a homeomorphism of X, μ is a regular Borel measure on X, and μ' is the set function defined by the formula*

$$\mu'(E) = \mu(T(E))$$

for every Borel set E, then μ' is also a regular Borel measure.

Proof. Since T preserves compact sets and (therefore) Borel sets, it is permissible to define μ' by the indicated formula, and clearly μ' is a Borel measure.

If E is any Borel set, by the regularity of μ there exists a sequence D_n of compact sets such that

$$D_n \subset T(E) \qquad \text{and} \qquad \mu(D_n) \uparrow \mu(T(E)).$$

Writing $D_n = T(C_n)$ with C_n compact, we have $C_n \subset E$ and

$$\mu'(C_n) = \mu(T(C_n)) = \mu(D_n) \uparrow \mu(T(E)) = \mu'(E),$$

hence μ' is regular by 61.5. ▮

Theorem 3. *If μ_1 and μ_2 are regular Borel measures on X, and T is a homeomorphism of X such that*

$$\mu_1(T(D)) = \mu_2(D)$$

for every compact G_δ D, then

$$\mu_1(T(E)) = \mu_2(E)$$

for every Borel set E.

Proof. Defining $\mu'(E) = \mu_1(T(E))$ for every Borel set E, we know from Theorem 2 that μ' is a regular Borel measure. Since $\mu'(D) = \mu_2(D)$ for every compact G_δ D, it follows from Theorem 1 that $\mu'(E) = \mu_2(E)$ for every Borel set E. Incidentally, the theorem and its proof remain valid if we replace "compact G_δ" by "bounded open Baire set." ▮

Corollary. *If μ is a regular Borel measure on X, T is a homeomorphism of X, and $c > 0$ is a constant such that*

$$\mu(T(D)) = c\mu(D)$$

for every compact G_δ D, then

$$\mu(T(E)) = c\mu(E)$$

for every Borel set E.

Proof. Apply Theorem 3 to the regular Borel measures $\mu_1 = \mu$ and $\mu_2 = c\mu$. Again, "compact G_δ" can be replaced by "bounded open Baire set." ∎

EXERCISES

1. If μ is a regular Borel measure on X, and U is an open Borel set, then there exists an open Baire set V such that $V \subset U$ and $\mu(U - V) = 0$.

2. If μ is a Borel measure on X, ν is the Baire restriction of μ, and ν^* is the outer measure induced by ν (on the class of all σ-bounded sets), then the following conditions are equivalent:

(a) μ is regular.

(b) $\mu(C) = \nu^*(C)$ for every compact set C.

(c) For every bounded open set U,

$$\mu(U) = \text{LUB}\,\{\mu(F) : F \subset U,\ F \text{ Baire set}\}.$$

(d) For every open Borel set U,

$$\mu(U) = \text{LUB}\,\{\mu(F) : F \subset U,\ F \text{ Baire set}\}.$$

3. If μ is a regular Borel measure on X, and f is a nonnegative Borel function which is integrable with respect to μ, then the indefinite integral μ_f is also a regular Borel measure.

4. If μ is a Borel measure on X, the *support* of μ, denoted $S(\mu)$, is defined to be the set $X - U$, where U is the union of the class of all open Borel sets of measure zero. Then:

(*i*) $S(\mu)$ is a closed set.

(*ii*) If C is any compact subset of $X - S(\mu)$, then $\mu(C) = 0$.

(*iii*) If μ is regular, and X is σ-bounded (equivalently, X is a Borel set), then $\mu(X - S(\mu)) = 0$. In this case $X - S(\mu)$ may be described as the largest open set of measure zero. If X is compact, then $S(\mu)$ is the smallest closed set whose measure equals that of X; it is not necessarily a Baire set.

5. If μ is a regular Borel measure on X, and E is any Borel set, then the contraction μ_E is also a regular Borel measure. Similarly if E is merely a weakly Borel set.

6. If μ is a Borel measure on X, then μ is regular if and only if

$$\mu(E) = \text{LUB}\,\{\mu(C) : C \subset E,\ C \text{ compact}\}$$

for every Borel set E of finite measure.

7. If μ is a regular Borel measure, and f is an integrable Borel function such that $\int_C f\,d\mu \geq 0$ for every compact set C, then $f \geq 0$ a.e. Similarly with "compact G_δ" instead of "compact."

8. If μ is a Borel measure whose Baire restriction is finite, then μ is also finite.

9. In order that a Borel measure μ be regular, it is necessary and sufficient that for each compact set C, there exist a compact G_δ D such that $C \subset D$ and $\mu(D - C) = 0$.

10. Suppose μ_i is a regular Borel measure on the locally compact space X_i $(i = 1, 2)$, and write $\mathscr{B}_i$ for the class of all Borel sets in X_i. Let $\mathscr{C}$ (respectively $\mathscr{U}$) be the class of all compact (respectively open) sets in $\mathscr{B}_1 \times \mathscr{B}_2$. Then the system

$$(\mathscr{B}_1 \times \mathscr{B}_2, \mu_1 \times \mu_2, \mathscr{C}, \mathscr{U})$$

is a regular system in the sense of Sect. 59. In particular, if it happens that $\mathscr{B}_1 \times \mathscr{B}_2$ is the class of all Borel sets in $X_1 \times X_2$, then $\mu_1 \times \mu_2$ is a regular Borel measure.

11. Let $(\mu_i)_{i \in I}$ be an increasingly directed family of Borel measures on X, and define $\mu = \text{LUB}\ \mu_i$ as in Sect. 10. Assume that for each compact set C, the family $\mu_i(C)$ is bounded; then μ is also a Borel measure. In order that μ be regular, it is necessary and sufficient that μ_i be regular for all i.

12. Let $(\mu_i)_{i \in I}$ be an arbitrary family of Borel measures on X, and define

$$\mu = \sum_{i \in I} \mu_i$$

as in Sect. 10. Assume that

$$\sum_{i \in I} \mu_i(C) < \infty$$

for each compact set C. Then the Borel measure μ is regular if and only if μ_i is regular for all i.

13. (Johnson) If μ_i is a regular Borel measure on the locally compact space X_i $(i = 1, 2)$, there exists a unique regular Borel measure ρ on $X_1 \times X_2$ which extends $\mu_1 \times \mu_2$; equivalently, there exists a unique regular Borel measure ρ on $X_1 \times X_2$ such that

$$\rho(E_1 \times E_2) = \mu_1(E_1)\mu_2(E_2)$$

for all Borel sets E_i in X_i. Specifically, if ν_i is the Baire restriction of μ_i, and ρ is the unique regular Borel measure which extends the Baire measure $\nu_1 \times \nu_2$, then ρ also extends $\mu_1 \times \mu_2$. (Cf. 65.1.)

14. (Johnson) Suppose X_i is a locally compact space $(i = 1, 2)$, and $M \subset X_1 \times X_2$. If M is an open (respectively closed, compact, Borel, compact G_δ, Baire) set in $X_1 \times X_2$, then every section of M is a set of the same type. If h is a Borel function on $X_1 \times X_2$, then every section of h is also a Borel function.

15. If μ is a regular Borel measure on X, and μ_w is the weakly Borel extension of μ described in Exercise 57.21, then

$$\mu_w(A) = \text{LUB}\ \{\mu(C) : C \subset A,\ C \text{ compact}\},$$

for every weakly Borel set A. It follows that $\mu_w(X - S(\mu)) = 0$, where $S(\mu)$ is the support of μ. If T is a homeomorphism of X, and $c > 0$ is a constant such that

$$\mu(T(E)) = c\mu(E)$$

for all Borel sets E, then

$$\mu_w(T(A)) = c\mu_w(A)$$

for all weakly Borel sets A (for this assertion, μ need not be regular).

16. (Johnson) Let μ_i be a regular Borel measure on the locally compact space X_i $(i = 1, 2)$, and let ρ be the unique regular Borel measure on $X_1 \times X_2$ which extends $\mu_1 \times \mu_2$ (Exercise 13). Let h be a Borel function on $X_1 \times X_2$ (see Exercise 14).

(*i*) If h is ρ-integrable, then the iterated integrals

$$\int\int h \, d\mu_2 \, d\mu_1 \qquad \text{and} \qquad \int\int h \, d\mu_1 \, d\mu_2$$

exist, and are equal to $\int h \, d\rho$.

(*ii*) If, conversely, $h \geq 0$ and the iterated integral

$$\int\int h \, d\mu_2 \, d\mu_1$$

exists, then h is ρ-integrable. (*Note:* It is not assumed that h is measurable with respect to the domain of $\mu_1 \times \mu_2$. Cf. Exercise 70.15.)

63. Contents

A **content** on the locally compact space X is a set function λ defined on the class of all compact sets, having the following properties:

(1) $0 \leq \lambda(C) < \infty$.

(2) $C \subset D$ implies $\lambda(C) \leq \lambda(D)$.

(3) $C \cap D = \varnothing$ implies $\lambda(C \cup D) = \lambda(C) + \lambda(D)$.

(4) $\lambda(C \cup D) \leq \lambda(C) + \lambda(D)$.

We emphasize that the values of a content are nonnegative real numbers. In words, a content is *monotone*, *additive*, and *subadditive*. It follows from (1) and (3) that $\lambda(\varnothing) = 0$. Observe that (2) cannot be deduced from (1) and (3) as in the case of measures, because $\lambda(D - C)$ may not be defined. A simple example of a content is the restriction of a Borel measure to the class of compact sets.

One motivation for discussing contents is as follows (see also Sect. 69). If a Baire measure ν on X is given, we wish to extend ν to a regular Borel measure μ on X. A natural place to begin is to try to define $\mu(C)$ for a compact set C. In any case $\nu^*(C)$ is defined, where ν^* is the outer measure induced by ν (Sect. 5); indeed, it is clear from 56.1 that the domain of definition of ν^*, namely the hereditary σ-ring generated by the class of Baire sets, is precisely the class of all σ-bounded sets. We shall see in the next section that the correspondence $C \to \nu^*(C)$ is a "regular" content. Now, it turns out that every "regular" content may be extended to a regular Borel measure, and this will yield the desired extension of ν (as we shall see in the next two sections). More generally, *every* content λ (not necessarily deduced from a Baire measure) leads to a regular Borel measure μ (which does not necessarily extend λ); the purpose of this section is to show how to construct μ from λ, whereas the question of whether or not μ extends λ is left to the next section.

If λ is any content on X, we define a set function λ_* on the class of all open Borel sets U by the formula

$$\lambda_*(U) = \text{LUB}\,\{\lambda(C): C \subset U, C \text{ compact}\};$$

λ_* is called the **inner content** induced by λ. We observe that the domain of definition of λ_* is closed under countable unions and finite intersections (whereas the domain of definition of λ is closed under finite unions and arbitrary intersections). The domain of λ_* may also be described as the class of all σ-bounded open sets (57.2).

Lemma. *If $C \subset U \cup V$, where C is compact and U, V are open, there exist compact sets D, E such that $C = D \cup E$, $D \subset U$, $E \subset V$.*

Proof. Since the compact sets $C - U$ and $C - V$ are disjoint, there exist disjoint open sets A, B such that $C - U \subset A$, $C - V \subset B$. Defining $D = C - A$ and $E = C - B$, it is easy to check that the sets D and E meet all requirements. ▮

Theorem 1. *If λ is a content on X, and λ_* is the inner content induced by λ, then:*

(1) $\lambda_*(\varnothing) = 0$.

(2) $\lambda_*(U) < \infty$ *for every bounded open set U.*

(3) λ_* *is monotone.*

(4) λ_* *is countably subadditive.*

(5) λ_* *is countably additive.*

Proof. (1) is immediate from $\lambda(\varnothing) = 0$.

(2) If U is a bounded open set, let D be any compact set containing U. For every compact subset C of U, we have $\lambda(C) \leq \lambda(D)$ by the monotonicity of λ; varying C, $\lambda_*(U) \leq \lambda(D) < \infty$.

(3) Let U and V be open Borel sets, such that $U \subset V$. If C is any compact subset of U, then also $C \subset V$, and so $\lambda(C) \leq \lambda_*(V)$; varying C, $\lambda_*(U) \leq \lambda_*(V)$.

(4) We first show that λ_* is subadditive. If U and V are open Borel sets, and C is a compact subset of $U \cup V$, then by the lemma there exist compact sets D and E such that $C = D \cup E$, $D \subset U$, $E \subset V$. Since λ is subadditive, it follows that

$$\lambda(C) \leq \lambda(D) + \lambda(E) \leq \lambda_*(U) + \lambda_*(V);$$

varying C,

$$\lambda_*(U \cup V) \leq \lambda_*(U) + \lambda_*(V).$$

By induction, λ_* is finitely subadditive. Finally, suppose U_k is a sequence of open Borel sets, and $U = \bigcup_1^\infty U_k$. If C is any compact subset of U, then $C \subset \bigcup_1^n U_k$ for a suitable index n, and so

$$\lambda(C) \leq \lambda_*\left(\bigcup_1^n U_k\right) \leq \sum_1^n \lambda_*(U_k) \leq \sum_1^\infty \lambda_*(U_k);$$

varying C,

$$\lambda_*(U) \leq \sum_1^\infty \lambda_*(U_k).$$

(5) We first show that λ_* is additive. Let U and V be disjoint open Borel sets. If C and D are compact sets such that $C \subset U$ and $D \subset V$, then $C \cup D$ is a compact subset of $U \cup V$; since C and D are necessarily disjoint, it follows from the additivity of λ that

$$\lambda_*(U \cup V) \geq \lambda(C \cup D) = \lambda(C) + \lambda(D).$$

Varying C,

$$\lambda_*(U \cup V) \geq \lambda_*(U) + \lambda(D);$$

varying D,

$$\lambda_*(U \cup V) \geq \lambda_*(U) + \lambda_*(V),$$

and this relation, combined with subadditivity, yields the desired relation

$$\lambda_*(U \cup V) = \lambda_*(U) + \lambda_*(V).$$

By induction, λ_* is finitely additive. Finally, suppose U_k is a sequence of mutually disjoint open Borel sets, and $U = \bigcup_1^\infty U_k$. Since $U \supset \bigcup_1^n U_k$ for each n, we have

$$\lambda_*(U) \geq \lambda_*\left(\bigcup_1^n U_k\right) = \sum_1^n \lambda_*(U_k)$$

by monotonicity and finite additivity; varying n, we have

$$\lambda_*(U) \geq \sum_1^\infty \lambda_*(U_k),$$

and this relation, combined with countable subadditivity, yields the desired relation

$$\lambda_*(U) = \sum_1^\infty \lambda_*(U_k). \quad \blacksquare$$

Let λ be a content on X, λ_* the inner content induced by λ, and $\mathscr{H}$ the class of all σ-bounded sets. Since every σ-bounded set A is contained in some open Borel set, we may define a set function λ^* on $\mathscr{H}$ by the formula

$$\lambda^*(A) = \text{GLB}\,\{\lambda_*(U) : A \subset U,\ U \text{ open Borel set}\}.$$

In view of the following theorem, we may refer to λ^* as the *outer measure induced by the content* λ:

Theorem 2. *If λ is a content on X, λ_* is the inner content induced by λ, and λ^* is the set function defined above on the class $\mathscr{H}$ of all σ-bounded sets, then:*

(1) *λ^* is an outer measure on the hereditary σ-ring $\mathscr{H}$.*

(2) *$\lambda^*(A) < \infty$ for every bounded set A.*

(3) *λ^* extends λ_*, that is, $\lambda^*(U) = \lambda_*(U)$ for every open Borel set U.*

(4) *If $U \subset D$, where U is open and D is compact, then*

$$\lambda_*(U) = \lambda^*(U) \le \lambda(D) \le \lambda^*(D).$$

Proof. (1) Obviously $\mathscr{H}$ is a hereditary σ-ring, and $0 \le \lambda^*(A) \le \infty$ for every A in $\mathscr{H}$. It is to be verified that (*i*) $\lambda^*(\varnothing) = 0$, (*ii*) λ^* is monotone, and (*iii*) λ^* is countably subadditive. The assertion (*i*) is immediate from $\lambda_*(\varnothing) = 0$. If $A \subset B$, and U is an open Borel set containing B, then also $A \subset U$, and so $\lambda^*(A) \le \lambda_*(U)$; varying U, $\lambda^*(A) \le \lambda^*(B)$. Finally, if A_k is a sequence in $\mathscr{H}$, and $A = \bigcup_1^\infty A_k$, the problem is to show that

$$\lambda^*(A) \le \sum_1^\infty \lambda^*(A_k);$$

this is trivial if one of the terms on the right is infinite, hence we may assume $\lambda^*(A_k) < \infty$ for all k. Given any $\varepsilon > 0$, we may choose for each k an open Borel set U_k such that

$$A_k \subset U_k \qquad \text{and} \qquad \lambda_*(U_k) \le \lambda^*(A_k) + \varepsilon/2^k.$$

Defining $U = \bigcup_1^\infty U_k$, we have $A \subset U$, and U is an open Borel set; since λ_* is countably subadditive by Theorem 1, it follows that

$$\lambda^*(A) \le \lambda_*(U) \le \sum_1^\infty \lambda_*(U_k) \le \sum_1^\infty \lambda^*(A_k) + \varepsilon,$$

and the desired inequality results from the arbitrariness of ε.

(2) If A is bounded, let C be a compact set containing A, and U a bounded open set containing C; then $A \subset U$, and

$$\lambda^*(A) \leq \lambda_*(U) < \infty$$

by Theorem 1.

(3) If U is an open Borel set, and V is any open Borel set containing U, then $\lambda_*(U) \leq \lambda_*(V)$ by Theorem 1; varying V, $\lambda_*(U) \leq \lambda^*(U)$. On the other hand $\lambda^*(U) \leq \lambda_*(U)$ results from $U \subset U$.

(4) Since U is a bounded open set, and therefore a Borel set (57.2), we have $\lambda_*(U) = \lambda^*(U)$ by (3).

If V is any open Borel set such that $D \subset V$, then $\lambda(D) \leq \lambda_*(V)$ by the definition of λ_*; varying V, $\lambda(D) \leq \lambda^*(D)$.

If C is any compact subset of U, then $\lambda(C) \leq \lambda(D)$ by the monotonicity of λ; varying C, $\lambda_*(U) \leq \lambda(D)$. ∎

If λ is a content on X, we may apply the general theory of outer measures, developed in Sect. 5, to the set function λ^*. Thus, a σ-bounded set A is λ^*-measurable in case

$$\lambda^*(B) = \lambda^*(B \cap A) + \lambda^*(B \cap A')$$

for every σ-bounded set B; equivalently (in view of the subadditivity of λ^*),

$$\lambda^*(B) \geq \lambda^*(B \cap A) + \lambda^*(B \cap A')$$

for every σ-bounded set B. According to 5.2, the class $\mathscr{M}$ of all λ^*-measurable sets is a σ-ring, and the restriction of λ^* to $\mathscr{M}$ is a (complete) measure. We are interested in the extent of the class $\mathscr{M}$; indeed, it will be shown in Theorem 3 that every Borel set belongs to $\mathscr{M}$, and that the restriction of λ^* to the class of Borel sets is a regular Borel measure. First we establish a convenient criterion for λ^*-measurability:

Lemma. *Let λ be a content on X, λ^* the outer measure induced by λ, and A a σ-bounded set. In order that A be λ^*-measurable, it is necessary and sufficient that*

$$\lambda^*(U) = \lambda^*(U \cap A) + \lambda^*(U \cap A')$$

for every open Borel set U.

Proof. Suppose the indicated relation holds for every open Borel set. Let B be an arbitrary σ-bounded set. If U is any open Borel set containing B, it follows from Theorem 2 that

$$\begin{aligned}\lambda_*(U) &= \lambda^*(U) \\ &= \lambda^*(U \cap A) + \lambda^*(U \cap A') \geq \lambda^*(B \cap A) + \lambda^*(B \cap A');\end{aligned}$$

varying U,

$$\lambda^*(B) \geq \lambda^*(B \cap A) + \lambda^*(B \cap A'). \quad ∎$$

Theorem 3. *If λ is a content on X, and λ^* is the outer measure induced by λ, then every Borel set is λ^*-measurable, and the restriction of λ^* to the class of Borel sets is a regular Borel measure.*

Proof. The class $\mathscr{M}$ of λ^*-measurable sets is a σ-ring (5.2); in proving that $\mathscr{M}$ contains every Borel set, it will therefore suffice to show that each compact set C is λ^*-measurable. We propose to apply the criterion of the lemma. Let U be an open Borel set; then $U \cap C'$ is also an open Borel set. Let D be a compact subset of $U \cap C'$; then $U \cap D'$ is also an open Borel set. If E is any compact subset of $U \cap D'$, then E and D are mutually disjoint compact subsets of U, hence

$$\lambda^*(U) = \lambda_*(U) \geq \lambda(D \cup E) = \lambda(D) + \lambda(E)$$

by Theorem 2 and the additivity of λ; varying E, we have

$$\lambda^*(U) \geq \lambda(D) + \lambda_*(U \cap D'),$$

that is,

$$\lambda^*(U) \geq \lambda(D) + \lambda^*(U \cap D').$$

Since $D \subset C'$, we have

$$\lambda^*(U \cap D') \geq \lambda^*(U \cap C)$$

by the monotonicity of λ^*. It follows that

$$\lambda^*(U) \geq \lambda(D) + \lambda^*(U \cap C);$$

varying D,

$$\lambda^*(U) \geq \lambda_*(U \cap C') + \lambda^*(U \cap C) = \lambda^*(U \cap C') + \lambda^*(U \cap C),$$

and so C is λ^*-measurable by the lemma.

Let μ be the restriction of λ^* to the class of all Borel sets. Since λ^* takes finite values on bounded sets (Theorem 2), it follows that μ is a Borel measure.

To show that μ is regular, it will suffice by 61.5 to show that each compact set C is outer regular. Indeed, it follows from the definitions that

$$\mu(C) = \lambda^*(C) = \text{GLB}\,\{\lambda_*(U) \colon C \subset U,\ U \text{ open Borel set}\},$$

and $\mu(U) = \lambda^*(U) = \lambda_*(U)$ for every open Borel set U (Theorem 2), thus

$$\mu(C) = \text{GLB}\,\{\mu(U) \colon C \subset U,\ U \text{ open Borel set}\}. \quad \blacksquare$$

If λ is a content on X, the restriction of λ^* to the class of Borel sets will be called the *regular Borel measure induced by* λ.

EXERCISES

1. If $C \subset U \cup V$, where C is a compact G_δ and U, V are open sets, there exist compact G_δ's D and E such that $C = D \cup E$, $D \subset U$, $E \subset V$.

2. If λ is a content on X, μ is the regular Borel measure induced by λ, and T is a homeomorphism of X such that

$$\lambda(T(C)) = \lambda(C)$$

for all compact sets C, then

$$\mu(T(E)) = \mu(E)$$

for all Borel sets E.

3. If λ is a content on X, and A is a σ-bounded set, then A is λ^*-measurable if and only if

$$\lambda^*(U) = \lambda^*(U \cap A) + \lambda^*(U \cap A')$$

for every bounded open set U.

64. Regular Contents

In the preceding section, it was shown that every content λ leads to a regular Borel measure μ in a natural way; namely, one defines μ to be the restriction of λ^* to the class of Borel sets (63.3). It will be shown in the next theorem that μ is an *extension* of λ if and only if λ is a "regular" content, in the sense to be defined below.

First we introduce a useful notation. If A and B are subsets of the locally compact space X, we shall say that A is **amply contained** in B in case A is contained in the interior of B; equivalently, there exists an open set U such that $A \subset U \subset B$. We shall write $A \prec B$ to indicate that A is amply contained in B. If $A \prec B$, we may also say that B *amply contains* A.

A content λ on X is said to be **regular** in case

$$\lambda(C) = \text{GLB}\,\{\lambda(D)\colon C \prec D,\ D \text{ compact}\},$$

for each compact set C; in other words, $\lambda(C)$ can be approximated by the numbers $\lambda(D)$, where D is a compact set amply containing C. The relationship of this concept with the concept of regular Borel measure is indicated by the following lemma:

Lemma. *If μ is a Borel measure on X, and λ is the restriction of μ to the class of compact sets, then the following conditions are equivalent:*

(a) *μ is a regular Borel measure.*

(b) *λ is a regular content.*

Proof. (a) *implies* (b): Let C be a compact set. Given any $\varepsilon > 0$, there exists, by the assumed regularity of μ, an open Borel set U such that

$$C \subset U \quad \text{and} \quad \mu(U) \leq \mu(C) + \varepsilon.$$

One has $C \subset V \subset D \subset U$ for a suitable compact set D and open set V (one could invoke 56.1, but there is also a simple argument based on the regularity of locally compact spaces). Then $C \prec D$, and

$$\lambda(D) = \mu(D) \leq \mu(U) \leq \mu(C) + \varepsilon = \lambda(C) + \varepsilon,$$

thus λ is regular.

(b) *implies* (a): To prove that μ is regular, it will suffice by 61.5 to show that each compact set C is outer regular. Indeed, given any $\varepsilon > 0$, let D be a compact set such that

$$C \prec D \quad \text{and} \quad \lambda(D) \leq \lambda(C) + \varepsilon$$

(this is possible by the assumed regularity of λ). If U is the interior of D, then $C \subset U \subset D$, and U is an open Borel set (57.2); since

$$\mu(U) \leq \mu(D) = \lambda(D) \leq \lambda(C) + \varepsilon = \mu(C) + \varepsilon,$$

we conclude that C is outer regular. ∎

Theorem 1. *Let λ be a content on X, and let μ be the regular Borel measure induced by λ (63.3). The following conditions are equivalent:*

(a) *λ is a regular content.*

(b) *μ extends λ, that is, $\lambda^*(C) = \lambda(C)$ for every compact set C.*

Proof. (b) *implies* (a): If μ extends λ, that is, if λ is the restriction of μ, then the regularity of μ implies that of λ by the lemma.

(a) *implies* (b): Let C be a compact set. Given any $\varepsilon > 0$, there exists, by assumption, a compact set D such that

$$C \prec D \quad \text{and} \quad \lambda(D) \leq \lambda(C) + \varepsilon.$$

Let U be any open set such that $C \subset U \subset D$; then $\lambda^*(C) \leq \lambda^*(U)$ by the monotonicity of λ^*. Since $\lambda^*(U) \leq \lambda(D)$ by 63.2, it follows that

$$\lambda^*(C) \leq \lambda(D) \leq \lambda(C) + \varepsilon;$$

varying ε, $\lambda^*(C) \leq \lambda(C)$. On the other hand $\lambda(C) \leq \lambda^*(C)$ by 63.2, and so $\lambda^*(C) = \lambda(C)$. ∎

The following theorem is the key to the extension of Baire measures, as we shall see in the next section:

Theorem 2. *If λ is a regular content on X, there exists a unique regular Borel measure μ such that*

$$\mu(C) = \lambda(C)$$

for every compact set C. Specifically, μ is the restriction of λ^* to the class of Borel sets.

Proof. If μ is the restriction of λ^* to the class of Borel sets, then μ is a regular Borel measure by 63.3, and μ extends λ by Theorem 1.

If μ' is *any* Borel measure on X such that $\mu'(C) = \lambda(C)$ for every compact set C, then $\mu' = \mu$ by 58.3. ▮

EXERCISES

1. If μ is a regular Borel measure, and λ is the restriction of μ to the class of compact sets, then the restriction of λ^* to the class of Borel sets is μ, that is, the regular Borel measure induced by λ is μ.

***2.** What about *Baire contents*, that is, nonnegative real-valued set functions defined on the class of compact G_δ's in X, having the formal properties of a content (monotonicity, additivity, subadditivity)? Presumably one could also develop the theory of contents along axiomatic lines (cf. Sects. 59, 61).

***3.** If λ is a regular content on X, then, given any compact set C, there exists a compact G_δ D such that $C \subset D$ and $\lambda(C) = \lambda(D)$. Does this property characterize regular contents?

65. The Regular Borel Extension of a Baire Measure

It will be shown in this section that a Baire measure may be extended to a regular Borel measure in one and only one way. The first step is to use the Baire measure to define a regular content:

Lemma. *If ν is a Baire measure on the locally compact space X, then the set function λ, defined for every compact set C by the formula*

$$\lambda(C) = \text{GLB}\,\{\nu(U) : C \subset U,\ U \textit{ open Baire set}\},$$

is a regular content on X. Moreover, $\lambda(D) = \nu(D)$ for every compact G_δ D.

Proof. If C is any compact set, then $C \subset U \subset D$ for a suitable open Baire set U and compact G_δ D (56.1), and so

$$\lambda(C) \leq \nu(U) \leq \nu(D) < \infty.$$

Obviously $\lambda(\varnothing) = 0$. It remains to show that λ is (*i*) monotone, (*ii*) subadditive, (*iii*) additive, and (*iv*) regular.

If C and D are compact sets such that $C \subset D$, and U is any open Baire set containing D, then also $C \subset U$, hence $\lambda(C) \leq \nu(U)$; varying U, $\lambda(C) \leq \lambda(D)$.

If C, D are compact sets, and U, V are open Baire sets such that $C \subset U$, $D \subset V$, then $U \cup V$ is an open Baire set containing $C \cup D$, and so

$$\lambda(C \cup D) \leq \nu(U \cup V) \leq \nu(U) + \nu(V);$$

varying U,

$$\lambda(C \cup D) \leq \lambda(C) + \nu(V);$$

varying V,

$$\lambda(C \cup D) \leq \lambda(C) + \lambda(D).$$

If C and D are disjoint compact sets, there exist disjoint open Baire sets U and V such that $C \subset U$ and $D \subset V$; this follows easily from the fact that the open Baire sets are basic (corollary of 56.1). If now W is any open Baire set containing $C \cup D$, then $C \subset U \cap W$ and $D \subset V \cap W$, hence

$$\begin{aligned}\nu(W) \geq \nu[(U \cap W) \cup (V \cap W)] &= \nu(U \cap W) + \nu(V \cap W) \\ &\geq \lambda(C) + \lambda(D);\end{aligned}$$

varying W, we have

$$\lambda(C \cup D) \geq \lambda(C) + \lambda(D).$$

Since λ is subadditive, we conclude that

$$\lambda(C \cup D) = \lambda(C) + \lambda(D).$$

Summarizing, λ is a content. To see that λ is regular, suppose C is a compact set, and $\varepsilon > 0$. By the definition of λ, we may choose an open Baire set U such that

$$C \subset U \qquad \text{and} \qquad \nu(U) \leq \lambda(C) + \varepsilon.$$

One has $C \subset V \subset D \subset U$ for a suitable open Baire set V and compact set D; since

$$C \prec D, \qquad \text{and} \qquad \lambda(D) \leq \nu(U) \leq \lambda(C) + \varepsilon,$$

we conclude that λ is regular.

Finally, let D be any compact G_δ. As shown in the proof of 62.1, there exists a sequence V_n of bounded open Baire sets such that $V_n \downarrow D$, and so

$$\nu(V_n) \downarrow \nu(D)$$

by finiteness. It is therefore clear that $\lambda(D) = \nu(D)$. (Incidentally, this is just the assertion that D is outer regular, and follows at once from 60.1.) ▮

The promised extension theorem is as follows:

Theorem 1. *If ν is a Baire measure on the locally compact space X, there exists a unique regular Borel measure μ such that*

$$\mu(F) = \nu(F)$$

for every Baire set F.

Proof. Define λ as in the lemma. Since λ is a content, the restriction of λ^* to the class of Borel sets is a regular Borel measure μ (63.3); since λ is regular, μ extends λ (64.1).

Let ν' be the Baire restriction of μ. For every compact G_δ D, we have $\nu'(D) = \mu(D) = \lambda(D)$, and $\lambda(D) = \nu(D)$ by the lemma, thus $\nu'(D) = \nu(D)$; it follows from 58.3 (or the regularity of Baire measures) that $\nu' = \nu$, that is, $\mu(F) = \nu(F)$ for every Baire set F.

Uniqueness follows from 62.1 (or 59.2). ∎

EXERCISES

1. The regular Borel measure induced by a finite Baire measure is finite.

2. With notation as in Theorem 1, one has $\mu(C) = \nu^*(C)$ for every compact set C. Compare with Exercise 70.12.

66. Integration of Continuous Functions with Compact Support

In this section we are interested primarily in the relationship between a regular Borel measure and the integral which it assigns to continuous functions with compact support. First we clear up a possible ambiguity:

Theorem 1. *Let μ be a Borel measure on X, and ν the Baire restriction of μ. If f is a Baire function, then f is μ-integrable if and only if it is ν-integrable, and in this case*

$$\int f\,d\mu = \int f\,d\nu.$$

Proof. We may assume without loss of generality that $f \geq 0$ (25.3). Let f_n be a sequence of simple Baire functions such that $0 \leq f_n \uparrow f$ (16.4). Of course these functions are also Borel functions. Since $\mu(N(f_n)) = \nu(N(f_n))$, it is clear that f_n is μ-integrable if and only if it is ν-integrable, and that in this case we have

$$\int f_n\,d\mu = \int f_n\,d\nu;$$

it is then immediate from the definitions in Sect. 24 that f is μ-integrable if and only if it is ν-integrable, and that in this case the values of the two integrals are equal. ∎

Recall that $\mathscr{L}$ denotes the class of all continuous real-valued functions on X with compact support. The functions in $\mathscr{L}$ are integrable with respect to all the measures on X we shall consider:

Theorem 2. *Every function in $\mathscr{L}$ is integrable with respect to any Baire measure or any Borel measure.*

Proof. Suppose ν is a Baire measure and $f \in \mathscr{L}$. Since f has compact support, it follows from 56.1 that there exists a compact G_δ D such that f vanishes outside D. By the continuity of f, and the compactness of D, there is a real number $c > 0$ such that $|f(x)| \leq c$ for all $x \in X$. Then

$$|f| \leq c\chi_D;$$

since χ_D is ν-integrable, and f is a Baire function (56.2), it follows from 25.7 that f is ν-integrable.

It follows that if μ is a Borel measure, and ν is the Baire restriction of μ, then every function in $\mathscr{L}$ is ν-integrable, and therefore μ-integrable by Theorem 1. ∎

Any two Baire measures which assign the same integral to each f in $\mathscr{L}$ are necessarily identical:

Theorem 3. *If ν_1 and ν_2 are Baire measures on X such that*

$$\int f\,d\nu_1 = \int f\,d\nu_2$$

for every f in $\mathscr{L}$, then $\nu_1 = \nu_2$.

Proof. Given any compact G_δ D, it will suffice by 58.3 (or regularity) to show that $\nu_1(D) = \nu_2(D)$. Choose any sequence f_n in $\mathscr{L}$ such that $f_n \downarrow \chi_D$ (55.2); since

$$\int f_n\,d\nu_1 = \int f_n\,d\nu_2$$

for all n, and

$$\int f_n\,d\nu_i \downarrow \nu_i(D)$$

by the (MCT), we conclude that $\nu_1(D) = \nu_2(D)$. ∎

There is a variant of Theorem 3 for regular Borel measures:

Theorem 4. *If μ_1 and μ_2 are regular Borel measures on X such that*

$$\int f\,d\mu_1 = \int f\,d\mu_2$$

for every f in $\mathscr{L}$, then $\mu_1 = \mu_2$.

Proof. Let ν_i be the Baire restriction of μ_i; by Theorems 1 and 2, we have

$$\int f\,d\nu_1 = \int f\,d\nu_2$$

for every f in $\mathscr{L}$; hence $\nu_1 = \nu_2$ by Theorem 3, and it follows from 62.1 that $\mu_1(E) = \mu_2(E)$ for every Borel set E. ∎

The remaining theorems in this section are for application in Chapter 9, and may be omitted until that time. If T is a homeomorphism of X and f is a real-valued function on X, we shall write

$$f^T(x) = f(Tx);$$

thus f^T is the composite function $f \circ T$. Since T is a homeomorphism, it is clear that f^T is continuous if and only if f is continuous; similarly for the property of being a Baire function, or a Borel function, or having compact support. In particular, $f \in \mathscr{L}$ implies $f^T \in \mathscr{L}$. Evidently

$$(f + g)^T = f^T + g^T,$$

and

$$(cf)^T = cf^T;$$

in particular, the correspondence $f \to f^T$ is a (bijective) linear mapping in $\mathscr{L}$.

The next results are concerned with the effect of a homeomorphism on a regular Borel measure.

Theorem 5. *If μ_1 and μ_2 are regular Borel measures on X, and T is a homeomorphism of X such that*

$$\mu_1(T^{-1}(D)) = \mu_2(D)$$

for every compact G_δ D, then

$$\int f^T\,d\mu_1 = \int f\,d\mu_2$$

for all f in $\mathscr{L}$.

Proof. If $f \in \mathscr{L}$, then also $f^T \in \mathscr{L}$; it follows from Theorem 2 that both functions are μ_i-integrable. It will suffice by linearity to consider $f \geq 0$. Let f_n be a sequence of simple Baire functions such that $0 \leq f_n \uparrow f$ (16.4). Then

$$0 \leq f_n^T \uparrow f^T;$$

since f^T is μ_1-integrable, and f_n^T is a Baire function, it follows from 25.7 that f_n^T is μ_1-integrable. Since

$$(\chi_E)^T = \chi_{T^{-1}(E)}$$

for every Borel set E, and since $\mu_1(T^{-1}(E)) = \mu_2(E)$ for all Borel sets E by 62.3, it follows that

$$\int f_n^T \, d\mu_1 = \int f_n \, d\mu_2$$

for all n; passing to the limit, our assertion follows from the (MCT) (or the definitions in Sect. 24). ▮

Corollary. *If μ is a regular Borel measure on X, T is a homeomorphism of X, and $c > 0$ is a constant such that*

$$\mu(T^{-1}(D)) = c\mu(D)$$

for every compact G_δ D, then

$$\int f^T \, d\mu = c \int f \, d\mu$$

for every f in $\mathscr{L}$.

Proof. Apply Theorem 5 to the regular Borel measures $\mu_1 = \mu$ and $\mu_2 = c\mu$. ▮

The converses of Theorem 5 and its Corollary also hold:

Theorem 6. *If μ_1 and μ_2 are regular Borel measures on X, and T is a homeomorphism of X such that*

$$\int f^T \, d\mu_1 = \int f \, d\mu_2$$

for all f in $\mathscr{L}$, then

$$\mu_1(T^{-1}(E)) = \mu_2(E)$$

for every Borel set E.

Proof. If D is any compact G_δ, it will suffice by 62.3 to show that $\mu_1(T^{-1}(D)) = \mu_2(D)$. Let f_n be a sequence in $\mathscr{L}$ such that $f_n \downarrow \chi_D$ (55.2). Then

$$f_n^T \downarrow (\chi_D)^T = \chi_A,$$

where $A = T^{-1}(D)$. For each n, we have

$$\int f_n^T \, d\mu_1 = \int f_n \, d\mu_2;$$

passing to the limit, it follows from the (MCT) that $\mu_1(A) = \mu_2(D)$. ▮

Corollary. *If μ is a regular Borel measure on X, T is a homeomorphism of X, and $c > 0$ is a constant such that*

$$\int f^T \, d\mu = c \int f \, d\mu$$

for all f in $\mathscr{L}$, then

$$\mu(T^{-1}(E)) = c\mu(E)$$

for every Borel set E.

Proof. Apply Theorem 6 to the regular Borel measures $\mu_1 = \mu$ and $\mu_2 = c\mu$. ∎

EXERCISES

1. If μ is a regular Borel measure on X, and C is any compact set, then

$$\mu(C) = \text{GLB} \int f \, d\mu,$$

where f varies over the class of all functions in $\mathscr{L}$ such that $\chi_C \leq f$.

2. Let $(X, \mathscr{S}, \mu)$ be an arbitrary measure space, $\mathscr{T}$ a σ-ring contained in $\mathscr{S}$, and ν the restriction of μ to $\mathscr{T}$. Let f be a real-valued function on X which is measurable with respect to $\mathscr{T}$. Then f is μ-integrable if and only if it is ν-integrable, and in this case

$$\int f \, d\mu = \int f \, d\nu.$$

3. If X is compact, μ is a regular Borel measure on X, and $S(\mu)$ is the support of μ, then $S(\mu)$ is the smallest closed set such that the relations $f \in \mathscr{L}$, $f = 0$ on $S(\mu)$, imply $\int f \, d\mu = 0$.

4. Assume the notations and hypothesis of Theorem 5. The conclusion may be strengthened as follows: if f is any μ_2-integrable Borel function, then the Borel function f^T is μ_1-integrable, and

$$\int f^T \, d\mu_1 = \int f \, d\mu_2.$$

67. Approximation of Baire Functions

Recall that $\mathscr{L}$ denotes the class of all continuous real-valued functions with compact support on the locally compact space X. According to 66.2, every f in $\mathscr{L}$ is integrable with respect to any Baire (or Borel) measure on X.

If ν is a Baire measure, it is clear from the proof of 66.3 that the characteristic function of a compact G_δ is the limit in mean $[\nu]$ of a suitable sequence of functions in $\mathscr{L}$. In the following theorem we show how to approximate any ν-integrable function (and in particular any function in $\mathscr{L}$) by linear combinations of characteristic functions of compact G_δ's:

Theorem 1. *Let ν be a Baire measure on X, and suppose f is a ν-integrable Baire function. Then, given any $\varepsilon > 0$, there exists a function g,*

$$g = \sum_1^n \alpha_i \chi_{D_i},$$

such that

$$\int |f - g| \, d\nu \leq \varepsilon,$$

where the D_i are mutually disjoint compact G_δ's, and the α_i are real numbers.

Proof. A function g of the indicated type is of course a ν-integrable simple Baire function.

Since f is the limit in mean $[\nu]$ of a sequence of integrable simple Baire functions (28.4), we may assume without loss of generality that f has the following form:

$$f = \sum_1^n \alpha_i \chi_{F_i},$$

where the α_i are real numbers, the F_i are mutually disjoint Baire sets, and $\nu(F_i) < \infty$ for all i. If

$$c = 1 + \max\{|\alpha_i| : 1 \leq i \leq n\},$$

then $c > 0$ and $|f| \leq c$.

Given any $\varepsilon > 0$; for each i there exists, by the regularity of ν (60.1), a compact G_δ D_i such that $D_i \subset F_i$ and

$$\nu(F_i - D_i) \leq \varepsilon/nc.$$

Defining

$$g = \sum_1^n \alpha_i \chi_{D_i},$$

we have

$$f - g = \sum_1^n \alpha_i(\chi_{F_i} - \chi_{D_i}) = \sum_1^n \alpha_i \chi_{F_i - D_i},$$

hence (by disjointness)

$$|f - g| = \sum_1^n |\alpha_i| \chi_{F_i - D_i}.$$

Then

$$\int |f - g| \, d\nu = \sum_1^n |\alpha_i| \nu(F_i - D_i) \leq \sum_1^n c(\varepsilon/nc) = \varepsilon. \quad \blacksquare$$

The next theorem asserts that if ν is any Baire measure on X, then $\mathscr{L}$ is "dense" in $\mathscr{L}^1(\nu)$ (with respect to convergence in mean); that is, every f in $\mathscr{L}^1(\nu)$ is the limit in mean of a suitable sequence of continuous functions with compact support:

Theorem 2. *Let ν be a Baire measure on X, and suppose f is a ν-integrable Baire function. Then, given any $\varepsilon > 0$, there exists a function h in $\mathscr{L}$ such that*

$$\int |f - h| \, d\nu \leq \varepsilon.$$

Proof. By Theorem 1, and the triangle inequality (Sect. 28), we may assume

$$f = \sum_1^n \alpha_i \chi_{D_i},$$

where the α_i are real numbers, and the D_i are compact G_δ's. In fact it is clearly sufficient to consider $f = \chi_D$, where D is a compact G_δ. By 55.2 there is a sequence h_n in $\mathscr{L}$ such that $h_n \downarrow f$, and so

$$\int h_n \, d\nu \downarrow \int f \, d\nu$$

by the (MCT). Then

$$\int |h_n - f| \, d\nu = \int (h_n - f) \, d\nu = \int h_n \, d\nu - \int f \, d\nu \to 0,$$

and so $h_n \to f$ in mean $[\nu]$. ∎

EXERCISE

1. If μ is a regular Borel measure on X, and f is a μ-integrable Borel function, then, given any $\varepsilon > 0$, there exists a function g such that

$$\int |f - g| \, d\mu \leq \varepsilon,$$

where g is a linear combination of mutually disjoint compact sets (or even compact G_δ's).

*68. Approximation of Borel Functions

The principal result of this section is as follows: if μ is a regular Borel measure on the locally compact space X, and f is any Borel function, then there exists a Baire function g (depending on μ) such that $f = g$ a.e. $[\mu]$. We begin with a special case:

Lemma. *If μ is a regular Borel measure on X, and E is any Borel set, then there exists a Baire set F such that*

$$\mu(E \triangle F) = 0,$$

that is, $\chi_E = \chi_F$ *a.e.* $[\mu]$.

Proof. Consider first the case that $\mu(E) < \infty$. By regularity, there is a sequence C_n of compact sets such that $C_n \subset E$ and $\mu(C_n) \uparrow \mu(E)$. Defining $G = \bigcup_1^\infty C_n$, we have $G \subset E$ and $\mu(G) = \mu(E)$; by finiteness, $\mu(E - G) = 0$. By 59.1 we may choose, for each n, a compact G_δ D_n such that $C_n \subset D_n$ and $\mu(D_n - C_n) = 0$; then $F = \bigcup_1^\infty D_n$ is a Baire set, $G \subset F$, and it follows from the relation

$$F - G \subset \bigcup_1^\infty (D_n - C_n)$$

that $\mu(F - G) = 0$. Since

$$\begin{aligned} E \triangle F &= (E \triangle G) \triangle (G \triangle F) \\ &= (E - G) \triangle (F - G) \subset (E - G) \cup (F - G), \end{aligned}$$

it is clear that $\mu(E \triangle F) = 0$.

In the general case, it follows from the σ-finiteness of μ that one can write $E = \bigcup_1^\infty E_n$, where E_n is a sequence of Borel sets of finite measure. It follows from the preceding case that for each n there is a Baire set F_n such that

$$\mu(E_n \triangle F_n) = 0;$$

then $F = \bigcup_1^\infty F_n$ is a Baire set, and it is clear from the relation

$$E \triangle F \subset \bigcup_1^\infty E_n \triangle F_n$$

that $\mu(E \triangle F) = 0$. ▮

Theorem 1. *Let μ be a regular Borel measure on X, and suppose f is any Borel function. Then there exists a Baire function g such that $f = g$ a.e. $[\mu]$.*

Proof. Let f_n be a sequence of simple Borel functions such that $f_n \to f$ pointwise on X (16.4). For each n, it is clear from the lemma that there exists a simple Baire function h_n such that

$$f_n = h_n \text{ a.e. } [\mu].$$

It follows from 19.4 that

$$h_n \to f \text{ a.e. } [\mu];$$

choose a Borel set E such that $\mu(E) = 0$ and $h_n \to f$ pointwise on $X - E$.

Consider the set F of all points x in X for which the sequence $h_n(x)$ is not Cauchy; thus,

$$F = \bigcup_{m=1}^{\infty} \bigcap_{n=1}^{\infty} \bigcup_{i,j \geq n} \{x: |h_i(x) - h_j(x)| \geq 1/m\}.$$

Clearly F is a Baire set. Since $h_n(x)$ is convergent whenever $x \in X - E$, we have $F \subset E$, and so $\mu(F) = 0$.

Define $g_n = \chi_{X-F} h_n$; then g_n is a Baire function (15.1), and

$$g_n = h_n \text{ a.e. } [\mu],$$

hence $g_n \to f$ a.e. $[\mu]$ by 19.4. Moreover, g_n converges pointwise on X to a real-valued function g, which is necessarily a Baire function by 14.3. Finally, $g = f$ a.e. $[\mu]$ by 19.2. ∎

We close with a result on the approximation (in mean) of Borel functions:

Theorem 2. *Let μ be a regular Borel measure on X, and suppose f is a μ-integrable Borel function. Then, given any $\varepsilon > 0$, there exists a function h in $\mathscr{L}$ such that*

$$\int |f - h| \, d\mu \leq \varepsilon.$$

Proof. By Theorem 1, there exists a Baire function g such that $g = f$ a.e. $[\mu]$; it follows from 25.5 that g is μ-integrable. Let ν be the Baire restriction of μ; then g is ν-integrable by 66.1. Given any $\varepsilon > 0$, it follows from 67.2 that there exists a function h in $\mathscr{L}$ such that

$$\int |g - h| \, d\nu \leq \varepsilon,$$

that is (66.1), $\int |g - h| \, d\mu \leq \varepsilon$; since

$$|f - h| = |g - h| \text{ a.e. } [\mu],$$

we conclude that $\int |f - h| \, d\mu \leq \varepsilon$. ∎

The applicability of the results of this section is sometimes limited on account of the following considerations. Suppose μ is a regular Borel measure, and E is a Borel set. According to the lemma of Theorem 1, there exists a Baire set F such that $\mu(E \triangle F) = 0$. If it happens that $E \subset F$, then this simply means $\mu(F - E) = 0$. The point we wish to make is that it is not always possible to arrange that F be a superset of E. Another way of looking at the matter is that in Theorem 1 we are able to find a Baire function g such that $g = f$ a.e. $[\mu]$, but we cannot always arrange to have $g \geq f$ (everywhere on X). We shall take up this question again in Sect. 70.

EXERCISES

1. Theorem 1 implies its lemma.

2. Let μ be a Borel measure on X, and ν the Baire restriction of μ. Suppose f and g are Baire functions such that $f = g$ a.e. $[\mu]$. Then $f = g$ a.e. $[\nu]$.

3. If μ is a Borel measure on X, ν is the Baire restriction of μ, and g_n is a sequence of Baire functions which is fundamental a.e. $[\mu]$, then the sequence is also fundamental a.e. $[\nu]$. Similarly with "a.e." replaced by "in measure."

4. Suppose X is a normal locally compact space, μ is a regular Borel measure on X, and $f \in \mathscr{L}^1(\mu)$. Given any $\varepsilon > 0$, there exists a continuous Baire function g on X such that

$$\mu(\{x: f(x) \neq g(x)\}) \leq \varepsilon,$$

that is, $\mu(N(f - g)) \leq \varepsilon$.

5. (Lusin's theorem) If μ is a regular Borel measure on X, f is a Borel function, and E is a Borel set of finite measure, then given any $\varepsilon > 0$ there exists a compact subset C of E such that

$$\mu(E - C) \leq \varepsilon$$

and such that the restriction of f to C is continuous. The proof can be based on the following series of remarks.

(*i*) If $\chi_E f$ is integrable, there exists a sequence f_n in $\mathscr{L}$ such that $f_n \to \chi_E f$ almost uniformly (Theorem 2, 28.3, and 21.4).

(*ii*) The uniform limit of continuous functions is continuous.

(*iii*) If E is any Borel set such that $\mu(E) > 0$, there exists a compact subset C of E such that $\mu(C) > 0$ and the restriction of f to C is continuous.

(*iv*) If C_1 and C_2 are compact (or merely closed) sets such that the restriction of f to each C_i is continuous, then the restriction of f to $C_1 \cup C_2$ is also continuous.

(*v*) If E is a Borel set of finite measure, there exists an increasing sequence C_n of compact subsets of E such that $\mu(C_n) \uparrow \mu(E)$ and the restriction of f to each C_n is continuous.

6. If μ is a finite regular Borel measure on X, and f is any Borel function on X, there exists a sequence f_n in $\mathscr{L}$ such that $f_n \to f$ almost uniformly.

69. The Riesz-Markoff Representation Theorem

If ν is a Baire measure on the locally compact space X, then the correspondence

$$f \to \int f \, d\nu$$

defines a "positive linear form" on the class $\mathscr{L}$ of continuous functions with compact support; that is,

$$\int (f + g)\, d\nu = \int f\, d\nu + \int g\, d\nu,$$

$$\int (cf)\, d\nu = c \int f\, d\nu,$$

and

$$\int f\, d\nu \geq 0 \qquad \text{when} \qquad f \geq 0.$$

In this section we are concerned with the converse problem. We assume that we are given a *positive linear form* φ on $\mathscr{L}$, that is, a real-valued function $f \to \varphi(f)$ defined on $\mathscr{L}$, such that

$$\varphi(f + g) = \varphi(f) + \varphi(g),$$

$$\varphi(cf) = c\varphi(f),$$

and

$$\varphi(f) \geq 0 \qquad \text{when} \qquad f \geq 0$$

(and hence $\varphi(f) \leq \varphi(g)$ when $f \leq g$). It will be shown (Riesz-Markoff theorem) that there exists a regular Borel measure μ such that

$$\varphi(f) = \int f\, d\mu$$

for all f in $\mathscr{L}$; such a measure is necessarily unique by 66.4. Our plan is as follows:

(1) The positive linear form φ will be used to define a regular content λ.

(2) The regular content λ will be extended to a regular Borel measure μ via 64.2.

(3) It will be verified that $\varphi(f) = \int f\, d\mu$ for every f in $\mathscr{L}$.

It will be convenient to make use of the following notation due to Halmos: if A is a subset of X and f is a function in $\mathscr{L}$, we shall write $A \subset f$ in case $\chi_A \leq f$. If $A \subset f$, obviously $f \geq 0$.

If C is a compact set, there exist functions f in $\mathscr{L}$ such that $C \subset f$. Indeed, by 54.3 there is an f in $\mathscr{L}$ such that $f \geq 0$, and $f = 1$ on C, and so $C \subset f$. We may therefore define a real-valued function λ on the class of all compact sets by the formula

$$\lambda(C) = \text{GLB}\, \{\varphi(f): C \subset f, f \in \mathscr{L}\}.$$

Evidently $0 \leq \lambda(C) < \infty$. Our first task is to show that λ is a regular content on X:

Lemma 1. *The set function λ defined above is a regular content.*

Proof. Since $\varnothing \subset 0$, and $\varphi(0) = 0$, it is clear that $\lambda(\varnothing) = 0$. If C and D are compact sets such that $C \subset D$, and if $D \subset f$, then also $C \subset f$, and so $\lambda(C) \leq \varphi(f)$; varying f, we have $\lambda(C) \leq \lambda(D)$, thus λ is *monotone.*

Let C and D be compact sets. If $C \subset f$ and $D \subset g$, then $C \cup D \subset f + g$ results from the relation $\chi_{C \cup D} \leq \chi_C + \chi_D$, and so

$$\lambda(C \cup D) \leq \varphi(f + g) = \varphi(f) + \varphi(g);$$

varying f and g (independently) we have

$$\lambda(C \cup D) \leq \lambda(C) + \lambda(D),$$

thus λ is *subadditive.*

Suppose C and D are disjoint compact sets. Let U and V be disjoint open sets such that $C \subset U$ and $D \subset V$. According to 54.3 there exist functions f, g in $\mathscr{L}$ such that $C \subset f$, $D \subset g$, $f = 0$ on $X - U$, $g = 0$ on $X - V$, $0 \leq f \leq 1$, $0 \leq g \leq 1$. It follows from disjointness that $f + g \leq 1$, and so $h(f + g) \leq h$ whenever $h \in \mathscr{L}$, $h \geq 0$. Suppose now $C \cup D \subset h$. Obviously $C \subset h$; since already $C \subset f$, it follows that $C \subset hf$. Similarly $D \subset hg$. Then

$$\begin{aligned} \lambda(C) + \lambda(D) \leq \varphi(hf) + \varphi(hg) &= \varphi(hf + hg) \\ &= \varphi(h(f + g)) \leq \varphi(h); \end{aligned}$$

varying h,

$$\lambda(C) + \lambda(D) \leq \lambda(C \cup D).$$

Since λ is subadditive, we conclude that

$$\lambda(C \cup D) = \lambda(C) + \lambda(D),$$

thus λ is *additive.*

It remains to show that λ is *regular*. Given a compact set C, and $\varepsilon > 0$, we seek a compact set D such that $C \prec D$ (in the notation of Sect. 64) and

$$\lambda(D) \leq \lambda(C) + 2\varepsilon.$$

By the definition of λ, there is an f in $\mathscr{L}$ such that

$$C \subset f \qquad \text{and} \qquad \varphi(f) \leq \lambda(C) + \varepsilon.$$

Let γ be a real parameter, $0 < \gamma < 1$ (the value to be fixed later), and for each γ define

$$U_\gamma = \{x : f(x) > \gamma\}, \qquad D_\gamma = \{x : f(x) \geq \gamma\};$$

evidently $C \subset U_\gamma \subset D_\gamma$, U_γ is open, and D_γ is compact. Thus $C \prec D_\gamma$, and it will suffice to choose γ so that

$$\lambda(D_\gamma) \le \lambda(C) + 2\varepsilon.$$

Now, it is clear from the definition of D_γ that $D_\gamma \subset (1/\gamma) f$, and so

$$\lambda(D_\gamma) \le \varphi((1/\gamma) f) = (1/\gamma)\varphi(f) \le (1/\gamma)[\lambda(C) + \varepsilon],$$

and the proof is concluded by choosing γ near enough to 1 so that

$$(1/\gamma)[\lambda(C) + \varepsilon] \le \lambda(C) + 2\varepsilon. \quad \blacksquare$$

According to 64.2, there exists a unique regular Borel measure μ which *extends* λ.

Lemma 2. *If C is a compact set and $\varepsilon > 0$, there exists an f in $\mathscr{L}$ such that $C \subset f$ and*

$$\varphi(f) \le \int f\, d\mu + \varepsilon.$$

Proof. By the definition of λ, we may choose f so that

$$C \subset f \qquad \text{and} \qquad \varphi(f) \le \lambda(C) + \varepsilon.$$

Since μ extends λ, and $\chi_C \le f$, we have

$$\varphi(f) \le \lambda(C) + \varepsilon = \mu(C) + \varepsilon = \int \chi_C\, d\mu + \varepsilon \le \int f\, d\mu + \varepsilon. \quad \blacksquare$$

Lemma 2 is a fragment in the direction of proving that $\varphi(f) \le \int f\, d\mu$. In the following lemma we prove an inequality in the reverse direction:

Lemma 3. *If $f \in \mathscr{L}$ and $f \ge 0$, then $\int f\, d\mu \le \varphi(f)$.*

Proof. It is clearly sufficient to find a sequence g_n of μ-integrable Borel functions such that

$$\int g_n\, d\mu \le \varphi(f) \qquad \text{for all } n,$$

and

$$\int g_n\, d\mu \to \int f\, d\mu.$$

Let f_n be a sequence of simple Baire functions such that $0 \le f_n \uparrow f$ (16.4). For each n, it is clear from the proof of 67.1 that there exists a simple Baire function g_n such that $0 \le g_n \le f_n$, g_n is a linear combination of characteristic functions of mutually disjoint compact G_δ's, and

$$\int |f_n - g_n|\, d\mu \le 1/n;$$

all that is involved here is that every Baire measure is regular. Now,

$$\int g_n\, d\mu = \int (g_n - f_n)\, d\mu + \int f_n\, d\mu,$$

and so

$$\int g_n\, d\mu \to \int f\, d\mu.$$

We are therefore reduced to the following situation: assuming $0 \leq g \leq f$, where g is a linear combination of characteristic functions of mutually disjoint compact G_δ's, it will suffice to show that

$$\int g \, d\mu \leq \varphi(f).$$

Say

$$g = \sum_1^m \alpha_i \chi_{C_i},$$

where $\alpha_i > 0$, and the C_i are mutually disjoint compact sets.

We may construct nonnegative functions $h_1, \ldots, h_m$ in $\mathscr{L}$ such that $h_i = 1$ on C_i, $h_i = 0$ on C_j when $j \neq i$, and $h_1 + \cdots + h_m \leq 1$. Indeed, let $U_1, \ldots, U_m$ be mutually disjoint open sets such that $C_i \subset U_i$, and choose h_i to be a function in $\mathscr{L}$ such that $0 \leq h_i \leq 1$, $h_i = 1$ on C_i, and $h_i = 0$ on $X - U_i$ (54.3).

For each i, clearly $\alpha_i \chi_{C_i} \leq h_i g$. On the other hand, $h_i g \leq h_i f$. Thus $\alpha_i \chi_{C_i} \leq h_i f$, that is,

$$C_i \subset (1/\alpha_i) h_i f.$$

By the definition of λ, we have

$$\lambda(C_i) \leq (1/\alpha_i)\varphi(h_i f),$$

thus (μ extends λ)

$$\alpha_i \mu(C_i) \leq \varphi(h_i f);$$

summing on i,

$$\int g \, d\mu = \sum_1^m \alpha_i \mu(C_i) \leq \sum_1^m \varphi(h_i f)$$

$$= \varphi\left(\left(\sum_1^m h_i\right) f\right) \leq \varphi(f). \quad \blacksquare$$

Theorem 1 (Riesz-Markoff). *If φ is a positive linear form on $\mathscr{L}$, there exists a unique regular Borel measure μ such that*

$$\varphi(f) = \int f \, d\mu$$

for all f in $\mathscr{L}$.

Proof. Let μ be the regular Borel measure constructed above. Given g in $\mathscr{L}$, let us show that $\varphi(g) = \int g \, d\mu$. By linearity, we may assume $0 \leq g \leq 1$. By Lemma 3 we have $\int g \, d\mu \leq \varphi(g)$, and so given any $\varepsilon > 0$, it will suffice to show that $\varphi(g) \leq \int g \, d\mu + \varepsilon$.

Let C be a compact set such that $g = 0$ on $X - C$. By Lemma 2, we may choose f in $\mathscr{L}$ so that $C \subset f$ and

$$(*) \qquad \varphi(f) \leq \int f \, d\mu + \varepsilon.$$

Observe that $g \leq f$; for, if $x \in C$ one has $g(x) \leq 1 = \chi_C(x) \leq f(x)$, whereas if $x \in X - C$ one has $g(x) = 0 \leq f(x)$. Thus $f - g \geq 0$, and it follows from Lemma 3 that

$$\int (f - g)\, d\mu \leq \varphi(f - g),$$

thus

$$\varphi(g) - \int g\, d\mu \leq \varphi(f) - \int f\, d\mu;$$

in view of relation (*), this implies

$$\varphi(g) - \int g\, d\mu \leq \varepsilon.$$

This concludes the proof of existence, and uniqueness follows from 66.4. ∎

With notation as in Theorem 1, we shall say that μ is the regular Borel measure *associated* with the positive linear form φ.

The rest of the material in this section is for application in Chapter 9. As in Sect. 66, we write $f^T(x) = f(Tx)$ when T is a homeomorphism of X and f is a real-valued function on X, and we observe that $f \in \mathscr{L}$ implies $f^T \in \mathscr{L}$.

Theorem 2. *Let φ be a positive linear form on $\mathscr{L}$, T a homeomorphism of X, and let φ' be the positive linear form on $\mathscr{L}$ defined by the formula*

$$\varphi'(f) = \varphi(f^T).$$

If μ (respectively μ') is the regular Borel measure associated with φ (respectively φ'), then

$$\mu'(E) = \mu(T^{-1}(E))$$

for every Borel set E.

Proof. Since $f \geq 0$ implies $f^T \geq 0$, it is clear that φ' is indeed a positive linear form. For all f in $\mathscr{L}$, we have

$$\int f^T\, d\mu = \varphi(f^T) = \varphi'(f) = \int f\, d\mu',$$

and so $\mu(T^{-1}(E)) = \mu'(E)$ for every Borel set E by 66.6. ∎

Corollary 1. *Let φ_1 and φ_2 be positive linear forms on $\mathscr{L}$, and let T be a homeomorphism of X such that*

$$\varphi_1(f^T) = \varphi_2(f)$$

for all f in $\mathscr{L}$. If μ_i is the regular Borel measure associated with φ_i, then

$$\mu_1(T^{-1}(E)) = \mu_2(E)$$

for every Borel set E.

Proof. Define $\varphi'(f) = \varphi_1(f^T)$ for all f in $\mathscr{L}$, and let μ' be the regular Borel measure associated with the positive linear form φ'; by Theorem 2, we have $\mu'(E) = \mu_1(T^{-1}(E))$ for every Borel set E.

By assumption, $\varphi'(f) = \varphi_2(f)$ for all f in $\mathscr{L}$, and so $\mu'(E) = \mu_2(E)$ for every Borel set E by 66.4. ▮

Corollary 2. *Let φ be a positive linear form on $\mathscr{L}$, T a homeomorphism of X, and $c > 0$ a constant such that*

$$\varphi(f^T) = c\varphi(f)$$

for all f in $\mathscr{L}$. If μ is the regular Borel measure associated with φ, then

$$\mu(T^{-1}(E)) = c\mu(E)$$

for every Borel set E.

Proof. It is clear from 55.2 and 59.2 that $c\mu$ is the regular Borel measure associated with the positive linear form $c\varphi$, and our assertion then follows on applying Corollary 1 to the forms $\varphi_1 = \varphi$ and $\varphi_2 = c\varphi$. ▮

EXERCISES

1. Granted the Riemann integral for continuous functions vanishing outside an interval, one can arrive at Lebesgue measure in the following way. If f is any continuous real-valued function defined on the real line R, which vanishes outside some interval (depending on the function), define $\varphi(f)$ to be the Riemann integral of f. Then φ is a positive linear form, and we may define Lebesgue measure to be the completion of the regular Borel measure μ associated with φ. To see that this definition is consistent with the one given in Sect. 7, it is sufficient to show that

$$\mu([a, b)) = b - a,$$

and this is easily shown by approximating the characteristic function of $[a, b)$ with a sequence of functions f_n in $\mathscr{L}$ whose Riemann integral tends to $b - a$ as $n \to \infty$.

2. If λ is any content on X, and μ is the regular Borel measure induced by λ, and if C is a compact set such that $C \subset f$, then $\lambda(C) \leq \int f\, d\mu$.

3. If φ_0 is a real-valued functional defined on the class of all nonnegative functions in $\mathscr{L}$, such that

$$\varphi_0(f + g) = \varphi_0(f) + \varphi_0(g) \qquad \text{and} \qquad \varphi_0(cf) = c\varphi_0(f)$$

(for $c \geq 0$), then φ_0 can be extended to a unique linear form φ on $\mathscr{L}$.

*70. Completion Regularity

A Baire measure ν on the locally compact space X will be called **completion regular** if it satisfies any one of the equivalent conditions of the following theorem:

Theorem 1. *If ν is a Baire measure on X, the following conditions on ν are equivalent:*

(a) *If f is any bounded Borel function, there exist bounded Baire functions g and h such that $g \geq f \geq h$ (everywhere on X) and $g = h$ a.e. $[\nu]$.*

(b) *If E is any Borel set, there exist Baire sets G and F such that $G \subset E \subset F$ and $\nu(F - G) = 0$.*

(c) *Every compact set (and hence every Borel set) is ν^*-measurable, where ν^* is the outer measure induced by ν.*

(d) *Every compact set (and hence every Borel set) is $\hat{\nu}$-measurable, where $\hat{\nu}$ is the completion of ν.*

Proof. (a) *implies* (b): If E is any Borel set, let $f = \chi_E$. By the assumption (a), there exist Baire functions g and h such that $g \geq f \geq h$ everywhere on X, and $g = h$ a.e. $[\nu]$. Let F_0 be a Baire set of measure zero such that $g = h$ on $X - F_0$.

Define $G = \{x: h(x) = 1\}$ and $F = \{x: g(x) \geq 1\}$; then G and F are Baire sets, and clearly $G \subset E \subset F$. If $x \in F - G$, then $g(x) \geq 1$ and $h(x) < 1$; thus $F - G \subset F_0$, and so $\nu(F - G) = 0$.

(b) *implies* (d): Suppose E is any Borel set. Choose Baire sets G and F such that $G \subset E \subset F$ and $\nu(F - G) = 0$. Then $E = G \cup (E - G)$, where G is a Baire set, and $E - G$ is a subset of a Baire set of measure zero; thus E is $\hat{\nu}$-measurable (see Sect. 9).

Since ν is σ-finite, (c) and (d) are equivalent by 9.2.

(d) *implies* (b): If E is a Borel set, by assumption one can write $E = G \cup S$, where G is a Baire set, and S is a subset of a Baire set F_0 such that $\nu(F_0) = 0$. Defining $F = G \cup F_0$, we have $G \subset E \subset F$, where G and F are Baire sets, and $\nu(F - G) = 0$ results from the relation $F - G \subset F_0$.

(b) *implies* (a): Let f be a bounded Borel function, say $|f(x)| \leq c < \infty$ for all x in X. Let μ be the unique regular Borel measure which extends ν (65.1). By 68.1, there exists a Baire function g_0 such that $g_0 = f$ a.e. $[\mu]$. Let E be a Borel set such that $\mu(E) = 0$ and $g_0 = f$ on $X - E$. By the assumption (b), there exist Baire sets G and F such that $G \subset E \subset F$ and $\nu(F - G) = 0$. Since

$$F = E \cup (F - E) \subset E \cup (F - G),$$

we have

$$\nu(F) = \mu(F) \leq \mu(E) + \mu(F - G) = \nu(F - G) = 0.$$

Since $g_0 = f$ on $X - F$, the functions

$$g = \chi_{X-F}g_0 + c\chi_F$$

and

$$h = \chi_{X-F}g_0 - c\chi_F$$

clearly meet all requirements (15.1). ∎

We shall say that a Baire measure ν is **monogenic** in case any two Borel measures extending ν are necessarily identical. In view of 65.1, this means that ν possesses a unique, and necessarily regular, Borel extension μ.

Theorem 2. *Every completion regular Baire measure is monogenic.*

Proof. Let ν be a completion regular Baire measure, and suppose μ_1 and μ_2 are any two Borel measures extending ν. Given any Borel set E, it is to be shown that $\mu_1(E) = \mu_2(E)$. Since ν is completion regular, we may choose Baire sets G and F such that $G \subset E \subset F$ and $\nu(F - G) = 0$. Then $\mu_i(F - G) = 0$, and so

$$\mu_i(G) = \mu_i(E) = \mu_i(F).$$

Thus,

$$\mu_1(E) = \mu_1(F) = \nu(F) = \mu_2(F) = \mu_2(E). \quad ∎$$

A Borel measure μ is said to be **completion regular** in case its Baire restriction ν is completion regular. It follows from the remarks preceding Theorem 2 that μ is then necessarily regular.

Theorem 3. *If μ is a regular Borel measure on X, the following conditions are equivalent:*

(a) *μ is completion regular.*

(b) *If E is any Borel set such that $\mu(E) = 0$, then there exists a Baire set F such that $E \subset F$ and $\mu(F) = 0$.*

Proof. Let ν be the Baire restriction of μ.

(a) *implies* (b): Let E be a Borel set such that $\mu(E) = 0$. Since ν is completion regular, we may choose Baire sets G and F such that $G \subset E \subset F$ and $\nu(F - G) = 0$. Thus $\mu(F - G) = 0$, and so

$$\mu(F) = \mu(E) = 0.$$

(b) *implies* (a): By Theorem 1, it will suffice to show that each compact set C is $\hat{\nu}$-measurable. Since μ is regular, there exists a compact G_δ D

such that $C \subset D$ and $\mu(D - C) = 0$ (59.1). By the assumption (b), there exists a Baire set F such that $D - C \subset F$ and $\mu(F) = 0$. Then

$$C = (D - F) \cup (C \cap F),$$

where $D - F$ is a Baire set and $C \cap F$ is a subset of a ν-null Baire set (namely F), and so C is $\hat{\nu}$-measurable (see Sect. 9). ▮

The next theorem is an indication of the superiority of completion regularity over regularity:

Theorem 4. *If μ_i is a completion regular Borel measure on the locally compact space X_i ($i = 1, 2$), and ν_i is the Baire restriction of μ_i, then every Borel measure on $X_1 \times X_2$ which extends $\nu_1 \times \nu_2$ is necessarily also an extension of $\mu_1 \times \mu_2$. In particular, $\mu_1 \times \mu_2$ may be extended to a (unique) regular Borel measure on $X_1 \times X_2$.*

Proof. In any case $\nu_1 \times \nu_2$ is a Baire measure on $X_1 \times X_2$ by 56.3.

Suppose τ is any Borel measure on $X_1 \times X_2$ which extends $\nu_1 \times \nu_2$. We assert that τ also extends $\mu_1 \times \mu_2$. Now, the domain of definition of $\mu_1 \times \mu_2$ is the Cartesian product σ-ring $\mathscr{S}_1 \times \mathscr{S}_2$, where $\mathscr{S}_i$ is the class of Borel sets of X_i. If $\mathscr{R}$ is the ring generated by the rectangles $C_1 \times C_2$, where C_i is a compact set in X_i, then the σ-ring generated by $\mathscr{R}$ is $\mathscr{S}_1 \times \mathscr{S}_2$ by 35.2; it will therefore suffice by 34.1 and the (UET) to show that

$$\tau(C_1 \times C_2) = (\mu_1 \times \mu_2)(C_1 \times C_2)$$

for all such rectangles. Assuming C_i is a compact set in X_i, it follows from 59.1 (and the regularity of μ_i) that there exists a compact G_δ D_i containing C_i such that $\mu_i(D_i - C_i) = 0$. Since μ_i is completion regular, by Theorem 3 there exists a Baire set F_i such that $D_i - C_i \subset F_i$ and $\mu_i(F_i) = 0$, that is, $\nu_i(F_i) = 0$. Now,

$$\begin{aligned} D_1 \times D_2 - C_1 \times C_2 &\subset [(D_1 - C_1) \times D_2] \cup [D_1 \times (D_2 - C_2)] \\ &\subset (F_1 \times D_2) \cup (D_1 \times F_2); \end{aligned}$$

since the D_i and F_i are Baire sets, and since τ extends $\nu_1 \times \nu_2$, we have

$$\begin{aligned} \tau(D_1 \times D_2 - C_1 \times C_2) &\leq \tau(F_1 \times D_2) + \tau(D_1 \times F_2) \\ &= (\nu_1 \times \nu_2)(F_1 \times D_2) + (\nu_1 \times \nu_2)(D_1 \times F_2) \\ &= \nu_1(F_1)\nu_2(D_2) + \nu_1(D_1)\nu_2(F_2) = 0. \end{aligned}$$

(Observe that the σ-finiteness of ν_i was used via 39.2). Thus

$$\tau(D_1 \times D_2 - C_1 \times C_2) = 0,$$

and so

$$\tau(C_1 \times C_2) = \tau(D_1 \times D_2) = (\nu_1 \times \nu_2)(D_1 \times D_2)$$
$$= \nu_1(D_1)\nu_2(D_2) = \mu_1(C_1)\mu_2(C_2) = (\mu_1 \times \mu_2)(C_1 \times C_2),$$

and our assertion is proved.

In particular, taking τ to be the unique regular Borel measure which extends $\nu_1 \times \nu_2$ (65.1), we have a regular Borel extension of $\mu_1 \times \mu_2$.

Observe that $\mu_1 \times \mu_2$ is an extension of $\nu_1 \times \nu_2$; indeed, if D_i is a compact G_δ in X_i, then

$$(\mu_1 \times \mu_2)(D_1 \times D_2) = \mu_1(D_1)\mu_2(D_2)$$
$$= \nu_1(D_1)\nu_2(D_2) = (\nu_1 \times \nu_2)(D_1 \times D_2),$$

and so by the familiar (UET) argument we see that $\mu_1 \times \mu_2 = \nu_1 \times \nu_2$ on the class of all Baire sets in $X_1 \times X_2$ (note 56.3).

Finally, suppose τ_1 and τ_2 are any two regular Borel measures on $X_1 \times X_2$ both of which extend $\mu_1 \times \mu_2$. Since the τ_i also extend $\nu_1 \times \nu_2$ by the preceding argument, we have $\tau_1 = \tau_2$ by 62.1. ∎

EXERCISES

1. The following is an example of a Baire measure ν which is not completion regular. Suppose x_0 is a point of X such that $\{x_0\}$ is not a G_δ. Define $\nu(F)$, for each Baire set F, to be 1 or 0 according as x_0 does or does not belong to F.

2. If x_0 is a point of X such that $\{x_0\}$ is not a G_δ, and μ is the Borel measure such that $\mu(E) \doteq 1$ or 0 according as x_0 does or does not belong to the Borel set E, then μ is regular, but not completion regular.

***3.** Is every monogenic Baire measure completion regular?

4. In order that every Baire measure on X be monogenic, it is necessary and sufficient that every Borel measure on X be regular.

***5.** If X is a locally compact space such that every Baire measure on X is monogenic, does it follow that every Borel set is a Baire set? What if every Baire measure is completion regular? In any case, it is easy to see that if every Baire measure on X is completion regular, then X is first countable.

6. (Johnson) If ν is a completion regular Baire measure, it does not follow that for each Borel function f there exist Baire functions g and h such that $g \geq f \geq h$ everywhere on X and $g = h$ a.e. $[\nu]$. That is, the word "bounded" cannot be omitted in condition (a) of Theorem 1.

***7.** If ν_i is a monogenic Baire measure on the locally compact space X_i $(i = 1, 2)$, is the Baire measure $\nu_1 \times \nu_2$ monogenic?

***8.** If ν_i is a completion regular Baire measure on the locally compact space X_i $(i = 1, 2)$, is $\nu_1 \times \nu_2$ completion regular? Is it even monogenic?

***9.** If μ_i is a Borel measure on X_i $(i = 1, 2)$, can $\mu_1 \times \mu_2$ be extended to a Borel measure on $X_1 \times X_2$?

***10.** If φ is a positive linear form on $\mathscr{L}$, and ν is the Baire measure associated with φ, what are necessary and sufficient conditions on φ such that ν is completion regular?

***11.** If ν is a Baire measure, the set $\mathscr{K}$ of all Borel measures which extend ν is a convex set. The condition that ν be monogenic is precisely that $\mathscr{K}$ consists of a single point. In the general case: (*i*) can $\mathscr{K}$ consist of a line segment? (*ii*) of what measure theoretic significance is an extremal point of $\mathscr{K}$?

12. If μ is a regular Borel measure, and ν is the Baire restriction of μ, then the following conditions are equivalent:

(a) μ is completion regular.

(b) The completions $\hat{\nu}$ and $\hat{\mu}$ have the same domain of definition, and $\hat{\nu} = \hat{\mu}$ on their common domain. Briefly, ν and μ have identical completions.

(c) $\nu^* = \mu^*$ (on their common domain, which is the class of all σ-bounded sets).

(d) $\mu(E) = \nu^*(E)$ for every Borel set E.

(e) every compact set of positive measure contains a Baire set of positive measure.

13. The "Haar measure" of a locally compact group is completion regular (see Sect. 90).

14. (Johnson) Theorem 4 can be generalized as follows. Suppose ν_i is a monogenic Baire measure on X_i $(i = 1, 2)$, and μ_i is the unique (regular) Borel measure which extends ν_i. If ρ is any Borel measure on $X_1 \times X_2$ which extends $\nu_1 \times \nu_2$, then ρ also extends $\mu_1 \times \mu_2$; in particular, there exists a (unique) regular Borel measure on $X_1 \times X_2$ which extends $\mu_1 \times \mu_2$. (See also Exercise 62.13.)

15. Let μ_i be a Borel measure on X_i $(i = 1, 2)$, and assume there exists a completion regular Borel measure ρ on $X_1 \times X_2$ which extends $\mu_1 \times \mu_2$. Let h be a Borel function on $X_1 \times X_2$.

(*i*) If h is ρ-integrable, then the iterated integrals

$$\int\int h \, d\mu_2 \, d\mu_1 \quad \text{and} \quad \int\int h \, d\mu_1 \, d\mu_2$$

exist, and are equal to $\int h \, d\rho$.

(*ii*) If, conversely, $h \geq 0$ and the iterated integral

$$\int\int h \, d\mu_2 \, d\mu_1$$

exists, then h is ρ-integrable. (*Note:* It is not assumed that h is measurable with respect to the domain of $\mu_1 \times \mu_2$.)

16. (Johnson) If μ_i is a completion regular Borel measure on the locally compact space X_i $(i = 1, 2)$, and if

$$\mathscr{B}(X_1 \times X_2) = \mathscr{B}(X_1) \times \mathscr{B}(X_2),$$

where $\mathscr{B}(X)$ denotes the class of Borel sets of a space X, then the Borel measure $\mu_1 \times \mu_2$ is also completion regular.

CHAPTER 9

Integration over Locally Compact Groups

71. Topological Groups

Let G be a group, with a multiplicatively notated (associative) internal law of composition. Thus we write xy for the product of x and y, e for the neutral element of G, and x^{-1} for the inverse of x. A topology on G is said to be **compatible** with the group structure in case $x \to x^{-1}$ is a continuous mapping in G, and $(x, y) \to xy$ is a continuous mapping of $G \times G$ into G (where it is understood that $G \times G$ bears the product topology). A **topological group** is a group G equipped with a compatible topology. In a topological group the continuous mapping $x \to x^{-1}$, being bijective and self-inverse, is a homeomorphism.

What we need to know about topological groups is easily stated in a paragraph. Suppose G is a topological group. For each s in G, the mapping $x \to sx$ is called *left translation* by s; it is a homeomorphism, with inverse mapping $x \to s^{-1}x$. Similarly for *right translations* $x \to xs$. It follows that a subset A of G is a compact (respectively open, closed, G_δ, F_σ, Baire, Borel, weakly Borel, bounded, σ-bounded) set if and only if sA has the same property. Similarly for As and A^{-1}. For the same reason, as U varies over all neighborhoods of e, the sets sU vary over all neighborhoods of $s = se$, and similarly for the sets Us; also, U^{-1} varies over all neighborhoods of $e = e^{-1}$. A subset A of G is called *symmetric* if $A^{-1} = A$. If U is any neighborhood of e, then $U \cap U^{-1}$ is a symmetric neighborhood of e, hence the symmetric neighborhoods of e are basic. If U is a neighborhood of e, there exists a neighborhood V of e such that $VV \subset U$; this follows at once from the continuity of $(x, y) \to xy$ at (e, e), and the fact that $ee = e$. While we are at it, we may take V to be symmetric.

For the rest of the chapter, G denotes a **locally compact** (Hausdorff) topological group. In particular, G is completely regular. The reader may be aware of the fact that G possesses (as does every topological group) two natural uniform structures, each of which yields the given topology (thus every topological group is a uniformizable topological space); this is still another source of the complete regularity of G. For our purposes, however, it will not be necessary to refer explicitly to the theory of uniform structures. (Nevertheless, a good deal of the theory of uniform structures is implicit in the fact, which we use freely, that every locally compact space is completely regular.) Our basic tools are local compactness, the complete regularity which it implies, and the topological group structure (to the extent described in the preceding paragraph). The one explicit piece of "uniformity" we need is as follows:

Theorem 1. *If f is a real-valued continuous function on G with compact support, then given any $\varepsilon > 0$ there exists a neighborhood V of e such that*

$$|f(x) - f(y)| < \varepsilon$$

whenever $x^{-1}y \in V$.

Proof. Writing $y = xs$, we have $x^{-1}y = s$. Thus, given $\varepsilon > 0$, we seek a neighborhood V of e such that $s \in V$ implies

$$|f(x) - f(xs)| < \varepsilon$$

for all x in G. Define

$$V = \{s \in G: |f(x) - f(xs)| < \varepsilon \text{ for all } x \text{ in } G\};$$

clearly $e \in V$, and the problem is to show that V is a neighborhood of e.

Let h be the function on $G \times G$ defined by

$$h(x, s) = |f(x) - f(xs)|;$$

clearly h is continuous, $h(x, e) = 0$ for all x in G, and

$$V = \{s \in G: h(x, s) < \varepsilon \text{ for all } x \text{ in } G\}.$$

By assumption there exists a compact set C such that $f = 0$ on $G - C$. Choose a compact and symmetric neighborhood W of e (for example if A is any compact neighborhood of e, we may take $W = A \cap A^{-1}$). Define

$$D = CW = \{xy: x \in C \text{ and } y \in W\};$$

since $(x, y) \to xy$ is continuous, and $C \times W$ is a compact subset of $G \times G$, it follows that D is a compact set in G. Clearly $C \subset D$.

We construct a finite covering of D in the following way. Given any y in D; since h is continuous at (y, e), and $h(y, e) = 0$, there exists a neighborhood N_y of (y, e) such that $h(x, s) < \varepsilon$ whenever $(x, s) \in N_y$. We may suppose N_y has the form $N_y = A_y \times U_y$, where A_y is a neighborhood of y, and U_y is a neighborhood of e. Since D is compact, we have

$$D \subset A_{y_1} \cup \cdots \cup A_{y_n}$$

for a suitable finite set of points $y_1, \ldots, y_n$. Define

$$U = U_{y_1} \cap \cdots \cap U_{y_n} \cap W;$$

then U is a neighborhood of e, and it will suffice to show that $U \subset V$.

Thus, given $s \in U$ and $x \in G$, it will suffice to show that

$$|f(x) - f(xs)| < \varepsilon.$$

We consider two cases, according as x does or does not belong to D. If $x \in D$, then $x \in A_{y_i}$ for some i; since also $s \in U_{y_i}$, it follows that $(x, s) \in N_{y_i}$, and so $h(x, s) < \varepsilon$, as we wished to show. If $x \notin D$, then also $x \notin C$, hence $f(x) = 0$; since $xs \notin C$ ($xs \in C$ would imply $x \in Cs^{-1} \subset CW = D$, contrary to assumption), we have also $f(xs) = 0$, and so $h(x, s) = 0 < \varepsilon$. ∎

EXERCISE

1. The class of Borel sets in G coincides with the class of Baire sets if and only if the topological group G is metrizable.

72. Translates, Haar Integral

If f is a real-valued function on G, and $s \in G$, the **left translate** of f by s is the function f_s on G defined by the formula

$$f_s(x) = f(s^{-1}x).$$

Since $x \to s^{-1}x$ is a homeomorphism of G, it is clear that f_s is continuous if and only if f is continuous. Similarly for the property of being a Baire function, or a Borel function, or having compact support.

In particular, if $f \in \mathscr{L}$ and $s \in G$, then $f_s \in \mathscr{L}$ (where $\mathscr{L}$ denotes, as in the preceding chapter, the class of all continuous real-valued functions on G with compact support). A (left) **Haar integral** for G is a positive linear form I on $\mathscr{L}$, not identically zero, such that

$$I(f_s) = I(f)$$

for all f in $\mathscr{L}$ and s in G; in other words, a Haar integral is a nonzero positive linear form on $\mathscr{L}$ which is invariant under left translations. If I is a Haar integral, and $c > 0$ is a real number, clearly cI is also a Haar integral. The fundamental existence theorem, to be proved in Sect. 74, is that *G possesses a Haar integral I.* The fundamental uniqueness theorem, to be proved in Sect 76, is that *the Haar integral is unique up to a factor of proportionality:* if J is another Haar integral for G, then $J = cI$ for a suitable real number $c > 0$. In both the existence and uniqueness proofs, we follow the exposition of A. Weil. The reader who already has these two theorems in his repertory, one way or another, may wish to pass directly to Sect. 77, for none of the (rather gory) details of their proofs will be needed from that point onward.

We now summarize the formal properties of left translation that will be used often in what follows. For each s in G, the correspondence $f \to f_s$ defines a bijective linear mapping in the class of all real-valued functions on G; specifically, $(f + g)_s = f_s + g_s$, $(cf)_s = cf_s$, and $f = (f_t)_s$ with $t = s^{-1}$. If $f \geq 0$, then $f_s \geq 0$. If f is bounded, then so is f_s, and $\|f_s\|_\infty = \|f\|_\infty$ inasmuch as f_s and f have the same range of values. One has $f_e = f$ for all f, and $(f_s)_t = f_{ts}$ for all $s, t \in G$; the latter identity is proved by the calculation

$$\begin{aligned}(f_s)_t(x) &= f_s(t^{-1}x) = f(s^{-1}(t^{-1}x)) \\ &= f((s^{-1}t^{-1})x) = f((ts)^{-1}x) = f_{ts}(x).\end{aligned}$$

73. Translation Ratios

Throughout Sects. 73–76 we shall write $\mathscr{P}$ for the class of all continuous real-valued functions f on G with compact support, such that $f \geq 0$ and f is not identically zero. Thus,

$$\mathscr{P} = \{f \in \mathscr{L} : f \geq 0, f \neq 0\}.$$

We note that if $f, g \in \mathscr{P}$, $c > 0$, and $s \in G$, then the functions $f + g$, cf, and f_s also belong to $\mathscr{P}$. In constructing a Haar integral for G, we may restrict attention to functions in $\mathscr{P}$ on account of the following lemma:

Lemma 1. *If $f \to I_0(f)$ is a real-valued function on $\mathscr{P}$ such that*

(1) $$I_0(f) > 0$$

(2) $$I_0(f + g) = I_0(f) + I_0(g)$$

(3) $$I_0(cf) = cI_0(f)$$

(4) $$I_0(f_s) = I_0(f),$$

for all $f, g \in \mathscr{P}$, $c > 0$, and $s \in G$, then I_0 may be extended to a Haar integral I.

Proof. Given $f \in \mathscr{L}$, the problem is to define $I(f)$. We may write $f = g - h$, with $g, h \in \mathscr{P}$ (for instance let f_0 be any function in $\mathscr{P}$, and define $g = f^+ + f_0$, $h = f^- + f_0$). We propose to define

$$I(f) = I_0(g) - I_0(h),$$

hence we must check that if also $f = g' - h'$ with $g', h' \in \mathscr{P}$, then

$$I_0(g') - I_0(h') = I_0(g) - I_0(h);$$

indeed, this follows from the relation $g + h' = h + g'$ and property (2). It is easy to see that I has the required properties. ∎

In this section it is the following properties of the functions f in $\mathscr{P}$ which are relevant:

(*i*) f is bounded,

(*ii*) $f \geq 0$,

(*iii*) f has compact support,

(*iv*) there exists a nonempty open set U (depending on f) and a real number $\eta > 0$ such that $f(x) \geq \eta$ for all x in U.

Thus, the functions under consideration are bounded, nonnegative, zero outside some compact set, and bounded away from zero on some nonempty open set; in this section we make no explicit use of continuity.

We now launch a series of lemmas which culminate (in the next section) in a proof of the existence of a functional I_0 having the properties (1) through (4) of Lemma 1.

Lemma 2. *If $f, g \in \mathscr{P}$, there exist elements $s_1, \ldots, s_n$ of G and positive real numbers $c_1, \ldots, c_n$ such that*

$$f \leq \sum_1^n c_i g_{s_i}.$$

Proof. Let C be a compact set such that $f = 0$ on $G - C$. Let U be a nonempty open set such that g is bounded below on U, say $g(x) \geq \eta > 0$ for all x in U. Since the sets sU $(s \in G)$ are clearly an open covering of G, by compactness we have

$$C \subset \bigcup_1^n s_i U$$

for suitable $s_1, \ldots, s_n$ in G. Let $M = \text{LUB} f$, and define $c_i = M/\eta$ for $i = 1, \ldots, n$. We assert that

(*) $$f(x) \leq \sum_1^n c_i g_{s_i}(x)$$

for all x in G. This is clear if $x \in G - C$. If $x \in C$, then $x \in s_j U$ for some j, and so

$$s_j^{-1}x \in U, \qquad g(s_j^{-1}x) \geq \eta, \qquad c_j g(s_j^{-1}x) \geq M;$$

thus $c_j g_{s_j}(x) \geq M \geq f(x)$, and the inequality $(*)$ follows at once. ∎

Under the circumstances of Lemma 2, we shall say that we have a **covering** of f by g. It follows from Lemma 2 that if I is a Haar integral and $g \in \mathscr{P}$, necessarily $I(g) > 0$. Indeed, let f be any function in $\mathscr{P}$ such that $I(f) > 0$ (such an f must exist since I is not identically zero), and consider a covering

$$f \leq \sum_1^n c_i g_{s_i}$$

of f by g (such a covering exists by Lemma 2); then

$$0 < I(f) \leq \sum_1^n c_i I(g_{s_i}) = \left(\sum_1^n c_i\right) I(g),$$

and so $I(g) > 0$.

Given any $f, g \in \mathscr{P}$, in view of Lemma 2 we may make the following definition:

$$(f:g) = \mathrm{GLB}\left\{\sum_1^n c_i : f \leq \sum_1^n c_i g_{s_i},\ c_i > 0,\ s_i \in G\right\}.$$

So to speak $(f:g)$ is the "ratio of f to g"; it is a nonnegative real number, indeed $(f:g) > 0$ by the following lemma:

Lemma 3. *Suppose $f, g \in \mathscr{P}$, and let $M = \mathrm{LUB}\, f$, $m = \mathrm{LUB}\, g$. Then $(f:g) \geq M/m$.*

Proof. Suppose $f \leq \sum_1^n c_i g_{s_i}$ is any covering of f by g. For any $x \in G$ one has

$$f(x) \leq \sum_1^n c_i g(s_i^{-1}x) \leq m \sum_1^n c_i,$$

hence $M \leq m \sum_1^n c_i$. Thus $\sum_1^n c_i \geq M/m$, and our assertion follows on taking GLB over all coverings of f by g. ∎

Lemma 4. *If $f, g, h \in \mathscr{P}$, $c > 0$, and $s \in G$, then:*

(1) $(f_s:g) = (f:g)$

(2) $(cf:g) = c(f:g)$

(3) $(f + g:h) \leq (f:h) + (g:h)$

(4) $f \leq g$ *implies* $(f:h) \leq (g:h)$.

Proof. (1) If $f \leq \sum_1^n c_i g_{s_i}$, then $f_s \leq \sum_1^n c_i g_{ss_i}$, hence

$$(f_s:g) \leq \sum_1^n c_i;$$

taking GLB over all coverings of f by g, we have $(f_s:g) \leq (f:g)$.

Applying this result to f_t in place of f, we have $(f_{st}:g) \le (f_t:g)$; in particular for $t = s^{-1}$, this yields $(f:g) \le (f_t:g)$, and $(f:g) \le (f_s:g)$ results on replacing s by s^{-1}.

(2) If $f \le \sum_1^n c_i g_{s_i}$, then $cf \le \sum_1^n cc_i g_{s_i}$, and so

$$(cf:g) \le \sum_1^n cc_i = c \sum_1^n c_i;$$

taking GLB over all coverings of f by g, we have $(cf:g) \le c(f:g)$. The reverse inequality results from replacing f by $(1/c)f$, and then c by $1/c$.

(3) If $f \le \sum_1^n c_i h_{s_i}$ and $g \le \sum_1^m d_j h_{t_j}$, then

$$f + g \le \sum_1^n c_i h_{s_i} + \sum_1^m d_j h_{t_j},$$

hence

$$(f + g:h) \le \sum_1^n c_i + \sum_1^m d_j;$$

our assertion follows on taking independent GLB's over all coverings of f by h and of g by h.

(4) This is clear. ∎

Lemma 5. *If $f, g, h \in \mathscr{P}$, then $(f:h) \le (f:g)(g:h)$.*

Proof. If $f \le \sum_1^n c_i g_{s_i}$ and $g \le \sum_1^m d_j h_{t_j}$, then

$$f \le \sum_i c_i\Big(\sum_j d_j (h_{t_j})_{s_i}\Big) = \sum_{i,j} c_i d_j h_{s_i t_j},$$

and so

$$(f:h) \le \sum_{i,j} c_i d_j = \Big(\sum_i c_i\Big)\Big(\sum_j d_j\Big);$$

our assertion follows on taking independent GLB's over all coverings of f by g and of g by h. ∎

Lemma 6. *If $f, g, h \in \mathscr{P}$, then*

$$\frac{1}{(h:f)} \le \frac{(f:g)}{(h:g)} \le (f:h).$$

Proof. By Lemma 5, we have

$$(h:g) \le (h:f)(f:g) \qquad \text{and} \qquad (f:g) \le (f:h)(h:g).$$

It is of future importance to note that the extreme members of the inequalities in the conclusion are independent of g. ∎

Finally, we observe that $(f:f) = 1$ for every f in $\mathscr{P}$. Indeed, $f \le 1 \cdot f$ shows that $(f:f) \le 1$; on the other hand, Lemma 5 implies $(f:f) \le (f:f)(f:f)$, and hence $(f:f) \ge 1$.

74. Existence of a Haar Integral

Fix $f_0 \in \mathscr{P}$, where $\mathscr{P}$ is the class of functions described in the preceding section; f_0 will play the role of a "reference function," and is to be fixed through Sect. 76, that is, during the proof of existence and uniqueness of the Haar integral.

For each pair of functions f, g in $\mathscr{P}$, we define

$$A_f(g) = \frac{(f:g)}{(f_0:g)}.$$

By Lemma 6 (in Sect. 73), we have

$$(f_0:f)^{-1} \le A_f(g) \le (f:f_0);$$

thus, if X_f is the closed interval

$$X_f = [(f_0:f)^{-1}, (f:f_0)],$$

we have $A_f(g) \in X_f$ for all $f, g \in \mathscr{P}$. Let X be the product topological space

$$X = \prod_{f \in \mathscr{P}} X_f;$$

then each $g \in \mathscr{P}$ determines a point of X, which we denote $A(g)$, whose fth coordinate is $A_f(g)$. The space X is compact, by Tychonoff's theorem.

A real-valued function g on G is said to be **symmetric** in case

$$g(x^{-1}) = g(x)$$

for all x in G. For each neighborhood V of the neutral element e, we denote by F_V the set of all points $A(g)$ of X for which $g \in \mathscr{P}$, $g = 0$ on $G - V$, and g is symmetric. Thus,

$$F_V = \{A(g): g \in \mathscr{P},\ g \text{ symmetric},\ g = 0 \text{ outside } V\}.$$

The reason for considering symmetric functions will not be clear until Sect. 76 (see the proof of Lemma 15).

Lemma 7. *For each neighborhood V of e, F_V is a nonempty subset of X.*

Proof. Let W be a symmetric neighborhood of e such that $W \subset V$, and let f be any function in $\mathscr{P}$ such that $f = 0$ on $G - W$. The function

$$g(x) = f(x) + f(x^{-1})$$

is symmetric, belongs to $\mathscr{P}$, and vanishes outside W, and therefore $A(g) \in F_V$. ∎

Lemma 8. *There exists a point $(I_f)_{f \in \mathscr{P}}$ of X with the following property: if $f_1, \ldots, f_n \in \mathscr{P}$, V is a neighborhood of e, and $\varepsilon > 0$, then there exists a symmetric function g in $\mathscr{P}$ such that g vanishes outside V and*

$$\left| \frac{(f_i : g)}{(f_0 : g)} - I_{f_i} \right| < \varepsilon$$

for $i = 1, \ldots, n$.

Proof. If U and V are any two neighborhoods of e, then

$$F_{U \cap V} \subset F_U \cap F_V$$

(a function which vanishes outside $U \cap V$ also vanishes outside U and outside V). It follows from Lemma 7 that the sets F_V have the "finite intersection property"; so, therefore, do their closures in the space X. Since X is compact, there exists a point

$$(I_f)_{f \in \mathscr{P}}$$

which is in the closure of every F_V. Thus, given $f_1, \ldots, f_n \in \mathscr{P}$, $\varepsilon > 0$, and a neighborhood V of e, there is a symmetric function g in $\mathscr{P}$ which vanishes outside V, such that $A(g)$ is within ε of $(I_f)_{f \in \mathscr{P}}$ in each of the coordinates $f_1, \ldots, f_n$, that is,

$$|A_{f_i}(g) - I_{f_i}| < \varepsilon$$

for $i = 1, \ldots, n$. ∎

Let us fix a point $(I_f)_{f \in \mathscr{P}}$ having the property in Lemma 8. (Actually such a point is necessarily unique, as a consequence of the uniqueness theorem in Sect. 76, but at the moment we settle for the existence of such a point.) The rest of the section is devoted to showing that the functional $f \to I_f$ satisfies the conditions of Lemma 1, and so may be extended to a Haar integral.

Lemma 9. *If $f, f' \in \mathscr{P}$, $c > 0$, and $s \in G$, then:*

(1) $I_f > 0$

(2) $I_{f+f'} \leq I_f + I_{f'}$

(3) $I_{cf} = cI_f$

(4) $I_{f_s} = I_f$.

Proof. (1) Since $I_f \in X_f$, it is clear that $I_f > 0$.

(2) Given any $\varepsilon > 0$, it will suffice to show that

$$I_{f+f'} < I_f + I_{f'} + 3\varepsilon.$$

Let $f_1 = f$, $f_2 = f'$, $f_3 = f + f'$, and let $V = G$; by Lemma 8 there exists a function g in $\mathscr{P}$ such that all three of the following inequalities hold:

$$\left|\frac{(f:g)}{(f_0:g)} - I_f\right| < \varepsilon,$$

$$\left|\frac{(f':g)}{(f_0:g)} - I_{f'}\right| < \varepsilon,$$

$$\left|\frac{(f+f':g)}{(f_0:g)} - I_{f+f'}\right| < \varepsilon.$$

Quoting part (3) of Lemma 4 at the appropriate step, we have

$$I_{f+f'} < \frac{(f+f':g)}{(f_0:g)} + \varepsilon \leq \frac{(f:g) + (f':g)}{(f_0:g)} + \varepsilon$$

$$= \frac{(f:g)}{(f_0:g)} + \frac{(f':g)}{(f_0:g)} + \varepsilon < (I_f + \varepsilon) + (I_{f'} + \varepsilon) + \varepsilon.$$

(3) In Lemma 8, take $f_1 = f$, $f_2 = cf$, and $V = G$; given $\varepsilon > 0$, let g be a function in $\mathscr{P}$ such that both of the following inequalities hold:

$$\left|\frac{(f:g)}{(f_0:g)} - I_f\right| < \varepsilon,$$

$$\left|\frac{(cf:g)}{(f_0:g)} - I_{cf}\right| < \varepsilon.$$

Since $(cf:g) = c(f:g)$ by Lemma 4, a simple calculation yields

$$|I_{cf} - cI_f| < \varepsilon + c\varepsilon,$$

and our assertion follows on letting $\varepsilon \to 0$.

(4) In Lemma 8, take $f_1 = f$, $f_2 = f_s$, and $V = G$; given $\varepsilon > 0$, choose $g \in \mathscr{P}$ so that both of the following inequalities hold:

$$\left|\frac{(f:g)}{(f_0:g)} - I_f\right| < \varepsilon,$$

$$\left|\frac{(f_s:g)}{(f_0:g)} - I_{f_s}\right| < \varepsilon.$$

Since $(f_s:g) = (f:g)$ by Lemma 4, it follows that $|I_{f_s} - I_f| < 2\varepsilon$. ∎

If the inequality

$$I_{f+f'} \leq I_f + I_{f'}$$

could be tightened to equality, the existence of a Haar integral would follow at once from Lemma 1. The reader will have observed, in the course of the proof of Lemma 9, that in our applications of Lemma 8 we have not made use of the possibility of

(*i*) taking smaller and smaller neighborhoods V;
(*ii*) choosing g to be symmetric.

The possibility (*ii*) will play no role until Sect. 76, in the proof of *uniqueness* of the Haar integral. In the next two lemmas we exploit the possibility (*i*) to show that in fact

$$I_{f+f'} = I_f + I_{f'},$$

and this is the heart of the proof of the *existence* of a Haar integral.

Lemma 10. *One has $I_{fh} + I_{fh'} \le I_f$, provided all of the functions f, h, h', fh, fh' belong to $\mathscr{P}$, and $h + h' \le 1$.*

Proof. Given any $\varepsilon > 0$, it will suffice to show that

$$I_{fh} + I_{fh'} \le (I_f + \varepsilon)(1 + 2\varepsilon) + 2\varepsilon.$$

By 71.1, there exists a neighborhood V of e such that both of the inequalities

$$|h(x) - h(s)| \le \varepsilon,$$

$$|h'(x) - h'(s)| \le \varepsilon$$

hold provided $x \in sV$. Let g be any function in $\mathscr{P}$ which vanishes outside V; evidently g_s vanishes outside sV, for each $s \in G$.

If $f \le \sum_1^n c_i g_{s_i}$ is any covering of f by g, we assert that

$$(*) \qquad fh \le \sum_1^n c_i[h(s_i) + \varepsilon]g_{s_i}.$$

The inequality (*) certainly holds at any point x such that $f(x) = 0$. If $f(x) > 0$, then $g_{s_j}(x) > 0$ for some j; for such a j, necessarily $x \in s_jV$, and so $h(x) \le h(s_j) + \varepsilon$ by the choice of V. Now,

$$f(x) \le \sum_1^n c_i g_{s_i}(x) = \sum{}' c_j g_{s_j}(x),$$

where the sum $\sum'$ is extended over all indices j for which $g_{s_j}(x) > 0$. Clearly

$$\begin{aligned} f(x)h(x) \le \sum{}' c_j g_{s_j}(x)h(x) &\le \sum{}' c_j[h(s_j) + \varepsilon]g_{s_j}(x) \\ &= \sum_1^n c_i[h(s_i) + \varepsilon]g_{s_i}(x), \end{aligned}$$

and the inequality is established. Similarly,

$$(**)\qquad fh' \leq \sum_1^n c_i[h'(s_i) + \varepsilon]g_{s_i}.$$

It follows from (*) and (**) that

$$(fh:g) \leq \sum_1^n c_i[h(s_i) + \varepsilon],$$

$$(fh':g) \leq \sum_1^n c_i[h'(s_i) + \varepsilon],$$

and so

$$(fh:g) + (fh':g) \leq \sum_1^n c_i[h(s_i) + h'(s_i) + 2\varepsilon];$$

since $h + h' \leq 1$, it follows that

$$(fh:g) + (fh':g) \leq \left(\sum_1^n c_i\right)(1 + 2\varepsilon),$$

and taking GLB over all coverings of f by g we have

$$(fh:g) + (fh':g) \leq (f:g)(1 + 2\varepsilon).$$

Dividing the latter inequality by $(f_0:g)$, we obtain

$$(***)\qquad A_{fh}(g) + A_{fh'}(g) \leq A_f(g)\cdot(1 + 2\varepsilon).$$

Summarizing: given any $\varepsilon > 0$, a neighborhood V of e has been found so that (***) holds for every g in $\mathscr{P}$ which vanishes outside V. According to Lemma 8 there is a g of this sort for which all three of the following inequalities hold:

$$|A_f(g) - I_f| \leq \varepsilon,$$

$$|A_{fh}(g) - I_{fh}| \leq \varepsilon,$$

$$|A_{fh'}(g) - I_{fh'}| \leq \varepsilon.$$

Combining these with (***), we have finally

$$\begin{aligned} I_{fh} + I_{fh'} &\leq A_{fh}(g) + A_{fh'}(g) + 2\varepsilon \\ &\leq A_f(g)\cdot(1 + 2\varepsilon) + 2\varepsilon \leq (I_f + \varepsilon)(1 + 2\varepsilon) + 2\varepsilon. \quad \blacksquare \end{aligned}$$

Lemma 11. *One has* $I_{f+f'} = I_f + I_{f'}$ *for all* f, f' *in* $\mathscr{P}$.

Proof. In view of part (2) of Lemma 9, it will suffice to show that $I_f + I_{f'} \leq I_{f+f'}$.

Let C be a compact set such that both f and f' vanish outside C. By 54.3 there exists an f'' in $\mathscr{P}$ such that $f''(x) = 1$ for all $x \in C$. Given any $\varepsilon > 0$, it is sufficient to show that

$$I_f + I_{f'} \leq I_{f+f'} + \varepsilon I_{f''}.$$

Defining $F = f + f' + \varepsilon f''$, we have $F \in \mathscr{P}$, and $F(x) \geq \varepsilon$ when $x \in C$. It follows that the function $1/F(x)$ is defined and continuous on C. Let h be the function defined by the formulas

$$h(x) = \begin{cases} \dfrac{f(x)}{F(x)} & \text{for } x \in C, \\ 0 & \text{for } x \in G - C. \end{cases}$$

If $A = \{x : f(x) = 0\}$, then A is a closed subset of G, and $A \cup C = G$; since h is continuous (and $h = 0$) on the closed set A, and h is continuous (and $h = f/F$) on the closed set C, it is an exercise in elementary topology that h is continuous on G, and so $h \in \mathscr{P}$. Similarly the formulas

$$h'(x) = \begin{cases} \dfrac{f'(x)}{F(x)} & \text{for } x \in C, \\ 0 & \text{for } x \in G - C, \end{cases}$$

define a function h' in $\mathscr{P}$.

Observe that $h + h' \leq 1$. Indeed, $h + h'$ vanishes on $G - C$, and for $x \in C$ one has

$$h(x) + h'(x) = \frac{f(x) + f'(x)}{f(x) + f'(x) + \varepsilon} < 1.$$

Since $Fh = f$ and $Fh' = f'$, it follows from Lemma 10 that

$$I_f + I_{f'} \leq I_F;$$

moreover,

$$I_F = I_{(f+f')+\varepsilon f''} \leq I_{f+f'} + \varepsilon I_{f''}$$

by Lemma 9, and so

$$I_f + I_{f'} \leq I_{f+f'} + \varepsilon I_{f''}. \quad \blacksquare$$

Theorem 1. *Every locally compact group G possesses a Haar integral.*

Indeed, given any f_0 in $\mathscr{P}$, there exists a Haar integral I with the following property: if $f_1, \ldots, f_n \in \mathscr{P}$, V is a neighborhood of e, and $\varepsilon > 0$, then there exists a symmetric function g in $\mathscr{P}$ such that g vanishes outside V and

$$\left| \frac{(f_i : g)}{(f_0 : g)} - I(f_i) \right| < \varepsilon$$

for $i = 1, \ldots, n$.

Proof. With notation as in Lemma 8, we define

$$I_0(f) = I_f$$

for all f in $\mathscr{P}$. By Lemmas 9 and 11, I_0 satisfies the hypothesis of Lemma 1, and so I_0 may be extended to a Haar integral I. ▮

Incidentally, it is clear from Theorem 1 that $I(f_0) = 1$. (Of course I depends on the reference function f_0.)

75. A Topological Lemma

The following result from general topology is needed for the proof of uniqueness of the Haar integral to be given in the next section:

Lemma 12. *If X is a locally compact space, C is a compact set, and $U_1, \ldots, U_n$ are open sets such that*

$$C \subset U_1 \cup \cdots \cup U_n,$$

then there exist continuous real-valued functions $f_1, \ldots, f_n$ on X with compact support, such that $0 \leq f_i \leq 1, f_i = 0$ on $X - U_i$, and

$$\sum_1^n f_i(x) = 1$$

for all x in C.

We shall approach the proof through two sublemmas, each of which is a special case of Lemma 12:

Sublemma 1. *If D is a compact space, and $V_1, \ldots, V_n$ are open sets in D such that*

$$D = V_1 \cup \cdots \cup V_n,$$

then there exist continuous real-valued functions $g_1, \ldots, g_n$ on D such that $0 \leq g_i \leq 1$, $g_i = 0$ on $D - V_i$, and

$$\sum_1^n g_i = 1.$$

Proof. By an easy induction on the Lemma to 63.1, there exist compact sets $D_1, \ldots, D_n$ such that

$$D = D_1 \cup \cdots \cup D_n$$

and $D_i \subset V_i$. By 54.3 there exists, for each i, a continuous real-valued function h_i on D such that $0 \le h_i \le 1$, $h_i = 1$ on D_i, and $h_i = 0$ on $D - V_i$. Define

$$h = \sum_1^n h_i;$$

since $h(x) > 0$ for all x in D, the function $1/h$ is defined and continuous on D. The functions $g_i = h_i/h$ clearly meet all requirements. ∎

Sublemma 2. *If D is a compact space, C is a closed subset of D, and $V_1, \ldots, V_n$ are open sets in D such that*

$$C \subset V_1 \cup \cdots \cup V_n,$$

then there exist continuous real-valued functions $g_1, \ldots, g_n$ on D such that $0 \le g_i \le 1$, $g_i = 0$ on $D - V_i$, and

$$\sum_1^n g_i(x) = 1$$

for all x in C.

Proof. Setting $V_{n+1} = D - C$, we have

$$D = V_1 \cup \cdots \cup V_n \cup V_{n+1},$$

with all the V_i open. By Sublemma 1, there exist continuous real-valued functions $g_1, \ldots, g_n, g_{n+1}$ on D such that $0 \le g_i \le 1$, $g_i = 0$ on $D - V_i$, and

$$\sum_1^{n+1} g_i(x) = 1$$

for all x in D. If in particular $x \in C = D - V_{n+1}$, then $g_{n+1}(x) = 0$, and so

$$\sum_1^n g_i(x) = 1. \quad ∎$$

Proof of Lemma 12. Let $C_1, \ldots, C_n$ be compact sets in X such that

$$C = C_1 \cup \cdots \cup C_n$$

and $C_i \subset U_i$. Since X is locally compact, it is easy to see that for each i there exists an open set V_i and a compact set D_i such that

$$C_i \subset V_i \subset D_i \subset U_i$$

(one could also quote 56.1). Let $D = D_1 \cup \cdots \cup D_n$.

Regard D as a compact space in the relative topology. Since the V_i are also open sets in D, and C is a closed set in D, and since

$$C \subset V_1 \cup \cdots \cup V_n,$$

it follows from Sublemma 2 that there exist continuous real-valued functions $g_1, \ldots, g_n$ on D such that $0 \leq g_i \leq 1$, $g_i = 0$ on $D - V_i$, and

$$\sum_1^n g_i(x) = 1$$

for all x in C. Define

$$f_i(x) = \begin{cases} g_i(x) & \text{for } x \in D, \\ 0 & \text{for } x \in X - D; \end{cases}$$

since f_i is continuous (indeed $f_i = 0$) on the closed set

$$X - V_i = (X - D) \cup (D - V_i),$$

and f_i is continuous (indeed $f_i = g_i$) on the closed set D, and since

$$(X - V_i) \cup D = X,$$

it follows that f_i is continuous on X. Obviously f_i has compact support, and $0 \leq f_i \leq 1$. Finally, if $x \in C$, then

$$\sum_1^n f_i(x) = \sum_1^n g_i(x) = 1. \quad \blacksquare$$

76. Uniqueness of the Haar Integral

We resume the notations in Sect. 74. Thus, a function f_0 in $\mathscr{P}$ is fixed, and I is the Haar integral whose existence is asserted in 74.1.

We suppose now that J is *any* Haar integral for G; we are to show that $J = cI$ for a suitable constant $c > 0$.

Lemma 13. *One has*

$$\frac{J(f)}{J(g)} \leq (f:g)$$

for all f, g in $\mathscr{P}$.

Proof. Note that $J(g) > 0$ by the remarks following Lemma 2. Let

$$f \leq \sum_1^n c_i g_{s_i}$$

be any covering of f by g. Then

$$J(f) \leq \sum_1^n c_i J(g_{s_i}) = \left(\sum_1^n c_i\right) J(g),$$

and our assertion follows on taking GLB over all coverings of f by g. $\blacksquare$

Lemma 14. *If C is a nonempty compact subset of G, and W is any open neighborhood of e, then there exist elements $s_1, \ldots, s_n$ of G and functions $f_1, \ldots, f_n$ in $\mathscr{P}$ such that*

(1) $C \subset s_1W \cup \cdots \cup s_nW.$

(2) $f_i = 0$ *on* $G - s_iW.$

(3) $\sum_1^n f_i(x) = 1$ *for all x in C.*

Proof. Since the sW are an open covering of G, it follows from the compactness of C that there exist elements $s_1, \ldots, s_n$ satisfying (1). Let us assume, moreover, that the s_iW are an irredundant covering of C, in other words that no s_iW can be omitted without invalidating the inclusion (1).

By Lemma 12, there exist continuous real-valued functions $f_1, \ldots, f_n$ on G with compact support, such that $0 \le f_i \le 1$, and conditions (2) and (3) are satisfied. It remains to show that no f_i is identically zero. Suppose to the contrary, say, that $f_1 = 0$. If $x \in C$, we have

$$\sum_2^n f_i(x) = 1$$

by (3), hence $f_j(x) > 0$ for some $j \ge 2$, and therefore $x \in s_jW$ by (2); thus

$$C \subset s_2W \cup \cdots \cup s_nW,$$

contrary to the assumed irredundancy. ∎

Lemma 15. *Suppose C is a compact set, and f, f' are functions in $\mathscr{P}$ such that $f = 0$ on $G - C$ and $f' = 1$ on C. Then, given any $\varepsilon > 0$, there exists a neighborhood U of e such that*

$$(f:g) \le \varepsilon(f':g) + \frac{J(f)}{J(g)}$$

for every symmetric g in $\mathscr{P}$ which vanishes outside U.

Proof. By 71.1 we may choose a neighborhood U of e so that

$$|f(x) - f(y)| \le \varepsilon$$

whenever $x^{-1}y \in U$. Let g be any symmetric function in $\mathscr{P}$ which vanishes outside U. We assert that

(1) $$[f(x) - \varepsilon]g_x \le fg_x$$

for all x in G. Indeed, suppose $x, y \in G$; if $x^{-1}y \in G - U$, then both sides of the desired inequality

$$[f(x) - \varepsilon]g(x^{-1}y) \le f(y)g(x^{-1}y)$$

are 0, whereas if $x^{-1}y \in U$ this inequality results from $|f(x) - f(y)| \le \varepsilon$.

Applying J to the inequality (1), and noting that $J(g_x) = J(g)$, we have

$$f(x) - \varepsilon \le \frac{J(fg_x)}{J(g)} \tag{2}$$

for all x in G.

Let $\eta > 0$ (ultimately we shall let $\eta \to 0$), and choose a symmetric open neighborhood W of e such that

$$|g(y) - g(z)| \le \eta$$

whenever $z^{-1}y \in W$ (71.1). By Lemma 14 there exist elements $s_1, \ldots, s_n$ of G, and functions $h_1, \ldots, h_n$ in $\mathscr{P}$ such that $h_i = 0$ on $G - s_iW$, and

$$\sum_1^n h_i(x) = 1$$

for all x in C. Since $f = 0$ on $G - C$, clearly

$$f(x) \sum_1^n h_i(x) = f(x)$$

for all x in G, thus

$$f = \sum_1^n fh_i.$$

Then for each x in G we have

$$fg_x = \sum_1^n fh_ig_x;$$

applying J,

$$J(fg_x) = \sum_1^n J(fh_ig_x) \tag{3}$$

for all x in G.

We assert that for each i,

$$h_ig_x \le [g_x(s_i) + \eta]h_i \tag{4}$$

for all x in G. Indeed, suppose $x, y \in G$. If $y \in G - s_iW$, then both sides of the desired inequality

$$h_i(y)g_x(y) \le [g_x(s_i) + \eta]h_i(y)$$

are 0. If instead $y \in s_iW$, that is if $s_i^{-1}y \in W$, then also

$$(x^{-1}s_i)^{-1}(x^{-1}y) = s_i^{-1}xx^{-1}y = s_i^{-1}y \in W,$$

and so

$$|g(x^{-1}y) - g(x^{-1}s_i)| \le \eta$$

by the choice of W; thus

$$|g_x(y) - g_x(s_i)| \leq \eta,$$

and the desired inequality follows at once.

Multiplying through (4) by f, and applying J, we have

$$J(fh_i g_x) \leq [g_x(s_i) + \eta]J(fh_i) \tag{5}$$

for all x in G, and $i = 1, \ldots, n$.

Combining (2), (3), and (5), we have

$$f(x) - \varepsilon \leq \frac{1}{J(g)} \sum_1^n [g_x(s_i) + \eta]J(fh_i)$$

for all x in G; defining

$$c_i = \frac{J(fh_i)}{J(g)},$$

it follows that

$$f(x) - \varepsilon \leq \sum_1^n c_i[g_x(s_i) + \eta], \tag{6}$$

for all x in G, and

$$\sum_1^n c_i = \frac{J(f)}{J(g)} \tag{7}$$

results from the relation $f = \sum_1^n fh_i$.

Since g is *symmetric*, evidently $g_x(s) = g_s(x)$ for all $x, s \in G$; it then follows from (6) and (7) that

$$f(x) - \varepsilon \leq \sum_1^n c_i[g_{s_i}(x) + \eta] = \sum_1^n c_i g_{s_i}(x) + \eta \frac{J(f)}{J(g)}$$

for all x in G, and so

$$f - \varepsilon \leq \sum_1^n c_i g_{s_i} + \eta \frac{J(f)}{J(g)}. \tag{8}$$

Multiplying through (8) by f', we have

$$f \leq \left[\varepsilon + \eta \frac{J(f)}{J(g)}\right] f' + \sum_1^n c_i g_{s_i}$$

(because $f'(x) = 1$ when $x \in C$ and $f(x) = 0$ when $x \in G - C$); therefore

$$f \leq \left[\varepsilon + \eta \frac{J(f)}{J(g)}\right] f' + \sum{}' c_i g_{s_i}, \tag{9}$$

where the sum $\sum'$ is extended over all indices i for which $c_i > 0$. It follows from (9) and Lemma 4 that

$$(f:g) \leq \left[\varepsilon + \eta \frac{J(f)}{J(g)}\right](f':g) + \sum' c_i(g_{s_i}:g);$$

since $(g_{s_i}:g) = (g:g) = 1$ (see the remark at the end of Sect. 73), and since

$$\sum' c_i = \sum_1^n c_i = J(f)/J(g),$$

it follows that

$$(f:g) \leq \left[\varepsilon + \eta \frac{J(f)}{J(g)}\right](f':g) + \frac{J(f)}{J(g)}; \tag{10}$$

letting $\eta \to 0$ in (10), the lemma is proved. ▮

Lemma 16. *Given any f in $\mathscr{P}$, there exists a neighborhood U of e and a constant $M > 0$ such that*

$$(f:g)J(g) \leq M$$

for every symmetric g in $\mathscr{P}$ which vanishes outside U.

Proof. Let C be a compact set such that f vanishes outside C. By 54.3, choose a function f' in $\mathscr{P}$ such that $f' = 1$ on C. Choose $\varepsilon > 0$ so that $\varepsilon(f':f) < 1$. Let U be the neighborhood of e given by Lemma 15; thus, for every symmetric g in $\mathscr{P}$ which vanishes outside U, we have

$$(f:g) \leq \varepsilon(f':g) + \frac{J(f)}{J(g)},$$

that is,

$$(f:g)J(g) \leq \varepsilon(f':g)J(g) + J(f). \tag{*}$$

But $(f':g) \leq (f':f)(f:g)$ by Lemma 5; substituting in (*), we have

$$(f:g)J(g) \leq \varepsilon(f':f)(f:g)J(g) + J(f),$$

that is,

$$(f:g)J(g)[1 - \varepsilon(f':f)] \leq J(f).$$

Since the quantity in brackets is > 0, and is independent of g, our assertion is clear. ▮

Theorem 1. *Every locally compact group G possesses a Haar integral I. Moreover, I is unique up to proportionality; that is, if J is any Haar integral for G, then $J = cI$ for a suitable constant $c > 0$.*

Proof. Let I be the Haar integral given by 74.1, where f_0 is the "reference function" fixed at the beginning of Sect. 74. Let J be an arbitrary Haar integral for G. Given any f in $\mathscr{P}$, it will suffice to show that $J(f) = J(f_0)I(f)$.

Let C be a compact set outside of which f vanishes, and choose (54.3) a function f' in $\mathscr{P}$ such that $f' = 1$ on C.

Similarly, let C_0 be a compact set outside of which f_0 vanishes, and choose f_0' in $\mathscr{P}$ so that $f_0' = 1$ on C_0.

By Lemma 16, there exists a neighborhood U of e and a constant $N > 0$ such that

$$(f':g)J(g) \leq N$$

for every symmetric g in $\mathscr{P}$ which vanishes outside U. Similarly there exists a neighborhood U_0 of e and a constant $N_0 > 0$ such that

$$(f_0':g)J(g) \leq N_0$$

for every symmetric g in $\mathscr{P}$ which vanishes outside U_0. Defining

$$V = U \cap U_0 \qquad \text{and} \qquad M = \max\{N, N_0\},$$

we have

$$(1) \qquad (f':g)J(g) \leq M \qquad \text{and} \qquad (f_0':g)J(g) \leq M$$

for every symmetric g in $\mathscr{P}$ which vanishes outside V.

Given any $\varepsilon > 0$. By Lemma 15 there exists a neighborhood W of e such that

$$(f:g) \leq \varepsilon(f':g) + \frac{J(f)}{J(g)}$$

for every symmetric g in $\mathscr{P}$ which vanishes outside W; since

$$\frac{J(f)}{J(g)} \leq (f:g)$$

by Lemma 13, it follows that

$$(2) \qquad J(f) \leq (f:g)J(g) \leq \varepsilon(f':g)J(g) + J(f)$$

for all symmetric $g \in \mathscr{P}$ which vanish outside W.

Similarly, there exists a neighborhood W_0 of e such that

$$(3) \qquad J(f_0) \leq (f_0:g)J(g) \leq \varepsilon(f_0':g)J(g) + J(f_0)$$

for all symmetric $g \in \mathscr{P}$ which vanish outside W_0.

Let $A = V \cap W \cap W_0$. Then if g is a symmetric function in $\mathscr{P}$ which vanishes outside A, all of the inequalities (1) through (3) hold. Dividing (2) by (3), we have

$$\frac{J(f)}{\varepsilon(f_0':g)J(g) + J(f_0)} \leq \frac{(f:g)}{(f_0:g)} \leq \frac{\varepsilon(f':g)J(g) + J(f)}{J(f_0)},$$

and it follows from (1) that

$$\frac{J(f)}{\varepsilon M + J(f_0)} \leq \frac{(f:g)}{(f_0:g)} \leq \frac{\varepsilon M + J(f)}{J(f_0)}. \tag{4}$$

It is important to note that ε was chosen after M.

It is clear from 74.1 that we may choose a sequence g_n of symmetric functions in $\mathscr{P}$ such that each g_n vanishes outside A, and

$$\frac{(f:g_n)}{(f_0:g_n)} \to I(f);$$

passing to the limit in (4) we obtain

$$\frac{J(f)}{\varepsilon M + J(f_0)} \leq I(f) \leq \frac{\varepsilon M + J(f)}{J(f_0)},$$

and the proof is concluded by letting $\varepsilon \to 0$. ∎

77. The Modular Function

For the rest of the chapter, I denotes a fixed Haar integral for the locally compact group G (74.1). Thus, I is a positive linear form, not identically zero, on the class $\mathscr{L}$ of continuous real-valued functions with compact support, such that $I(f_s) = I(f)$ for all f in $\mathscr{L}$ and s in G. We know further that every Haar integral for G is proportional to I (76.1). *This is all we need to extract from Sects.* 74–76, *the supporting details being immaterial from now on.*

If f is a real-valued function on G, and $t \in G$, the **right translate** of f by t is the function f^t on G defined by

$$f^t(x) = f(xt).$$

Since $x \to xt$ is a homeomorphism of G, it is clear that f^t is continuous if and only if f is continuous. Similarly for the property of being a Baire function, or a Borel function, or having compact support. In particular, $f \in \mathscr{L}$ implies $f^t \in \mathscr{L}$.

For each $t \in G$, the correspondence $f \to f^t$ defines a bijective linear mapping in the class of all real-valued functions on G; specifically, $(f + g)^t = f^t + g^t$, $(cf)^t = cf^t$, and $f = (f^s)^t$ with $s = t^{-1}$. If $f \geq 0$, then $f^t \geq 0$. If f is bounded, then so is f^t, and $\|f^t\|_\infty = \|f\|_\infty$. One has $f^e = f$ for all f, and $(f^s)^t = f^{ts}$ for all $s, t \in G$.

Finally, we have the following relationship between left and right translation: $(f_s)^t = (f^t)_s$. This is immediate from the associative law in G.

Theorem 1. *For each t in G, there exists a unique real number $\Delta(t) > 0$ such that*

$$I(f^t) = \Delta(t)I(f)$$

for all f in $\mathscr{L}$.

Proof. Defining $J(f) = I(f^t)$ for all f in $\mathscr{L}$, it is clear that J is a positive linear form on $\mathscr{L}$, not identically zero. Moreover,

$$J(f_s) = I((f_s)^t) = I((f^t)_s) = I(f^t) = J(f)$$

for all f in $\mathscr{L}$ and s in G; thus J is a Haar integral, and so must be proportional to I. ∎

The function $t \to \Delta(t)$ defined by Theorem 1 is called the **modular function** for G. The next lemmas lead up to the assertion that Δ is a continuous homomorphism of G into the group of positive real numbers.

Lemma 1. *The modular function is a homomorphism of G into the group of positive real numbers; that is,*

$$\Delta(st) = \Delta(s)\,\Delta(t)$$

for all s, t in G. In particular, $\Delta(e) = 1$ and $\Delta(t^{-1}) = 1/\Delta(t)$.

Proof. Fix any f in $\mathscr{L}$ such that $I(f) \neq 0$, and calculate:

$$\Delta(st)I(f) = I(f^{st}) = I((f^t)^s) = \Delta(s)I(f^t) = \Delta(s)\,\Delta(t)I(f). \quad ∎$$

Lemma 2. *Let $f \in \mathscr{L}$. Given any $\varepsilon > 0$, there exists a neighborhood U of e such that*

$$\|f^t - f\|_\infty \leq \varepsilon$$

whenever $t \in U$. Briefly, $f^t \to f$ uniformly as $t \to e$.

Proof. By 71.1 there is a neighborhood U of e such that

$$|f(y) - f(x)| \leq \varepsilon$$

whenever $x^{-1}y \in U$. If $t \in U$, then for all $x \in G$ one has

$$x^{-1}(xt) = t \in U,$$

hence $|f(xt) - f(x)| \leq \varepsilon$, that is, $|f^t(x) - f(x)| \leq \varepsilon$. ∎

Lemma 3. *If φ is any positive linear form on $\mathscr{L}$, and C is a compact set, there exists a constant $M \geq 0$ such that*

$$|\varphi(f)| \leq M\|f\|_\infty$$

for every f in $\mathscr{L}$ which vanishes outside C.

Proof. Choose a function g in $\mathscr{L}$ such that $g \geq 0$, and $g = 1$ on C (54.3). If $f \in \mathscr{L}$ vanishes outside C, clearly $fg = f$; multiplying through the inequality

$$-\|f\|_\infty \leq f \leq \|f\|_\infty$$

by g, we have

$$-\|f\|_\infty g \leq f \leq \|f\|_\infty g,$$

hence

$$-\|f\|_\infty \varphi(g) \leq \varphi(f) \leq \|f\|_\infty \varphi(g).$$

The constant $M = \varphi(g)$ meets the requirements of the lemma. Incidentally, this is a lemma about locally compact spaces; the group structure is irrelevant here. ▮

Theorem 2. *The modular function $t \to \Delta(t)$ is a continuous homomorphism of G into the group of positive real numbers.*

Proof. In view of Lemma 1, it remains to prove that for each fixed $t_0 \in G$, we have $\Delta(t) \to \Delta(t_0)$ as $t \to t_0$. Since $\Delta(t) = \Delta(tt_0^{-1})\Delta(t_0)$ and $\Delta(e) = 1$, it is clearly sufficient to prove that $\Delta(s) \to 1$ as $s \to e$. Thus, given any $\varepsilon > 0$, the problem is to find a neighborhood U of e such that

$$|\Delta(s) - 1| \leq \varepsilon$$

whenever $s \in U$.

Fix any $f \in \mathscr{L}$ such that $I(f) = 1$, and let C be a compact set outside of which f vanishes. Let V be a fixed compact symmetric neighborhood of e, and define $D = CV$. Evidently D is compact.

We assert that if $t \in V$, then $f^t - f$ vanishes outside D. Indeed, if $f^t(x) - f(x) \neq 0$, that is if $f(xt) \neq f(x)$, then not both of $f(xt)$ and $f(x)$ can be 0; hence either $xt \in C$ or $x \in C$, thus $x \in Ct^{-1} \subset CV = D$ or $x \in C \subset D$.

Now, by Lemma 3 there is a constant $M \geq 0$ such that

$$|I(g)| \leq M\|g\|_\infty$$

for every g in $\mathscr{L}$ which vanishes outside D. In particular, we have

$$|I(f^t - f)| \leq M\|f^t - f\|_\infty$$

whenever $t \in V$. Since

$$I(f^t - f) = I(f^t) - I(f) = \Delta(t)I(f) - I(f) = \Delta(t) - 1,$$

we have

$$(1) \qquad |\Delta(t) - 1| \leq M \| f^t - f \|_\infty$$

whenever $t \in V$.

Given any $\varepsilon > 0$, there exists, by Lemma 2, a neighborhood W of e such that

$$(2) \qquad M \| f^t - f \|_\infty \leq \varepsilon$$

whenever $t \in W$. The neighborhood $U = V \cap W$ clearly meets our requirements. ∎

EXERCISES

1. If $f \in \mathscr{L}$, then $f_s \to f_t$ uniformly and $f^s \to f^t$ uniformly, as $s \to t$.

2. The locally compact group G is called *unimodular* in case $\Delta(t) = 1$ for all t in G. In order that G be unimodular, it is necessary and sufficient that $I(f') = I(f)$ for all f in $\mathscr{L}$, where $f'(x) = f(x^{-1})$. If G is Abelian, or compact, or discrete, then G is necessarily unimodular.

78. Haar Measure

For the rest of the chapter, we shall write μ for the unique regular Borel measure on G such that

$$I(f) = \int f \, d\mu$$

for all f in $\mathscr{L}$ (69.1), where I denotes the Haar integral which was fixed at the beginning of Sect. 77. In other words, μ is the regular Borel measure associated with the positive linear form I on $\mathscr{L}$.

A **Haar measure** on G is a regular Borel measure ρ such that (1) $\rho(sE) = \rho(E)$ for all Borel sets E and all s in G, and (2) ρ is not identically zero.

Theorem 1. *If E is any Borel set, and $s \in G$, then*

$$\mu(sE) = \mu(E),$$

thus μ is a Haar measure on G.

Proof. Fix $s \in G$, and let T be the homeomorphism $x \to s^{-1}x$ of G. If $f \in \mathscr{L}$, then $f(Tx) = f(s^{-1}x) = f_s(x)$; thus $f^T = f_s$, in the notation of Sect. 66. Since $I(f^T) = I(f_s) = I(f)$ for all f in $\mathscr{L}$, we conclude from the corollary of 66.6 that

$$\mu(T^{-1}(E)) = \mu(E)$$

for all Borel sets E; that is, $\mu(sE) = \mu(E)$ for all Borel sets E. ▮

Haar measure is unique up to a factor of proportionality:

Theorem 2. *If ρ is any Haar measure on G, there exists a constant $c > 0$ such that $\rho = c\mu$, that is,*

$$\rho(E) = c\mu(E)$$

for every Borel set E.

Proof. Let J be the positive linear form on $\mathscr{L}$ defined by

$$J(f) = \int f\,d\rho.$$

We assert that J is a Haar integral for G. Indeed, if $s \in G$, let T be the homeomorphism $x \to s^{-1}x$; thus $f^T = f_s$ for all $f \in \mathscr{L}$, as noted in the proof of Theorem 1. For every Borel set E, we have

$$\rho(T^{-1}(E)) = \rho(sE) = \rho(E);$$

it follows from the corollary of 66.5 that

$$\int f^T\,d\rho = \int f\,d\rho,$$

in other words $J(f_s) = J(f)$, for all f in $\mathscr{L}$.

Let $c > 0$ be a constant such that $J = cI$. Thus

$$\int f\,d\rho = c\int f\,d\mu$$

for all f in $\mathscr{L}$, and so $\rho = c\mu$ by 66.4 (with $\mu_1 = \rho$ and $\mu_2 = c\mu$). ▮

In view of Theorem 2, we shall say that "μ is Haar measure on G" instead of "μ is *a* Haar measure on G." The behavior of Haar measure under right translation is as follows:

Theorem 3. *If μ is Haar measure on G, then*

$$\mu(Es) = \Delta(s^{-1})\mu(E)$$

for every Borel set E, and all $s \in G$.

Proof. Fix $s \in G$, and let T be the homeomorphism $x \to xs$. Then $f^T(x) = f(Tx) = f(xs) = f^s(x)$, thus $f^T = f^s$. For all f in $\mathscr{L}$, we have

$$I(f^T) = I(f^s) = \Delta(s)I(f)$$

by 77.1, and so

$$\mu(T^{-1}(E)) = \Delta(s)\mu(E)$$

for every Borel set E by Corollary 2 of 69.2. Thus

$$\mu(Es^{-1}) = \Delta(s)\mu(E)$$

for all Borel sets E, and our assertion follows on replacing s by s^{-1}. ▌

For the rest of the chapter, we shall write ν for the Baire restriction of μ; thus, ν is the unique Baire measure on G such that

$$I(f) = \int f\,d\nu$$

for all f in $\mathscr{L}$ (66.3). We shall refer to ν as **restricted Haar measure** on G. Restricted Haar measure is uniquely determined, up to proportionality, by the property of being invariant under left translations:

Theorem 4. *If ν' is any Baire measure on G such that*

$$\nu'(sF) = \nu'(F)$$

for all Baire sets F and all $s \in G$, then ν' is proportional to ν.

Proof. If μ' is a (the) regular Borel measure on G which extends ν' (65.1), then $\mu'(sE) = \mu'(E)$ for all Borel sets E and all $s \in G$ (corollary of 62.3), and so μ' is proportional to μ by Theorem 2. ▌

Since the σ-ring of Baire sets in the product topological space $G \times G$ is precisely the Cartesian product of the σ-ring of Baire sets in G with itself (Lemma 2 of 56.3), since we are unable to make the analogous statement for Borel sets, and since Fubini's theorem is an indispensable measure theoretic tool, it is not surprising that on certain key occasions we shall be doing business with the Baire measure ν rather than the Borel measure μ (see the discussion of convolution in Sects. 81 and 84). In particular, 56.3 yields the following result (which really has nothing to do with the group structure):

Theorem 5. *If ν is restricted Haar measure on G, then $\nu \times \nu$ is a Baire measure on the product topological space $G \times G$.*

Consistent with the notations in Sect. 25, we write $\mathscr{L}^1(\mu)$ for the class of all μ-integrable Borel functions, and $\mathscr{L}^1(\nu)$ for the class of all

ν-integrable Baire functions. According to 66.1, a Baire function f is μ-integrable if and only if it is ν-integrable, and in this case

$$\int f\,d\mu = \int f\,d\nu.$$

In place of the phrase "μ-integrable Borel function" we shall prefer simply **integrable Borel function.** Similarly, in place of "ν-integrable Baire function," we shall prefer **integrable Baire function.** If f is an integrable Baire function, we may write either $\int f\,d\mu$ or $\int f\,d\nu$ depending on the context.

It will occasionally be suggestive (especially in contexts involving two or more variables) to write dx in place of $d\mu$ or $d\nu$. Thus we shall write

$$\int f(x)\,dx = \int f\,d\mu$$

when f is an integrable Borel function, and

$$\int f(x)\,dx = \int f\,d\nu$$

when f is an integrable Baire function. Whether dx stands for Haar measure or restricted Haar measure depends on whether the function being integrated is a Borel function or a Baire function, and this will always be made clear in the context.

EXERCISES

1. It can be shown that if ρ is any Borel measure on G such that $\rho(sE) = \rho(E)$ for all Borel sets E and all $s \in G$, then ρ is automatically regular, and hence is proportional to Haar measure μ. This follows from the completion regularity of the restricted Haar measure ν (see Sect. 90).

2. Strictly speaking, the terms Haar integral, Haar measure, restricted Haar measure, should be qualified by the word "left." Thus, a *right Haar integral* for G is a positive linear form K on $\mathscr{L}$, not identically zero, such that $K(f^t) = K(f)$ for all f in $\mathscr{L}$ and t in G. Every locally compact group possesses a right Haar integral, and this is unique up to a factor of proportionality; the "right" theory is trivially deducible from the "left" theory by considering the mapping $x \to x^{-1}$. In order that G be unimodular, it is necessary and sufficient that every (some) left Haar integral be a right Haar integral.

3. If μ is Haar measure for G, then G is unimodular if and only if

$$\mu(D) = \mu(D^{-1})$$

for every compact G_δ D, and in this case

$$\mu(E) = \mu(E^{-1})$$

for all Borel sets E.

4. In order that G be discrete (as a topological space), it is necessary and sufficient that $\mu(\{e\}) > 0$.

5. If ν_i is restricted Haar measure on the locally compact group G_i $(i = 1, 2)$, then $\nu_1 \times \nu_2$ is restricted Haar measure for the product topological group $G_1 \times G_2$. Moreover, if μ_i is Haar measure on G_i, then $\mu_1 \times \mu_2$ is an extension of $\nu_1 \times \nu_2$. It is noted in Sect. 90 that Haar measure on $G_1 \times G_2$ extends $\mu_1 \times \mu_2$.

6. If μ is Haar measure for G, and $\rho = \mu_w$ is the weakly Borel extension of μ described in Exercise 57.21, then:

(1) $$\rho(A) = \text{LUB}\,\{\rho(C) : C \subset A,\, C \text{ compact}\},$$

(2) $$\rho(sA) = \rho(A),$$

(3) $$\rho(As) = \Delta(s^{-1})\rho(A),$$

for all weakly Borel sets A and all s in G (cf. Exercise 62.15). The properties (1) and (2) determine a weakly Borel measure on G uniquely up to a factor of proportionality.

7. (Johnson) There exists a compact group G for which the inclusion

$$\mathscr{B}(G) \times \mathscr{B}(G) \subset \mathscr{B}(G \times G)$$

is proper. (Recall that $\mathscr{B}(X)$ denotes the class of all Borel sets in the locally compact space X.)

79. Translates of Integrable Functions

The relations $I(f_s) = I(f)$ and $I(f^s) = \Delta(s)I(f)$, valid for every f in $\mathscr{L}$ and all $s \in G$, are easily generalized for integrable Borel functions:

Theorem 1. *If f is an integrable Borel function, and $s \in G$, then f_s is also an integrable Borel function, and*

$$\int f_s \, d\mu = \int f \, d\mu.$$

Proof. Since $x \to s^{-1}x$ is a homeomorphism of G, it is clear that f_s is a Borel function.

Writing $f = f^+ - f^-$, we may assume by linearity that $f \geq 0$. Let f_n be a sequence of simple Borel functions such that $0 \leq f_n \uparrow f$. Then $0 \leq (f_n)_s \uparrow f_s$.

Suppose E is any Borel set of finite measure. Since $(\chi_E)_s = \chi_{sE}$, and since $\mu(sE) = \mu(E) < \infty$ by 78.1, it follows that $(\chi_E)_s$ is an integrable Borel function, and

$$\int (\chi_E)_s \, d\mu = \int \chi_E \, d\mu.$$

It is now clear that each $(f_n)_s$ is an integrable simple Borel function, and

$$\int (f_n)_s \, d\mu = \int f_n \, d\mu.$$

In particular, since $\int f_n\,d\mu \uparrow \int f\,d\mu$ (see Sect. 24), the sequence $\int (f_n)_s\,d\mu$ is bounded; therefore (Sect. 24) f_s is integrable, and

$$\int (f_n)_s\,d\mu \uparrow \int f_s\,d\mu. \quad \blacksquare$$

After replacing s by s^{-1}, the result of Theorem 1 may be expressed in dx notation as follows:

$$\int f(x)\,dx = \int f(sx)\,dx;$$

in other words, it is permissible to make the substitution $x \to sx$ under the integral sign.

Theorem 2. *If f is an integrable Borel function, and $s \in G$, then f^s is also an integrable Borel function, and*

$$\int f^s\,d\mu = \Delta(s) \int f\,d\mu.$$

Proof. Clearly one can paraphrase the proof of Theorem 1 by citing, at the appropriate point, 78.3 in place of 78.1. $\blacksquare$

In dx notation, the result of Theorem 2 assumes the following form:

$$\int f(xs)\,dx = \Delta(s) \int f(x)\,dx,$$

that is,

$$\int f(x)\,dx = \int \Delta(s^{-1}) f(xs)\,dx;$$

in other words, it is permissible to make the substitution $x \to xs$ under the integral sign, provided the "compensating" factor $\Delta(s^{-1})$ is inserted.

Since homeomorphisms preserve also Baire functions, it is immediate from 66.1 that Theorems 1 and 2 imply their Baire variants.

EXERCISES

1. If f is an integrable Borel function, then $f_s \to f_t$ and $f^s \to f^t$ in mean $[\mu]$ as $s \to t$.

2. If f is an integrable Borel function, by 68.2 we may choose a sequence f_n in $\mathscr{L}$ such that $\int |f_n - f|\,d\mu \to 0$. One can then base the proofs of Theorems 1 and 2 on the corresponding properties of the Haar integral I on $\mathscr{L}$ (and the $\mathscr{L}^1$ completeness theorem 31.1).

80. Adjoints of Continuous Functions with Compact Support

We may motivate the definition of adjunction by examining the effect of the substitution $x \to x^{-1}$ under the integral sign. For the moment we need consider only functions in $\mathscr{L}$. For each f in $\mathscr{L}$, let us write $f'(x) = f(x^{-1})$. Defining

$$J(f) = I(f')$$

for all f in $\mathscr{L}$, the problem of studying the correspondence $f \to I(f')$ amounts to studying the positive linear form J.

Is J a Haar integral? If $f \in \mathscr{L}$ and $s \in G$, one has $(f_s)' = (f')^s$, and so

$$J(f_s) = I((f_s)') = I((f')^s) = \Delta(s)I(f') = \Delta(s)J(f).$$

So to speak, J fails to be a Haar integral by the (function) factor Δ (unless, of course, Δ is identically 1). In other words, I fails to be invariant under the substitution $x \to x^{-1}$ to the extent that Δ fails to be identically 1.

The remedy for this is as follows. For each f in $\mathscr{L}$, let us write

$$\tilde{f}(x) = \Delta(x)f(x^{-1});$$

thus, $\tilde{f} = \Delta f'$. The function $\tilde{f}$ is called the **adjoint** of f. Since Δ is continuous (77.2), and $f' \in \mathscr{L}$, it follows that $\tilde{f} \in \mathscr{L}$. It is easily checked that the correspondence $f \to \tilde{f}$ is an involutive linear mapping in $\mathscr{L}$, that is,

$$(\tilde{f})^{\sim} = f, \ (f + g)^{\sim} = \tilde{f} + \tilde{g}, \quad \text{and} \quad (cf)^{\sim} = c\tilde{f}.$$

Moreover, $f \geq 0$ implies $\tilde{f} \geq 0$. The following relation between adjunction and translation is easily verified:

$$(f_s)^{\sim} = \Delta(s^{-1})(\tilde{f})^s.$$

Theorem 1. *One has $I(\tilde{f}) = I(f)$ for all f in $\mathscr{L}$.*

Proof. Defining $J(f) = I(\tilde{f})$ for all f in $\mathscr{L}$, it is clear that J is a positive linear form, not identically zero, such that

$$J(f_s) = I((f_s)^{\sim}) = \Delta(s^{-1})I((\tilde{f})^s) = \Delta(s^{-1})\,\Delta(s)I(\tilde{f}) = J(f).$$

Thus J is a Haar integral, and so $J = cI$ for a suitable constant $c > 0$. Then

$$I(f) = I((\tilde{f})^{\sim}) = J(\tilde{f}) = cI(\tilde{f}) = cJ(f) = c^2I(f),$$

and so $c = 1$. ∎

In dx notation, Theorem 1 reads as follows:

$$\int f(x)\,dx = \int \Delta(x) f(x^{-1})\,dx$$

for each f in $\mathscr{L}$; in other words, one can make the substitution $x \to x^{-1}$ under the integral sign, provided the compensating (function) factor $x \to \Delta(x)$ is inserted. This will be generalized for integrable Baire and Borel functions in Sect. 82.

81. Convolution of Continuous Functions with Compact Support

If f and g are functions in $\mathscr{L}$, we shall define a sort of group theoretic product of f and g, called the **convolution** of f and g; this will also be a function in $\mathscr{L}$, to be denoted $f * g$. The defining formula is

$$(f * g)(x) = \int f(xy) g(y^{-1})\,dy,$$

where dy denotes, in this section, *restricted* Haar measure. To justify the definition, we observe that for each fixed x, the integrand is a continuous function with compact support. Indeed, if C and D are compact sets outside of which f and g vanish, respectively, then the continuous function

$$y \to f(xy) g(y^{-1})$$

vanishes outside the compact set $(x^{-1}C) \cap D^{-1}$, and so may be integrated (simply by applying to it the positive linear form I).

Theorem 1. *If $f, g \in \mathscr{L}$, then $f * g \in \mathscr{L}$.*

Proof. Let C and D be as above. Defining

$$h(x, y) = f(xy) g(y^{-1}),$$

it is easy to see that $h(x, y) \neq 0$ implies $x \in CD$; it follows that if $x \notin CD$, the function $y \to h(x, y)$ is identically zero, and so

$$(f * g)(x) = \int h(x, y)\,dy = 0.$$

Thus $f * g$ vanishes outside the compact set CD.

It remains to show that $f * g$ is continuous. Given any $\varepsilon > 0$, let V be a neighborhood of e such that

$$|f(x) - f(s)| \leq \varepsilon$$

whenever $xs^{-1} \in V$ (apply 71.1 to the function $x \to f(x^{-1})$). If $x \in Vs$, then $(xy)(sy)^{-1} = xs^{-1} \in V$ for all $y \in G$, hence

$$|(f * g)(x) - (f * g)(s)| = \left|\int f(xy)g(y^{-1})\,dy - \int f(sy)g(y^{-1})\,dy\right|$$

$$\leq \int |f(xy) - f(sy)| \cdot |g(y^{-1})|\,dy$$

$$\leq \varepsilon \int |g(y^{-1})|\,dy. \quad \blacksquare$$

The following properties of convolution are immediate from the definition, and the linearity of integration:

Theorem 2. *If $f, g, h \in \mathscr{L}$, and c is a real number, then:*

(1) $f * (g + h) = f * g + f * h$

(2) $(f + g) * h = f * h + g * h$

(3) $(cf) * g = c(f * g) = f * (cg)$.

From an algebraic point of view, a key property of convolution is associativity:

Theorem 3 (Associative law). *If $f, g, h \in \mathscr{L}$, then*

$$f * (g * h) = (f * g) * h.$$

Proof. This is one of the three crucial points at which it will be essential to work with *restricted* Haar measure (the others occur in Theorem 4, and in the definition of convolution of integrable Baire functions in Sect. 84).

Let $x \in G$ be fixed. Then

$$[f * (g * h)](x) = \int f(xz)(g * h)(z^{-1})\,dz$$

$$= \int f(xz)\left(\int g(z^{-1}y)h(y^{-1})\,dy\right)dz$$

$$= \int \left(\int f(xz)g(z^{-1}y)h(y^{-1})\,dy\right)dz;$$

defining

$$F(y, z) = f(xz)g(z^{-1}y)h(y^{-1}),$$

we have in effect shown that the iterated integral

$$\int\int F(y, z)\,dy\,dz$$

exists (in the sense of Sect. 40), and is equal to $[f*(g*h)](x)$. On the other hand,

$$\begin{aligned}[(f*g)*h](x) &= \int (f*g)(xy)h(y^{-1})\,dy \\ &= \int\Big(\int f(xyz)g(z^{-1})\,dz\Big)h(y^{-1})\,dy \\ &= \int\Big(\int f(xz)g(z^{-1}y)\,dz\Big)h(y^{-1})\,dy \\ &= \int\Big(\int f(xz)g(z^{-1}y)h(y^{-1})\,dz\Big)\,dy,\end{aligned}$$

where the left invariance of integration was used in making the substitution $z \to y^{-1}z$. We have in effect shown that the iterated integral

$$\int\int F(y, z)\,dz\,dy$$

exists, and is equal to $[(f*g)*h](x)$.

It is easy to see that F is a continuous function on the product topological space $G \times G$, with compact support. Therefore F is integrable with respect to any Baire measure on $G \times G$ (66.2), and in particular with respect to $\nu \times \nu$ (78.5); then

$$\int\int F(y, z)\,dy\,dz = \int\int F(y, z)\,dz\,dy$$

by Fubini's theorem (41.1), that is,

$$[f*(g*h)](x) = [(f*g)*h](x). \quad \blacksquare$$

Theorems 1–3 may be summarized in algebraic language as follows: $\mathscr{L}$ is an associative algebra over the real numbers, with respect to the pointwise linear operations, and convolution products. Moreover, $f \to I(f)$ is a homomorphism of this algebra:

Theorem 4. *If $f, g \in \mathscr{L}$, then $I(f*g) = I(f)I(g)$.*

Proof. Since the function $F(x, y) = f(xy)g(y^{-1})$ is continuous on $G \times G$, and has compact support, it is integrable with respect to the

Baire measure $\nu \times \nu$ (78.5 and 66.2). Quoting Fubini's theorem, 77.1, and 80.1 at the appropriate steps, we have

$$\begin{aligned}\int (f * g)(x)\, dx &= \int \Big(\int f(xy) g(y^{-1})\, dy\Big)\, dx \\ &= \int g(y^{-1}) \Big(\int f(xy)\, dx\Big)\, dy \\ &= \int g(y^{-1}) I(f^y)\, dy \\ &= \int g(y^{-1})\, \Delta(y) I(f)\, dy \\ &= I(f) I(\tilde{g}) \\ &= I(f) I(g). \quad \blacksquare\end{aligned}$$

The relation between convolution and adjunction in $\mathscr{L}$ is as follows:

Theorem 5. *If $f, g \in \mathscr{L}$, then $(f * g)^\sim = \tilde{g} * \tilde{f}$.*

Proof. Substituting $y \to x^{-1}y$ at the appropriate step, we have

$$\begin{aligned}(\tilde{g} * \tilde{f})(x) &= \int \tilde{g}(xy) \tilde{f}(y^{-1})\, dy = \int \Delta(xy) g(y^{-1}x^{-1})\, \Delta(y^{-1}) f(y)\, dy \\ &= \Delta(x) \int f(y) g(y^{-1}x^{-1})\, dy = \Delta(x) \int f(x^{-1}y) g(y^{-1})\, dy \\ &= \Delta(x)(f * g)(x^{-1}) = (f * g)^\sim(x). \quad \blacksquare\end{aligned}$$

Thus, $f \to \tilde{f}$ is an involution in the associative algebra $\mathscr{L}$: $(\tilde{f})^\sim = f$, $(f + g)^\sim = \tilde{f} + \tilde{g}$, $(cf)^\sim = c\tilde{f}$, and $(f * g)^\sim = \tilde{g} * \tilde{f}$.

The results of this section will be suitably generalized for integrable Baire functions, in Sects. 84 and 85, but only after a considerable amount of technical preparation.

EXERCISES

1. In order that G be Abelian, it is necessary and sufficient that $f * g = g * f$ for all f and g in $\mathscr{L}$.

2. One has $(f * g)_s = f_s * g$ and $(f * g)^s = f * g^s$ for all f, g in $\mathscr{L}$ and $s \in G$. If moreover G is unimodular, then $f * g_s = f^t * g$ and $f^s * g = f * g_t$, where $t = s^{-1}$.

82. Adjoints of Integrable Functions

If f is an arbitrary real-valued function on G, we may define the **adjoint** of f to be the function

$$\tilde{f}(x) = \Delta(x) f(x^{-1}).$$

The following properties of adjunction are easily checked: $(\tilde{f})^{\sim} = f$, $(f + g)^{\sim} = \tilde{f} + \tilde{g}$, $(cf)^{\sim} = c\tilde{f}$; moreover, $f \geq 0$ implies $\tilde{f} \geq 0$. It will be shown in Theorem 1 that adjunction preserves measurability in each of the two competing senses. The way for this is prepared by two lemmas of a general character:

Lemma 1. *If f is measurable with respect to a σ-ring $\mathscr{S}$, and g is measurable with respect to the σ-ring $\mathscr{S}_\lambda$, then fg is measurable with respect to $\mathscr{S}$.*

Proof. We shall use the terms "measurable" and "locally measurable" with respect to $\mathscr{S}$ as in Sect. 11. Thus, $\mathscr{S}_\lambda$ is the class of all locally measurable sets. Let $h = fg$.

Consider first the case that $f = \chi_E$ with $E \in \mathscr{S}$. If M is any Borel set of real numbers, then

$$N(h) \cap h^{-1}(M) = E \cap N(g) \cap g^{-1}(M);$$

since E is measurable and $N(g) \cap g^{-1}(M)$ is locally measurable, it follows that $N(h) \cap h^{-1}(M)$ is measurable, and so h is measurable.

In the general case, let $E = N(f)$. Then $h = (\chi_E f)g = f(\chi_E g)$ is the product of two measurable functions, and is therefore measurable (13.6). ∎

Lemma 2. *If f is a continuous real-valued function on a locally compact space, and $\mathscr{S}$ is the class of all Baire (respectively Borel) sets, then f is measurable with respect to $\mathscr{S}_\lambda$.*

Proof. If c is any real number, the set $A = \{x: f(x) \geq c\}$ is a closed G_δ (55.1), hence its intersection with any compact G_δ is a compact G_δ. It follows that $A \cap F$ is a Baire set, for every Baire set F (Corollary of 1.6), and so $A \in \mathscr{S}_\lambda$. Since $\mathscr{S}_\lambda$ is a σ-algebra, it follows (as remarked in Sect. 13) that f is measurable with respect to $\mathscr{S}_\lambda$.

Replacing "Baire" by "Borel," and "G_δ" by "set," the same proof works for the Borel case (it is not sufficient merely to argue that every Baire set is a Borel set). ∎

Theorem 1. *If f is a Baire (respectively Borel) function on G, then $\tilde{f}$ is also a Baire (respectively Borel) function.*

Proof. Suppose f is a Baire function. Since $x \to x^{-1}$ is a homeomorphism, it follows that $x \to f(x^{-1})$ is also a Baire function. Moreover, $x \to \Delta(x)$ is continuous (77.2), and so

$$x \to \Delta(x) f(x^{-1})$$

is a Baire function by Lemmas 1 and 2.

Replacing "Baire" by "Borel," the same proof works for the Borel case. ∎

The next three lemmas lead up to a proof of the highly useful fact that if E is a Borel set of (Haar) measure zero, then E^{-1} also has measure zero.

Lemma 3. *If D is a compact G_δ, and $f = \chi_D$, then $\tilde{f}$ is an integrable Baire function, and*

$$\int \tilde{f}\, d\nu = \int f\, d\nu = \nu(D).$$

Proof. Choose a sequence f_n in $\mathscr{L}$ such that $f_n \downarrow f$ (55.2). Then $\tilde{f}_n \downarrow \tilde{f}$, and so $\tilde{f}$ is a Baire function (either by Theorem 1, or a combination of 56.2 and 14.3). Since the $\tilde{f}_n$ are integrable (66.2) and $\tilde{f}_n \geq \tilde{f} \geq 0$, it follows from 25.7 that $\tilde{f}$ is integrable. By the (MCT),

$$I(\tilde{f}_n) \downarrow \int \tilde{f}\, d\nu \qquad \text{and} \qquad I(f_n) \downarrow \int f\, d\nu = \nu(D);$$

since $I(\tilde{f}_n) = I(f_n)$ for all n (80.1), we conclude that

$$\int \tilde{f}\, d\nu = \int f\, d\nu. \quad ∎$$

Lemma 4. *If D is a compact G_δ such that $\nu(D) = 0$, then also $\nu(D^{-1}) = 0$.*

Proof. The set $A = D^{-1}$ is also a compact G_δ. Defining $f = \chi_A$, we have

$$\tilde{f}(x) = \Delta(x)\chi_A(x^{-1}) = \Delta(x)\chi_D(x),$$

thus $\tilde{f} = \Delta\chi_D$. Obviously $\tilde{f} = 0$ a.e. $[\nu]$. Quoting Lemma 3, we have

$$\nu(A) = \int f\, d\nu = \int \tilde{f}\, d\nu = 0. \quad ∎$$

Lemma 5. *If C is a compact set such that $\mu(C) = 0$, then also $\mu(C^{-1}) = 0$.*

Proof. Since μ is regular, there exists a compact G_δ D such that $C \subset D$ and $\mu(D - C) = 0$ (59.1). Then $\mu(D) = \mu(C) = 0$. Since $C^{-1} \subset D^{-1}$, it follows that $\mu(C^{-1}) \leq \mu(D^{-1}) = 0$ by Lemma 4. ∎

Theorem 2. *If E is a Borel set such that $\mu(E) = 0$, then also $\mu(E^{-1}) = 0$.*

Proof. If C is any compact subset of E^{-1}, it will suffice by regularity to show that $\mu(C) = 0$. Since $C^{-1} \subset E$, we have $\mu(C^{-1}) \leq \mu(E) = 0$; since C^{-1} is also compact, it follows from Lemma 5 that $\mu(C) = 0$. ∎

A typical application of Theorem 2 is as follows:

Corollary. *If f and g are real-valued functions on G such that $f = g$ a.e. $[\nu]$ (respectively a.e. $[\mu]$), then also $\tilde{f} = \tilde{g}$ a.e. $[\nu]$ (respectively a.e. $[\mu]$).*

Proof. If F is a Baire (respectively Borel) set of measure zero, on whose complement $f = g$, then F^{-1} is also a Baire (respectively Borel) set of measure zero (Theorem 2), and it is clear that $\tilde{f} = \tilde{g}$ on the complement of F^{-1}. ∎

In the next theorem it will be shown that adjunction preserves integrability, and the value of the integral.

Lemma 6. *If C is a compact set, and $f = \chi_C$, then $\tilde{f}$ is an integrable Borel function, and*

$$\int \tilde{f}\,d\mu = \int f\,d\mu = \mu(C).$$

Proof. Let D be a compact G_δ such that $C \subset D$ and $\mu(D - C) = 0$ (59.1), and let $g = \chi_D$. Evidently $f = g$ a.e. $[\mu]$, and so $\tilde{f} = \tilde{g}$ a.e. $[\mu]$ by the above corollary. Since $\tilde{f}$ is a Borel function (Theorem 1), and $\tilde{g}$ is integrable (Lemma 3), it follows from 25.5 that $\tilde{f}$ is integrable, and

$$\int \tilde{f}\,d\mu = \int \tilde{g}\,d\mu = \int g\,d\mu = \int f\,d\mu. \quad ∎$$

Lemma 7. *If E is a Borel set such that $\mu(E) < \infty$, and $f = \chi_E$, then $\tilde{f}$ is an integrable Borel function, and*

$$\int \tilde{f}\,d\mu = \int f\,d\mu = \mu(E).$$

Proof. Since μ is regular, there exists a sequence C_n of compact sets such that $C_n \subset E$ and $\mu(C_n) \uparrow \mu(E)$. Replacing C_n by $C_1 \cup \cdots \cup C_n$, we may assume $C_n\uparrow$. Defining $E_0 = \bigcup_1^\infty C_n$, we have $C_n \uparrow E_0$, $E_0 \subset E$, and $\mu(E - E_0) = 0$.

Let $f_n = \chi_{C_n}$ and $f_0 = \chi_{E_0}$. Evidently $f_0 = f$ a.e. $[\mu]$, and $f_n \uparrow f_0$. Then $\tilde{f}_n \uparrow \tilde{f}_0$, and by the corollary of Theorem 2, $\tilde{f}_0 = \tilde{f}$ a.e. $[\mu]$. By Lemma 6, each $\tilde{f}_n$ is integrable, and

$$\int \tilde{f}_n\,d\mu = \int f_n\,d\mu = \mu(C_n) \leq \mu(E) < \infty$$

for all n; it follows from the (MCT) that the Borel function $\tilde{f}_0$ is integrable, and

$$\int \tilde{f}_0\,d\mu = \text{LUB} \int \tilde{f}_n\,d\mu = \text{LUB}\,\mu(C_n) = \mu(E_0) = \mu(E) = \int f\,d\mu.$$

Finally, since $\tilde{f}$ is a Borel function (Theorem 1), and $\tilde{f} = \tilde{f}_0$ a.e. $[\mu]$, it follows from 25.5 that $\tilde{f}$ is integrable, and

$$\int \tilde{f}\,d\mu = \int \tilde{f}_0\,d\mu = \int f\,d\mu. \quad ∎$$

Theorem 3. *If f is an integrable Baire (respectively Borel) function on G, then $\tilde{f}$ is also an integrable Baire (respectively Borel) function, and*

$$\int \tilde{f}\, d\mu = \int f\, d\mu.$$

Proof. Suppose f is an integrable Baire function. By linearity, we may clearly assume $f \geq 0$. In any case $\tilde{f}$ is a Baire function by Theorem 1.

Let f_n be a sequence of simple Baire functions such that $0 \leq f_n \uparrow f$. Then $0 \leq \tilde{f}_n \uparrow \tilde{f}$; since it is clear from Lemma 7 that $\tilde{f}_n$ is integrable, and that

$$\int \tilde{f}_n\, d\mu = \int f_n\, d\mu \leq \int f\, d\mu < \infty$$

for all n, it follows from the (MCT) that $\tilde{f}$ is integrable, and

$$\int \tilde{f}\, d\mu = \text{LUB} \int \tilde{f}_n\, d\mu = \text{LUB} \int f_n\, d\mu = \int f\, d\mu.$$

Replacing "Baire" by "Borel," the same proof works for the Borel case. ∎

In dx notation, Theorem 3 reads as follows:

$$\int f(x)\, dx = \int \Delta(x) f(x^{-1})\, dx$$

for every integrable Borel function.

It follows from Theorem 3 that the correspondence $f \to \tilde{f}$ is an involutive linear mapping in $\mathscr{L}^1(\mu)$, that is, $(\tilde{f})^{\sim} = f$, $(f + g)^{\sim} = \tilde{f} + \tilde{g}$, and $(cf)^{\sim} = c\tilde{f}$. Moreover $\int \tilde{f}\, d\mu = \int f\, d\mu$, and $f \in \mathscr{L}^1(\nu)$ implies $\tilde{f} \in \mathscr{L}^1(\nu)$.

83. The Operation $f \triangledown g$

The present section is technical preparation for the next, in which the concept of convolution will be extended from the class of continuous functions with compact support to the class of integrable Baire functions. The main difference is that the convolution of two functions in $\mathscr{L}$ is itself a function in $\mathscr{L}$, and in particular is defined everywhere on G, whereas the convolution of two functions in $\mathscr{L}^1(\nu)$ will in general be defined only on a subset of G. This is a natural and unavoidable circumstance; it requires some extra care, but causes no real difficulty, as we hope to make clear in the sequel.

If f and g are any two real-valued functions defined on G, we shall write $f \triangledown g$ for the function on $G \times G$ defined by the formula

$$(f \triangledown g)(x, y) = f(xy)g(y^{-1});$$

this is precisely the function of two variables which occurs in the definition of convolution for functions in $\mathscr{L}$. The operation

$$(f, g) \to f \triangledown g$$

is easily seen to be bilinear, that is,

$$(f_1 + f_2) \triangledown g = f_1 \triangledown g + f_2 \triangledown g,$$

$$f \triangledown (g_1 + g_2) = f \triangledown g_1 + f \triangledown g_2,$$

$$(cf) \triangledown g = c(f \triangledown g) = f \triangledown (cg).$$

It is important to have explicit formulas for the sections of $f \triangledown g$:

Lemma 1. *If f and g are real-valued functions on G, then*

$$(f \triangledown g)^y = g(y^{-1}) \cdot f^y,$$

$$(f \triangledown g)_x = f_{x^{-1}} \cdot g'$$

for all $x, y \in G$, where $g'(y) = g(y^{-1})$.

Proof. Observe that $(f \triangledown g)^y$ is the y-section of $f \triangledown g$ as defined in Sect. 36, whereas f^y is the right translate of f by y. There should be no difficulty in keeping the notations straight: since we ignore the group structure of $G \times G$, functions of two variables will never be translated; on the other hand, one cannot take sections of a function of one variable.

The assertion of the lemma is immediate from the formulas

$$(f \triangledown g)^y(x) = (f \triangledown g)_x(y) = (f \triangledown g)(x, y) = f(xy)g(y^{-1}). \quad ▮$$

If in particular f and g are functions in $\mathscr{L}$, we have

$$(f * g)(x) = \int f(xy)g(y^{-1})\, dy = \int (f \triangledown g)_x\, d\nu = I((f \triangledown g)_x),$$

and so it is quite natural to consider the combination $f \triangledown g$ in seeking to extend the notion of convolution beyond $\mathscr{L}$.

Since homeomorphisms preserve Baire and Borel sets, Lemma 1 implies at once:

Lemma 2. *If f and g are Baire (respectively Borel) functions on G, then every section of $f \triangledown g$ is a Baire (respectively Borel) function.*

The following result is the lock on the theory of convolution of integrable functions, and the key which turns it is Fubini's theorem, as we shall see in the next section:

Theorem 1. *If f and g are integrable Baire functions, then:*

(1) *For each $y \in G$, $(f \triangledown g)^y$ is an integrable Baire function.*

(2) *Indeed, $(f \triangledown g)^y = g(y^{-1}) f^y$, and*

$$\int (f \triangledown g)^y \, d\nu = \left(\int f \, d\nu\right) \tilde{g}(y).$$

(3) *The iterated integral $\iint (f \triangledown g)(x, y) \, dx \, dy$ exists, and is equal to $\left(\int f \, d\nu\right)\left(\int g \, d\nu\right)$.*

Proof. Assertions (1) and (2) are immediate from the lemmas and 79.2. Assertion (3) then follows from 82.3, and the definition of iterated integral (Sect. 40). ∎

In view of Lemma 2, it is clear that the statement and proof of Theorem 1 are valid with "Baire" replaced by "Borel." Since this fact will not be needed, we let it go at that.

84. Convolution of Integrable Baire Functions

Up to this point we have developed the Baire and Borel theories in parallel, but we now come to a parting of ways. We shall make the next strides with the Baire theory alone. (It is possible, and quite unnecessary, to adapt these results to the Borel case, as we shall see in Sect. 89.)

It will be convenient to introduce the mapping S defined in $G \times G$ by the formula

$$S(x, y) = (xy, y^{-1}).$$

In view of the continuity of the mappings $(x, y) \to xy$ and $y \to y^{-1}$, it is clear that S is continuous; moreover, S is bijective and self-inverse, and so is a homeomorphism of $G \times G$. In particular, S preserves the domain of definition of the product measure $\nu \times \nu$, namely, the class of Baire sets in $G \times G$. If f and g are real-valued functions on G, and one defines

$$H(x, y) = f(x) g(y),$$

evidently $f \triangledown g$ is the composite function $H \circ S$.

Lemma. *If f and g are Baire functions on G, then $f \triangledown g$ is a Baire function on $G \times G$.*

Proof. Defining $H(x, y) = f(x) g(y)$, it is clear from the preliminary remarks that it will suffice to show that H is a Baire function. Since each of f and g is the pointwise limit of a sequence of simple Baire functions (16.4), it is clearly sufficient (14.3, and bilinearity) to consider the case that $f = \chi_E$ and $g = \chi_F$, where E and F are Baire sets. In this case, $H = \chi_{E \times F}$ is the characteristic function of a Baire set. ▮

It will be convenient from now on to write $\pi = \nu \times \nu$. Thus, π is a Baire measure on $G \times G$ (78.5).

Theorem 1. *If f and g are integrable Baire functions on G, then $f \triangledown g$ is a π-integrable Baire function on $G \times G$, and*

$$\int (f \triangledown g)\, d\pi = \left(\int f\, d\nu\right)\left(\int g\, d\nu\right).$$

Proof. It is clear from bilinearity that it is sufficient to consider $f \geq 0$, $g \geq 0$. Then also $f \triangledown g \geq 0$. Since $f \triangledown g$ is a Baire function by the lemma, and since the iterated integral

$$\iint (f \triangledown g)(x, y)\, dx\, dy$$

exists by 83.1, it follows from the converse of Fubini's theorem (41.2) that $f \triangledown g$ is π-integrable; then by Fubini's theorem (41.1), and 83.1, we have

$$\int (f \triangledown g)\, d\pi = \iint (f \triangledown g)(x, y)\, dx\, dy = \left(\int f\, d\nu\right)\left(\int g\, d\nu\right). \quad ▮$$

As far as convolution is concerned, our object in all of this is the following corollary:

Corollary. *If f and g are integrable Baire functions on G, then $(f \triangledown g)_x$ is an integrable Baire function, for almost all x. Indeed, there exists a Baire set F and an integrable Baire function h such that:*

(1) $\nu(F) = 0$;

(2) *for each $x \in G - F$, $(f \triangledown g)_x$ is an integrable Baire function, and*

$$\int (f \triangledown g)_x\, d\nu = h(x).$$

Necessarily

$$\int h\, d\nu = \left(\int f\, d\nu\right)\left(\int g\, d\nu\right).$$

Proof. In any case, $(f \triangledown g)_x$ is a Baire function, for each $x \in G$ (Lemma 2 of 83.1). Since $f \triangledown g$ is π-integrable by Theorem 1, it follows from Fubini's theorem that the iterated integral

$$\iint (f \triangledown g)(x, y)\, dy\, dx$$

exists; this means, according to the definition of iterated integral (Sect.

40), that the desired F and h exist, the value of the iterated integral being $\int h\,d\nu$. Then

$$\int h\,d\nu = \int\int (f \triangledown g)(x, y)\,dy\,dx$$
$$= \int (f \triangledown g)\,d\pi = \left(\int f\,d\nu\right)\left(\int g\,d\nu\right)$$

by Fubini's theorem and Theorem 1. ▌

As threatened in the preceding section, we are on the verge of considering functions which are only partially defined on G, that is, functions whose domain of definition is some subset of G. If f is such a function, we shall write D_f for its domain of definition. We shall say that f is **defined a.e.** $[\nu]$ in case there exists a Baire set F such that $\nu(F) = 0$ and $G - F \subset D_f$; in other words, the domain of definition of f includes the complement of a ν-null Baire set.

If f and g are real-valued functions defined on G, the **convolution** of f and g (in that order) is the function $f * g$ defined as follows. The domain of definition D_{f*g} is the set of all x in G for which $(f \triangledown g)_x$ is an integrable Baire function, that is,

$$(f \triangledown g)_x \in \mathscr{L}^1(\nu);$$

for such an x, the value of $f * g$ is defined by the formula

$$(f * g)(x) = \int (f \triangledown g)_x\,d\nu = \int (f \triangledown g)(x, y)\,dy = \int f(xy)g(y^{-1})\,dy.$$

In the text we are concerned only with the case that f and g are integrable; other possibilities are considered in the exercises.

It will be convenient in this and the next section to abbreviate "a.e. $[\nu]$" to "a.e."

Theorem 2. *If f and g are integrable Baire functions on G, then $f * g$ is defined a.e. Indeed, there exists an integrable Baire function h such that*

$$h = f * g \text{ a.e.};$$

necessarily

$$\int h\,d\nu = \left(\int f\,d\nu\right)\left(\int g\,d\nu\right).$$

Proof. The existence of an h having all of the required properties is immediate from the corollary of Theorem 1, and the definition of $f * g$. If k is any other integrable Baire function such that $f * g = k$ a.e., then also $k = h$ a.e., and so

$$\int k\,d\nu = \int h\,d\nu = \left(\int f\,d\nu\right)\left(\int g\,d\nu\right). \quad ▌$$

If f is any integrable Baire function, it will be convenient to write

$$\|f\|_1 = \int |f| \, d\nu.$$

The number $\|f\|_1$ is called the $\mathscr{L}^1$-*norm* of f, and evidently has the following properties:

(*i*) $\|f\|_1 \geq 0$
(*ii*) $\|f + g\|_1 \leq \|f\|_1 + \|g\|_1$
(*iii*) $\|cf\|_1 = |c| \, \|f\|_1$
(*iv*) $\|f\|_1 = 0$ if and only if $f = 0$ a.e.

In the notation of Sect. 28,

$$\rho(f, g) = \|f - g\|_1,$$

thus "$f_n \to f$ in mean" if and only if

$$\|f_n - f\|_1 \to 0,$$

and "f_n is MF" if and only if

$$\|f_m - f_n\|_1 \to 0$$

as $m, n \to \infty$. (The concept of $\mathscr{L}^1$-norm can of course be defined in the context of any measure space, and has nothing in particular to do with locally compact spaces, let alone locally compact groups.)

Corollary. *If f, g, h are integrable Baire functions, and $h = f * g$ a.e., then*

$$\|h\|_1 \leq \|f\|_1 \|g\|_1.$$

If moreover $f \geq 0$ and $g \geq 0$, then $h \geq 0$ a.e. and

$$\|h\|_1 = \|f\|_1 \|g\|_1.$$

Proof. Observe that the functions $f * g$ and $|f| * |g|$ have the same domain of definition; for, $(f \triangledown g)_x$ is a Baire function on G (for each x), and the relation

$$|(f \triangledown g)_x| = |f \triangledown g|_x = (|f| \triangledown |g|)_x$$

shows that $(f \triangledown g)_x$ is integrable if and only if $(|f| \triangledown |g|)_x$ is integrable (25.3).

If $x \in D_{f*g}$, then by 25.3 we have

$$|(f * g)(x)| = \left| \int (f \triangledown g)_x \, d\nu \right| \leq \int |(f \triangledown g)_x| \, d\nu$$

$$= \int (|f| \triangledown |g|)_x \, d\nu = (|f| * |g|)(x);$$

thus $|f * g| \le |f| * |g|$ (on their common domain). Let h and k be integrable Baire functions such that $h = f * g$ a.e. and $k = |f| * |g|$ a.e. (Theorem 2); then $|h| \le |k|$ a.e. Evidently $|f| * |g| \ge 0$, hence $k \ge 0$ a.e.; then $|h| \le |k| = k$ a.e. It follows from Theorem 2 that

$$\int |h| \, d\nu \le \int k \, d\nu = \left(\int |f| \, d\nu\right)\left(\int |g| \, d\nu\right),$$

that is, $\|h\|_1 \le \|f\|_1 \|g\|_1$.

If in particular $f \ge 0$ and $g \ge 0$, then $f * g \ge 0$, and so $h \ge 0$ a.e.; thus $h = |h|$ a.e., and it follows that

$$\begin{aligned} \|h\|_1 &= \int |h| \, d\nu = \int h \, d\nu \\ &= \left(\int f \, d\nu\right)\left(\int g \, d\nu\right) = \|f\|_1 \|g\|_1. \quad \blacksquare \end{aligned}$$

The relation between convolution and adjunction is as follows:

Theorem 3. *If f, g, h are integrable Baire functions such that $h = f * g$ a.e., then $\tilde{h} = \tilde{g} * \tilde{f}$ a.e.*

Proof. In any case, $\tilde{f}$, $\tilde{g}$, and $\tilde{h}$ are integrable Baire functions (82.3). Suppose x is in the domain of definition of $\tilde{g} * \tilde{f}$, in other words, $(\tilde{g} \triangledown \tilde{f})_x$ is integrable. Since

$$\begin{aligned} (\tilde{g} \triangledown \tilde{f})_x(y) &= \tilde{g}(xy)\tilde{f}(y^{-1}) = \Delta(xy) g(y^{-1}x^{-1}) \, \Delta(y^{-1}) f(y) \\ &= \Delta(x) f(y) g(y^{-1}x^{-1}), \end{aligned}$$

it follows that the (Baire) function $k(y) = f(y) g(y^{-1}x^{-1})$ is integrable. Now

$$k_x(y) = k(x^{-1}y) = f(x^{-1}y) g(y^{-1}) = (f \triangledown g)_{x^{-1}}(y),$$

and so $(f \triangledown g)_{x^{-1}} = k_x$ is integrable; thus x^{-1} belongs to the domain of definition of $f * g$, and moreover

$$\begin{aligned} (\tilde{g} * \tilde{f})(x) &= \int (\tilde{g} \triangledown \tilde{f})_x \, d\nu = \Delta(x) \int k \, d\nu = \Delta(x) \int k_x \, d\nu \\ &= \Delta(x) \int (f \triangledown g)_{x^{-1}} \, d\nu = \Delta(x)(f * g)(x^{-1}). \end{aligned}$$

The steps of the argument are reversible: if $(f * g)(x^{-1})$ is defined, then so is $(\tilde{g} * \tilde{f})(x)$; the point is that $y \to k(y)$ is integrable if and only if $y \to \Delta(x)k(y)$ is integrable.

By assumption there exists a Baire set F such that $\nu(F) = 0$, and such that $x \notin F$ implies $(f * g)(x)$ is defined and is equal to $h(x)$. If $x \notin F^{-1}$, then $x^{-1} \notin F$, and so $(f * g)(x^{-1})$ is defined and is equal to $h(x^{-1})$; that is, $(\tilde{g} * \tilde{f})(x)$ is defined and is equal to

$$\Delta(x)(f * g)(x^{-1}) = \Delta(x)h(x^{-1}) = \tilde{h}(x).$$

Since $\nu(F^{-1}) = 0$ by 82.2, we conclude that $\tilde{g} * \tilde{f} = \tilde{h}$ a.e. $\blacksquare$

EXERCISES

1. If f and g are Baire functions, and $f = 0$ a.e. $[\nu]$, then $f * g$ and $g * f$ are everywhere defined on G, and are identically zero.

2. If f and g are integrable Baire functions, then

$$\|f \triangledown g\|_1 = \|f\|_1 \|g\|_1.$$

3. If $f, g \in \mathscr{L}^2(\nu)$, then $f \triangledown g \in \mathscr{L}^2(\pi)$, and

$$\|f \triangledown g\|_2 = \|f\|_2 \|g\|_2.$$

4. If f is a real-valued function defined on a subset D_f of G, we may define a function $\tilde{f}$ on $(D_f)^{-1}$ by the formula

$$\tilde{f}(x) = \Delta(x) f(x^{-1}).$$

Then, if f and g are integrable Baire functions, the functions $\tilde{g} * \tilde{f}$ and $(f * g)^{\sim}$ have identical domains of definition, and $\tilde{g} * \tilde{f} = (f * g)^{\sim}$ on their common domain.

5. If f is an integrable Baire function, and g is an essentially bounded Baire function, then the convolution $f * g$ is everywhere defined on G, and is a uniformly continuous function. If moreover G is unimodular, then the convolution $g * f$ is also everywhere defined.

6. If G is unimodular, and f, g are Baire functions such that $f \in \mathscr{L}^p$ and $g \in \mathscr{L}^q$ (where $p > 1$, $q > 1$, and $p + q = pq$), then the convolution $f * g$ is defined everywhere on G, and is a bounded continuous function on G,

$$\|f * g\|_\infty \leq \|f\|_p \|g\|_q.$$

7. If f, g are Baire functions such that $f \in \mathscr{L}^1$ and $g \in \mathscr{L}^p$ (where $p > 1$), then $f * g$ is defined a.e. (See also Exercise 88.3.)

8. If G is unimodular, and f is a Baire function on G, then $(f * \tilde{f})(e)$ is defined if and only if $f \in \mathscr{L}^2(\nu)$. In the notation of Sect. 33, one has

$$(f|g) = (f * \tilde{g})(e)$$

for all f, g in $\mathscr{L}^2(\nu)$.

85. Associativity of Convolution

As in the preceding section, we abbreviate "a.e. $[\nu]$" to "a.e."

If f, g, h are functions in $\mathscr{L}$, then $f * (g * h) = (f * g) * h$ by 81.3; all of the indicated convolution products are themselves functions in $\mathscr{L}$, and in particular they are defined everywhere on G.

What if f, g, h are integrable Baire functions? The convolutions $g * h$ and $f * g$ can at any rate be defined, but they are in general defined only partially on G; it follows that we cannot in general *define* $f * (g * h)$ and $(f * g) * h$, let alone compare them. However, we know from 84.2 that there exist integrable Baire functions k_1 and k_2 such that $g * h = k_1$ a.e. and $f * g = k_2$ a.e.; as we shall see in Theorem 2, the associative law for the convolution of integrable functions takes the form $f * k_1 = k_2 * h$ a.e. The preparatory definitions and lemmas are concerned with the "bilinearity" of convolution.

If f and g are real-valued functions defined on the subsets D_f and D_g of G, respectively, then $f + g$ denotes the function defined on $D_f \cap D_g$ by the formula

$$(f + g)(x) = f(x) + g(x).$$

If c is a real number, then cf denotes the function defined on D_f by the formula

$$(cf)(x) = cf(x).$$

We write $f \supset g$, and say that f is an *extension* of g, in case $D_f \supset D_g$ and $f(x) = g(x)$ for all x in D_g. We write $f = g$ in case both $f \supset g$ and $g \supset f$.

Lemma 1. *If f, g_1, g_2 are integrable Baire functions, then the function $f * g_1 + f * g_2$ is defined a.e., and*

$$f * (g_1 + g_2) \supset f * g_1 + f * g_2.$$

In particular, $f * (g_1 + g_2) = f * g_1 + f * g_2$ a.e.

Proof. According to 84.2, there exists a Baire set F_i of measure zero such that $G - F_i$ is included in the domain of $f * g_i$. Then $F = F_1 \cup F_2$ is a Baire set of measure zero, on whose complement the function $f * g_1 + f * g_2$ is defined, and so this function is defined a.e.

Suppose x is a point of G such that both $(f * g_1)(x)$ and $(f * g_2)(x)$ are defined, in other words both $(f \triangledown g_1)_x$ and $(f \triangledown g_2)_x$ are integrable. Taking x-sections in the relation

$$f \triangledown (g_1 + g_2) = f \triangledown g_1 + f \triangledown g_2,$$

we conclude that $(f \triangledown (g_1 + g_2))_x$ is integrable; indeed,

$$(f \triangledown (g_1 + g_2))_x = (f \triangledown g_1)_x + (f \triangledown g_2)_x,$$

and so integration yields

$$[f * (g_1 + g_2)](x) = (f * g_1)(x) + (f * g_2)(x),$$

and we have shown that $f * (g_1 + g_2) \supset f * g_1 + f * g_2$. ▮

Arguing similarly with the relation

$$(f_1 + f_2) \triangledown g = f_1 \triangledown g + f_2 \triangledown g,$$

we have:

Lemma 2. *If f_1, f_2, g are integrable Baire functions, then the function $f_1 * g + f_2 * g$ is defined a.e., and*

$$(f_1 + f_2) * g \supset f_1 * g + f_2 * g.$$

In particular, $(f_1 + f_2) * g = f_1 * g + f_2 * g$ *a.e.*

Similarly the relation $(cf) \triangledown g = f \triangledown (cg) = c(f \triangledown g)$ yields the following:

Lemma 3. *If f and g are integrable Baire functions, and c is a nonzero real number, then the functions $(cf) * g, f * (cg)$, and $c(f * g)$ all have the same domain of definition (namely, that of $f * g$), and*

$$(cf) * g = f * (cg) = c(f * g).$$

The situation for $c = 0$ is as follows. Writing θ for the function identically zero on G, it is clear that $\theta * g$ and $f * \theta$ are defined everywhere on G, and $\theta * g = f * \theta = \theta$. Thus $(0f) * g = f * (0g) = \theta$. On the other hand, the function $0(f * g)$ has the same domain of definition as $f * g$, and its values are all zero, and so $0(f * g) \subset \theta$. Thus,

$$(0f) * g = f * (0g) = 0(f * g) = \theta \text{ a.e.}$$

Combining this with Lemma 3, we may say that

$$(cf) * g = f * (cg) = c(f * g) \text{ a.e.}$$

for all real numbers c, and integrable Baire functions f and g.

Lemma 4. *If f_1, g_1, f, g are integrable Baire functions, then the following relation holds a.e.:*

$$f_1 * g_1 - f * g = (f_1 - f) * g_1 + f * (g_1 - g).$$

Proof. For each of the four convolutions, there is a Baire set of measure zero whose complement is contained in the domain of definition of the convolution. If F is the union of these four null sets, then F is a Baire set of measure zero on whose complement all four of the convolutions are defined. If $x \in G - F$, our assertion follows on taking x-sections in the relation

$$f_1 \triangledown g_1 - f \triangledown g = (f_1 - f) \triangledown g_1 + f \triangledown (g_1 - g)$$

and integrating. ∎

The relation between convolution and convergence in mean is given by the following key "continuity theorem":

Theorem 1. *If f_n and g_n are mean fundamental sequences of integrable Baire functions, and if, for each n, h_n is an integrable Baire function such that $h_n = f_n * g_n$ a.e., then the sequence h_n is also mean fundamental. If, moreover, f and g are integrable Baire functions to which the sequences f_n and g_n converge in mean, respectively, and if h is an integrable Baire function such that $f * g = h$ a.e.* (84.2), *then $h_n \to h$ in mean.*

Proof. According to Lemma 4, we have, almost everywhere,

$$\begin{aligned} h_m - h_n &= f_m * g_m - f_n * g_n \\ &= (f_m - f_n) * g_m + f_n * (g_m - g_n). \end{aligned}$$

By 84.2, we may choose integrable Baire functions A_{mn}, B_{mn} such that the following relations hold almost everywhere:

$$\begin{aligned} A_{mn} &= (f_m - f_n) * g_m, \\ B_{mn} &= f_n * (g_m - g_n). \end{aligned}$$

Evidently $|h_m - h_n| \le |A_{mn}| + |B_{mn}|$ a.e.; integrating this relation, and quoting the corollary of 84.2, we have

$$\begin{aligned} \|h_m - h_n\|_1 &\le \|A_{mn}\|_1 + \|B_{mn}\|_1 \\ &\le \|f_m - f_n\|_1 \|g_m\|_1 + \|f_n\|_1 \|g_m - g_n\|_1, \end{aligned}$$

hence $\|h_m - h_n\|_1 \to 0$ as $m, n \to \infty$. In other words, the sequence h_n is mean fundamental.

Assume now that f and g are integrable Baire functions such that $f_n \to f$ and $g_n \to g$, in mean. (We are not quoting 31.1 here, but are simply assuming f and g to be given.) Choose an integrable Baire function h so that $h = f * g$ a.e. (84.2). We have, almost everywhere,

$$\begin{aligned} h_n - h &= f_n * g_n - f * g \\ &= (f_n - f) * g_n + f * (g_n - g); \end{aligned}$$

then $\|h_n - h\|_1 \le \|f_n - f\|_1 \|g_n\|_1 + \|f\|_1 \|g_n - g\|_1$ by a repetition of the earlier argument, and so $\|h_n - h\|_1 \to 0$. ∎

Since convolution in $\mathscr{L}$ is associative, and $\mathscr{L}$ is dense in $\mathscr{L}^1(\nu)$, we may deduce the associativity of convolution for integrable functions:

Theorem 2 (Associative law). *Let f, g, h be integrable Baire functions, and choose* (84.2) *integrable Baire functions k_1, k_2 so that $g * h = k_1$ a.e. and $f * g = k_2$ a.e. Then $f * k_1 = k_2 * h$ a.e.*

Proof. According to 67.2, we may choose sequences f_n, g_n, h_n in $\mathscr{L}$, so that $f_n \to f$, $g_n \to g$, and $h_n \to h$, in mean. By 84.2, there exist integrable Baire functions A and B such that

$$f * k_1 = A \text{ a.e.} \qquad \text{and} \qquad k_2 * h = B \text{ a.e.}$$

It will suffice to show that $A = B$ a.e.

Since $g_n \to g$ and $h_n \to h$ in mean, and since $g_n * h_n \in \mathscr{L}$, it follows from Theorem 1 that

$$g_n * h_n \to k_1 \qquad \text{in mean.}$$

Since $f_n \to f$ and $g_n * h_n \to k_1$ in mean, and since

$$f_n * (g_n * h_n) \in \mathscr{L},$$

another application of Theorem 1 yields

$$f_n * (g_n * h_n) \to A \text{ in mean.}$$

Similarly $(f_n * g_n) * h_n \to B$ in mean. Since

$$f_n * (g_n * h_n) = (f_n * g_n) * h_n$$

by 81.3, we conclude that $A = B$ a.e. ▮

EXERCISE

1. If f, g_1, g_2 are nonnegative Baire functions, then

$$f * (g_1 + g_2) = f * g_1 + f * g_2.$$

*86. The Group Algebra

The foregoing results on $\mathscr{L}^1(\nu)$ may be recast in a slightly different form so as to take advantage of the terminology and theory of normed algebras. The present section is devoted to the necessary reformulation; there are no new theorems here.

As in the preceding two sections, it will be convenient to abbreviate "a.e. $[\nu]$" to "a.e." Let us denote by $\mathscr{N}$ the class of all f in $\mathscr{L}^1(\nu)$ such that $\|f\|_1 = 0$ (equivalently, f is a Baire function such that $f = 0$ a.e.). Evidently $\mathscr{N}$ is a linear subspace of $\mathscr{L}^1(\nu)$, and so we may form the quotient vector space

$$\mathscr{L}^1(\nu)/\mathscr{N},$$

which we denote by $L^1(\nu)$. Thus, $L^1(\nu)$ is the class of all cosets $f + \mathcal{N}$, where $f \in \mathcal{L}^1(\nu)$, with the following linear operations:

$$(f + \mathcal{N}) + (g + \mathcal{N}) = (f + g) + \mathcal{N},$$
$$c(f + \mathcal{N}) = cf + \mathcal{N}.$$

We shall also write $[f]$ for the coset $f + \mathcal{N}$. Thus

$$[f] + [g] = [f + g] \qquad \text{and} \qquad c[f] = [cf].$$

Evidently $[f] = [g]$ if and only if $f - g \in \mathcal{N}$, that is, $f = g$ a.e.

If $u = [f] = [g]$, then $\|f\|_1 = \|g\|_1$ results from the relation $|f| = |g|$ a.e.; it follows that the *norm* of u may be unambiguously defined by the formula

$$\|u\| = \|f\|_1.$$

The properties of the $\mathcal{L}^1$-norm $\| \ \|_1$ yield the following properties of $\| \ \|$:

(1) $$\|u\| \geq 0$$

(2) $$\|u + v\| \leq \|u\| + \|v\|$$

(3) $$\|cu\| = |c| \, \|u\|$$

(4) $$\|u\| = 0 \text{ if and only if } u = 0,$$

where $0 = [\theta]$. Thus, the vector space $L^1(\nu)$, together with the function $u \to \|u\|$, satisfies the axioms for a *normed space*. In fact, $L^1(\nu)$ is a *Banach space*, that is:

(5) if u_n is a sequence in $L^1(\nu)$ such that $\|u_m - u_n\| \to 0$ as $m, n \to \infty$, then there exists an element u of $L^1(\nu)$ such that $\|u_n - u\| \to 0$; this is immediate from the $\mathcal{L}^1$ completeness theorem (31.1).

All of the above is valid in the context of an arbitrary measure space. The underlying group structure, as reflected in the convolution product, has yet to be exploited. If u and v are any elements in L^1, we may define their *product* uv as follows. Choose $f, g \in \mathcal{L}^1$ so that $u = [f]$ and $v = [g]$. By 84.2, there exists an h in $\mathcal{L}^1$ such that

$$h = f * g \text{ a.e.}$$

We propose to define

$$uv = [h],$$

and so must verify that if $f_0, g_0, h_0 \in \mathcal{L}^1$ are such that $f = f_0$ a.e., $g = g_0$ a.e., and $h_0 = f_0 * g_0$ a.e., then $h = h_0$ a.e. Indeed, by Lemma 4 in Sect. 85, we have

$$h_0 - h = (f_0 - f) * g_0 + f * (g_0 - g) \text{ a.e.};$$

arguing as in the proof of 85.1, we have

$$\begin{aligned}\|h_0 - h\|_1 &\leq \|f_0 - f\|_1 \|g_0\|_1 + \|f\|_1 \|g_0 - g\|_1 \\ &= 0 \cdot \|g_0\|_1 + \|f\|_1 \cdot 0,\end{aligned}$$

thus $h_0 = h$ a.e. Summarizing: if $f, g, h \in \mathscr{L}^1$ and $f * g = h$ a.e., then

$$[f][g] = [h].$$

It is easy to see that L^1 is an *algebra* over the real numbers; that is, in addition to the vector space structure, we have:

$$u(v + w) = uv + uw \tag{6}$$

$$(u + v)w = uw + vw \tag{7}$$

$$(cu)v = u(cv) = c(uv); \tag{8}$$

these relations are immediate from Lemmas 1, 2, 3, respectively, of Sect. 85. Moreover, the algebra L^1 is *associative*, that is,

$$u(vw) = (uv)w; \tag{9}$$

this is immediate from 85.2. If $u, v \in L^1$, then

$$\|uv\| \leq \|u\| \, \|v\|; \tag{10}$$

this follows at once from the fact that if $f, g, h \in \mathscr{L}^1$ satisfy $h = f * g$ a.e., then $\|h\|_1 \leq \|f\|_1 \|g\|_1$ (corollary of 84.2). We may summarize (1) through (10) by saying that the vector space L^1, together with the norm and multiplication defined above, satisfies the axioms for a *Banach algebra* over the real numbers.

The **group algebra** of the locally compact group G is defined to be the Banach algebra L^1 just constructed. A suggestive notation for the group algebra is $L^1(G)$.

If $u = [f] = [g]$, then $\int f \, d\nu = \int g \, d\nu$, hence we may unambiguously define

$$\int u = \int f \, d\nu.$$

It follows that $u \to \int u$ is a *homomorphism* of the group algebra $L^1(G)$ onto the algebra of real numbers; that is,

$$\int (u + v) = \int u + \int v \tag{11}$$

$$\int (cu) = c \int u \tag{12}$$

$$\int (uv) = \left(\int u\right)\left(\int v\right) \tag{13}$$

(the latter equality follows from 84.2). Moreover, the relation

$$\left|\int f\,d\nu\right| \le \int |f|\,d\nu$$

implies that

$$\left|\int u\right| \le \|u\|, \tag{14}$$

and so

$$\left|\int u - \int v\right| \le \|u - v\|;$$

this shows that the correspondence $u \to \int u$ is a *continuous* homomorphism of the group algebra.

If $u = [f] = [g]$, so that $f = g$ a.e., then $\tilde{f} = \tilde{g}$ a.e. (corollary of 82.2); it follows from 82.3 that we may unambiguously define

$$\tilde{u} = [\tilde{f}],$$

and the result is an *involution* of the group algebra, that is,

$$(\tilde{u})^{\sim} = u \tag{15}$$

$$(u + v)^{\sim} = \tilde{u} + \tilde{v} \tag{16}$$

$$(cu)^{\sim} = c\tilde{u} \tag{17}$$

$$(uv)^{\sim} = \tilde{v}\tilde{u} \tag{18}$$

(the latter equality follows from 84.3). Moreover,

$$\|\tilde{u}\| = \|u\| \tag{19}$$

by 82.3.

Summarizing: relative to the operations introduced above, $L^1(G)$ is an (associative) Banach algebra, with an involution $u \to \tilde{u}$ satisfying $\|\tilde{u}\| = \|u\|$; moreover, $u \to \int u$ is a homomorphism of $L^1(G)$ onto the algebra of real numbers, and is continuous in virtue of the relation $\left|\int u\right| \le \|u\|$.

It may appear at first glance that the essential structure of the group algebra is its associative algebra structure, and that the involution is rather a curiosity thrown in free of charge. However, it should be observed that the possibility of integrating adjoints was crucial for the very definition of convolution (see 83.1). Thus, convolution and adjunction are intimately related, and not just incidental cohabitants of $\mathscr{L}^1(\nu)$.

EXERCISES

1. In order that G be Abelian, it is necessary and sufficient that $uv = vu$ for all u and v in $L^1(G)$.

2. In order that the group algebra of G possess a unity element, it is necessary and sufficient that G be discrete.

3. The class $\mathscr{N}$ is an "ideal" in $\mathscr{L}^1$ in the sense that if $f \in \mathscr{L}^1$ and $g \in \mathscr{N}$, then the functions $f * g$ and $g * f$ are everywhere defined on G and belong to $\mathscr{N}$ (indeed, they are identically zero).

4. Strictly speaking, we have constructed the *real group algebra* of G. In many applications, for instance the theory of the Fourier transformation, one requires the *complex group algebra;* this can be constructed either by paraphrasing the foregoing results on convolution for complex-valued integrable functions (Exercise 25.5), or by "complexifying" the real group algebra. The first of these procedures turns out to be the easiest approach to the complex analogue of the crucial formula (10). Incidentally, the adjoint of a complex-valued function f is defined by the formula

$$\tilde{f}(x) = \Delta(x)\overline{f(x^{-1})},$$

where the bar denotes complex conjugation.

*87. Convolution of Integrable Simple Baire Functions

In some arguments it is convenient to approximate an integrable function by continuous functions with compact support (see, for example, the proof of 85.2); at other times, approximation by integrable simple functions is more convenient (see, for example, the proof of 82.3). This depends, roughly speaking, on which aspect of the Riesz-Markoff theorem one is emphasizing—the linear form aspect, or the measure aspect.

In the present context of a locally compact group, an advantage of approximation by functions in $\mathscr{L}$ is that the convolution of any two such functions is again a function in $\mathscr{L}$. It is therefore of some interest to know that the convolution of any two integrable simple Baire functions, though usually not simple, is defined everywhere on G and is itself an integrable Baire function. To prove this, it will of course suffice, by bilinearity, to consider convolutions $\chi_E * \chi_F$, where E and F are Baire sets of finite measure.

As in Sect. 84, we write $\pi = \nu \times \nu$ for the product (Baire) measure on $G \times G$, and S for the homeomorphism $(x,y) \to (xy,y^{-1})$ of $G \times G$. Recall that S is self-inverse.

Lemma 1. *If E and F are Baire sets in G, then $\chi_E \triangledown \chi_F$ is a Baire function on $G \times G$; indeed,*

$$\chi_E \triangledown \chi_F = \chi_M,$$

where $M = S(E \times F)$.

Proof. Writing $h(x, y) = \chi_E(x)\chi_F(y) = \chi_{E \times F}(x, y)$, we have $\chi_E \triangledown \chi_F = h \circ S$. Clearly $(h \circ S)(x, y) = 1$ if and only if $S(x, y) \in E \times F$, that is, $(x, y) \in S(E \times F) = M$; otherwise, $(h \circ S)(x, y) = 0$. This shows that $\chi_E \triangledown \chi_F = \chi_M$; since S is a homeomorphism, M is a Baire set, and so χ_M is a Baire function (cf. the lemma to 84.1). ∎

If A and B are arbitrary subsets of G, and $x \in G$, we have the following formula for the x-section of $S(A \times B)$:

$$(S(A \times B))_x = (x^{-1}A) \cap B^{-1};$$

indeed, since S is self-inverse, the following relations imply one another: $y \in (S(A \times B))_x$, $(x, y) \in S(A \times B)$, $S(x, y) \in A \times B$, $(xy, y^{-1}) \in A \times B$, $xy \in A$ and $y^{-1} \in B$, $y \in (x^{-1}A) \cap B^{-1}$.

Lemma 2. *If E and F are Baire sets of finite measure, then $\chi_E * \chi_F$ is defined everywhere on G, and*

$$(\chi_E * \chi_F)(x) = \nu((S(E \times F))_x)$$

for all $x \in G$.

Proof. Let $f = \chi_E$, $g = \chi_F$, and let x be any element of G. According to Lemma 1, we have $(f \triangledown g)_x = \chi_{M_x}$, where $M = S(E \times F)$. Since

$$M_x = (x^{-1}E) \cap F^{-1}$$

is a Baire set, since $M_x \subset x^{-1}E$, and so

$$\nu(M_x) \leq \nu(x^{-1}E) = \nu(E) < \infty,$$

we conclude that $(f \triangledown g)_x$ is an integrable Baire function, and that

$$\int (f \triangledown g)_x \, d\nu = \nu(M_x).$$

Thus, $(f * g)(x)$ is defined, and is equal to $\nu(M_x)$. Incidentally, $0 \leq (f * g)(x) \leq \nu(E)$ for all $x \in G$, and so $f * g$ is bounded. ∎

For the next theorem, recall the notations f_M and g^M introduced in Sect. 37:

Theorem 1. *Let E and F be Baire sets of finite measure, and let* $h = \chi_E \triangledown \chi_F$ *and* $M = S(E \times F)$. *Then:*

(1) *Every section of h is an integrable Baire function.*

(2) $g^M = \nu(E)(\chi_F)^\sim$.

(3) $f_M = \chi_E * \chi_F$.

(4) $\pi(M) = \pi(E \times F) = \nu(E)\nu(F) < \infty$.

(5) *The functions* f_M *and* g^M *are integrable Baire functions.*

(6) *In particular,* $\chi_E * \chi_F$ *is an integrable Baire function.*

(7) *The transformation S is measure preserving, that is,* $\pi(S(N)) = \pi(N)$ *for every Baire set N in* $G \times G$.

Proof. (1) Let $x, y \in G$. Since h is the characteristic function of M, it follows that h_x is the characteristic function of the Baire set

$$M_x = (x^{-1}E) \cap F^{-1}$$

(cf. Lemma 2 to 83.1, and 36.8). On the other hand, by Lemma 1 to 83.1 we have

$$h^y = \chi_F(y^{-1})\chi_A,$$

where $A = Ey^{-1}$.

(2) Let $y \in G$. Since $h^y = \chi_F(y^{-1})\chi_A$, where $A = Ey^{-1}$, it follows from 78.3 that h^y is integrable, and

$$\int h^y \, d\nu = \chi_F(y^{-1})\nu(A) = \chi_F(y^{-1})\,\Delta(y)\nu(E) = \nu(E)(\chi_F)^\sim(y).$$

Since h is the characteristic function of M, it follows that h^y is the characteristic function of M^y, and so

$$g^M(y) = \nu(M^y) = \int h^y \, d\nu = \nu(E)(\chi_F)^\sim(y).$$

(3) Quoting Lemma 2, we have

$$f_M(x) = \nu(M_x) = (\chi_E * \chi_F)(x)$$

for all $x \in G$, and in particular f_M is real-valued.

(4) *and* (5): Since $g^M = \nu(E)(\chi_F)^\sim$, it follows from 82.3 that g^M is an integrable Baire function, and

$$\int g^M \, d\nu = \nu(E)\nu(F) = \pi(E \times F).$$

Moreover, by 40.1 we have $\pi(M) < \infty$, and

$$\pi(M) = \int g^M \, d\nu = \pi(E \times F).$$

Also by 40.1, we know there exists an integrable Baire function f such that $f_M = f$ a.e. $[\nu]$; to show that the real-valued function f_M is an integrable Baire function, it will therefore suffice (25.5) to show that it is a Baire function. Since ν is σ-finite, we may choose an increasing sequence of rectangles $E_n \times F_n$, where the E_n and F_n are Baire sets of finite measure, such that

$$M \subset \bigcup_1^\infty E_n \times F_n$$

(see the proof of 40.1). Defining $M_n = M \cap (E_n \times F_n)$, we have $M_n \uparrow M$, and so $f_{M_n} \uparrow f_M$ pointwise on G (37.3). Since each f_{M_n} is a Baire function by Lemma 2 to 40.1, it follows that f_M is a Baire function (14.3).

(6) This is immediate from (3) and (5).

(7) Define $\rho(N) = \pi(S(N))$ for every Baire set N in $G \times G$; since S is a homeomorphism, it is clear that ρ is a Baire measure. According to (4), we have $\rho(E \times F) = \pi(E \times F)$ for every rectangle whose sides are Baire sets of finite measure, hence $\rho = \pi$ on the ring $\mathscr{R}$ generated by all such rectangles (34.1). Since ρ and π are finite on $\mathscr{R}$, it follows from the (UET) that $\rho = \pi$ on the σ-ring $\mathfrak{S}(\mathscr{R})$ generated by $\mathscr{R}$; in other words (35.2 and 56.3), $\rho = \pi$ (on the class of all Baire sets in $G \times G$). ▮

In view of the bilinearity of $(f, g) \to f \triangledown g$, we have at once:

Corollary. *If f and g are integrable simple Baire functions on G, then $f * g$ is defined everywhere on G; indeed, $f * g$ is an integrable Baire function.*

EXERCISES

1. If f and g are integrable simple Baire functions, then $f * g$ is bounded.

2. If ρ is Haar measure for the product topological group $G \times G$, then $\rho(S(M)) = \rho(M)$ for all Borel sets M in $G \times G$.

***3.** Let H be a locally compact group with restricted Haar measure τ, and suppose S is a homeomorphism of H such that $\tau(S(N)) = \tau(N)$ for all Baire sets N in H. What assumptions on S would ensure the existence of a group G such that $H = G \times G$?

4. If f and g are nonnegative integrable Baire functions on G such that $f * g$ is defined everywhere on G, then $f * g$ is an integrable Baire function. The assumption of nonnegativity can be dropped.

5. If f and g are Baire functions on G, then $f \triangledown g$ is a Baire function on $G \times G$.

88. The Domain of $f * g$

If f and g are integrable Baire functions, it was shown in 84.2 that the domain D_{f*g} of $f * g$ contains the complement of a Baire set of measure zero. We shall show in this section that D_{f*g} is in fact itself the complement of a Baire set of measure zero (and so D_{f*g} is locally measurable with respect to the σ-ring of Baire sets). The proof will be based on a general result concerning the product of σ-finite measure spaces, a result which could already have been proved in Sect. 40:

Lemma 1. *Suppose $(X, \mathscr{S}, \mu_1)$ and $(Y, \mathscr{T}, \mu_2)$ are σ-finite measure spaces, $(X \times Y, \mathscr{S} \times \mathscr{T}, \mu_1 \times \mu_2)$ is the product measure space, and h is a real-valued function on $X \times Y$ which is measurable with respect to the σ-ring $\mathscr{S} \times \mathscr{T}$. Then the set*

$$A = \{x \in X: h_x \in \mathscr{L}^1(\mu_2)\}$$

is locally measurable with respect to $\mathscr{S}$; indeed, $X - A \in \mathscr{S}$. If moreover the iterated integral $\iint h \, d\mu_2 \, d\mu_1$ exists, then $\mu_1(X - A) = 0$.

Proof. In any case, we know from 36.8 that every section of h is measurable. Since h_x is measurable, and since $|h_x| = |h|_x$, it follows from 25.3 that h_x is integrable if and only if $|h|_x$ is integrable. Thus, in proving that $X - A$ is measurable, we may assume without loss of generality that $h \geq 0$.

As shown in the proof of 40.2, there exists an increasing sequence of measurable rectangles $P_n \times Q_n$ with sides of finite measure, and a sequence h_n of $(\mu_1 \times \mu_2)$-integrable simple functions such that $0 \leq h_n \uparrow h$ and $N(h_n) \subset P_n \times Q_n$.

It follows from Lemma 5 to 40.2 that for each n, every section of h_n is integrable, and defining

$$f_n(x) = \int (h_n)_x \, d\mu_2$$

for all $x \in X$, we have $f_n \in \mathscr{L}^1(\mu_1)$. Now, for each $x \in X$, we have $0 \leq (h_n)_x \uparrow h_x$; it follows from the (MCT) that h_x is μ_2-integrable if and only if the sequence $\int (h_n)_x \, d\mu_2$ is bounded, that is, the sequence $f_n(x)$ is bounded. Thus we have

$$A = \bigcup_{m=1}^{\infty} \bigcap_{n=1}^{\infty} \{x \in X: f_n(x) \leq m\},$$

and so

$$X - A = \bigcap_{m=1}^{\infty} \bigcup_{n=1}^{\infty} \{x \in X: f_n(x) > m\};$$

it follows from the remarks at the end of Sect. 12 that $X - A \in \mathscr{S}$.

Suppose now that h is a measurable function on $X \times Y$ such that the iterated integral

$$\iint h \, d\mu_2 \, d\mu_1$$

exists (we no longer assume that $h \geq 0$), and let

$$A = \{x \colon h_x \in \mathscr{L}^1(\mu_2)\}.$$

By the definition of iterated integral, there is a set E in $\mathscr{S}$ such that $\mu_1(E) = 0$ and $X - E \subset A$. Then $X - A \subset E$, and so

$$\mu_1(X - A) \leq \mu_1(E) = 0. \quad \blacksquare$$

Theorem 1. *If f and g are integrable Baire functions on G, then the domain of definition of $f * g$ is the complement of a Baire set of measure zero.*

Proof. Writing $h = f \triangledown g$, the domain of definition of $f * g$ is the set

$$A = \{x \colon h_x \in \mathscr{L}^1(\nu)\};$$

since h is a Baire function on $G \times G$ (lemma to 84.1), and since the iterated integral

$$\iint h(x, y) \, dy \, dx$$

exists by 84.1 and Fubini's theorem (41.1), our assertion follows from the lemma. $\blacksquare$

Assume again the notations of Lemma 1 and its proof. Let us assume further that $h \geq 0$, and that the iterated integral

$$\iint h \, d\mu_2 \, d\mu_1$$

exists (equivalently, h is $(\mu_1 \times \mu_2)$-integrable; see 41.1 and 41.2). Thus we have $0 \leq h_n \uparrow h$,

$$f_n(x) = \int (h_n)_x \, d\mu_2$$

for all $x \in X$, and $A = \{x \colon h_x \in \mathscr{L}^1(\mu_2)\}$. Writing $\rho = \mu_1 \times \mu_2$, it follows from Lemma 5 to 40.2 that

$$\int f_n \, d\mu_1 = \int h_n \, d\rho$$

for all n. Let us define a function f on X by the formulas

$$f(x) = \begin{cases} \int h_x \, d\mu_2 & \text{when } x \in A, \\ 0 & \text{when } x \in X - A. \end{cases}$$

By the (MCT) for μ_2, we have $\chi_A f_n \uparrow f$ pointwise on X, and in particular f is measurable with respect to $\mathscr{S}$ (15.1 and 14.3). Since $\mu_1(X - A) = 0$, we have

$$\int \chi_A f_n \, d\mu_1 = \int f_n \, d\mu_1 = \int h_n \, d\rho \leq \int h \, d\rho$$

for all n; since $\chi_A f_n \uparrow f$, it follows from the (MCT) for μ_1 that f is μ_1-integrable.

More generally, if h is any $(\mu_1 \times \mu_2)$-integrable function, we may apply the above reasoning to h^+ and h^-; since

$$h_x = (h^+)_x - (h^-)_x = (h_x)^+ - (h_x)^-$$

is μ_2-integrable if and only if both $(h_x)^+$ and $(h_x)^-$ are μ_2-integrable, we have at once:

Lemma 2. *If $(X, \mathscr{S}, \mu_1)$ and $(Y, \mathscr{T}, \mu_2)$ are σ-finite measure spaces, h is a $(\mu_1 \times \mu_2)$-integrable function on $X \times Y$, and if*

$$A = \{x \in X\colon h_x \in \mathscr{L}^1(\mu_2)\},$$

then the function f on X defined by the formulas

$$f(x) = \begin{cases} \int h_x \, d\mu_2 & \text{when } x \in A, \\ 0 & \text{when } x \in X - A, \end{cases}$$

is μ_1-integrable.

Combining Lemma 2 with Theorem 1, we have at once:

Theorem 2. *If f and g are integrable Baire functions on G, then the function k defined on G by the formulas*

$$k(x) = \begin{cases} (f * g)(x) & \text{when } x \in D_{f*g}, \\ 0 & \text{when } x \in G - D_{f*g}, \end{cases}$$

is an integrable Baire function, and $k = f * g$ a.e. $[\nu]$.

We close with an observation on the group algebra $L^1(G)$ defined in Sect. 86. Suppose $u, v \in L^1(G)$; say $u = [f]$ and $v = [g]$. The observation is that the function $f * g$ is uniquely determined by the cosets u and v, and does not depend on the particular choice of representatives f and g; indeed:

Theorem 3. *If f, g, f_1, g_1 are Baire functions on G, such that $f = f_1$ a.e. $[\nu]$ and $g = g_1$ a.e. $[\nu]$, then the functions $f * g$ and $f_1 * g_1$ have identical domains of definition, and*

$$(f * g)(x) = (f_1 * g_1)(x)$$

for all x in their common domain; briefly, $f * g = f_1 * g_1$.

Proof. Suppose first that either $f = 0$ a.e. $[\nu]$ or $g = 0$ a.e. $[\nu]$. For each x in G, we have

$$(f \triangledown g)_x(y) = f(xy)g(y^{-1})$$

for all $y \in G$, hence it follows from either 78.1 or 82.2, as the case may be, that $(f \triangledown g)_x = 0$ a.e. $[\nu]$. Since $(f \triangledown g)_x$ is a Baire function (Lemma 2 to 83.1), we conclude that $(f \triangledown g)_x$ is integrable (25.5), and that

$$\int (f \triangledown g)_x \, d\nu = 0.$$

Summarizing: if either $f = 0$ a.e. $[\nu]$ or $g = 0$ a.e. $[\nu]$, then $f * g$ is defined everywhere on G, and $(f * g)(x) = 0$ for all $x \in G$.

The general assertion of the theorem then follows from the relation

$$(f_1 \triangledown g_1)_x - (f \triangledown g)_x = [(f_1 - f) \triangledown g_1]_x + [f \triangledown (g_1 - g)]_x;$$

indeed, the right hand member is an integrable Baire function with integral 0 (for each x in G), hence $(f_1 \triangledown g_1)_x$ is integrable if and only if $(f \triangledown g)_x$ is integrable, and the values of their integrals are then equal. ∎

EXERCISES

1. If G is Abelian, then $f * g = g * f$ for all integrable Baire functions f and g.

2. If G is unimodular, f and g are integrable Baire functions, and g is *central* in the sense that $g(yx) = g(xy)$ for all $x, y \in G$, then $f * g = g * f$.

3. If f and g are Baire functions such that $f \in \mathscr{L}^1$ and $g \in \mathscr{L}^p$, where $p > 1$, then there exists a function k in $\mathscr{L}^p$ such that $k = f * g$ a.e., and one has

$$\|k\|_p \leq \|f\|_1 \|g\|_p.$$

(See also Exercise 84.7.)

*89. Convolution of Integrable Borel Functions

It will be convenient in this section to abbreviate "a.e. $[\mu]$" to "a.e."

Until Sect. 84, the theory of Baire and Borel functions on G had been developed in parallel. For technical reasons involving Fubini's theorem, it was necessary in Sect. 84 to restrict attention to Baire functions in order to show that convolutions of integrable functions are defined on "substantial" subsets of G. In the ensuing discussion in Sects. 85–88, again only Baire functions were considered; this was largely a matter of convenience, for we shall now see that virtually all of the results of Sects. 84–88 may be adapted so as to yield the analogous results for Borel functions (the exceptions being 84.1 and certain parts of 87.1). As might be expected, this will be done with the aid of 68.1: every Borel function f is equal, a.e., to a suitable Baire function g.

First, the definition of convolution given in Sect. 84 must be extended. If f and g are real-valued functions defined on G, the **convolution** of f and g (in that order) is the function $f * g$ defined as follows: the domain of definition of $f * g$ is the set of all x in G for which $(f \triangledown g)_x$ is an integrable Borel function, and for such an x the value of $f * g$ is defined by

$$(f * g)(x) = \int (f \triangledown g)_x \, d\mu = \int f(xy) g(y^{-1}) \, dy.$$

If in particular f and g are Baire functions, so that every section of $f \triangledown g$ is a Baire function (Lemma 2 to 83.1), it follows from 66.1 that the definition of convolution just given is consistent with the definition in Sect. 84.

When f and g are arbitrary real-valued functions on G, a pertinent question is how extensive is the domain of definition of $f * g$? We shall see in Theorem 1 that if f and g are integrable Borel functions, then $f * g$ is defined a.e.

Lemma. *If f, $\hat{f}$, g, $\hat{g}$ are real-valued functions on G such that $f = \hat{f}$ a.e. and $g = \hat{g}$ a.e., then for each x in G one has*

$$(f \triangledown g)_x = (\hat{f} \triangledown \hat{g})_x \text{ a.e.}$$

Proof. Fix any $x \in G$. Since

$$(f \triangledown g)_x(y) = f(xy) g(y^{-1})$$

and

$$(\hat{f} \triangledown \hat{g})_x(y) = \hat{f}(xy) \hat{g}(y^{-1})$$

for all $y \in G$, it is clear from 78.1 and 82.2 that $(f \triangledown g)_x = (\hat{f} \triangledown \hat{g})_x$ a.e. ∎

Theorem 1. *If f and g are integrable Borel functions, then $f * g$ is defined a.e. Indeed, there exists an integrable Borel function h such that $f * g = h$ a.e.*

Proof. According to 68.1, we may choose Baire functions $\hat{f}$ and $\hat{g}$ such that $f = \hat{f}$ a.e. and $g = \hat{g}$ a.e. Of course $\hat{f}$ and $\hat{g}$ are also integrable (25.5). It follows from 84.2 that $\hat{f} * \hat{g}$ is defined a.e., indeed there exists an integrable Baire function h such that $h = \hat{f} * \hat{g}$ a.e. Let E be a Borel set of measure zero such that for each x in $G - E$, $(\hat{f} \triangledown \hat{g})_x$ is integrable, and

$$\int (\hat{f} \triangledown \hat{g})_x \, d\mu = h(x)$$

(if we like, E can be taken to be a Baire set).

Fixing any $x \in G - E$, it will suffice to show that $(f * g)(x)$ is defined and is equal to $h(x)$. Indeed, $(f \triangledown g)_x = (\hat{f} \triangledown \hat{g})_x$ a.e. by the lemma, $(f \triangledown g)_x$ is a Borel function (Lemma 2 to 83.1), and $(\hat{f} \triangledown \hat{g})_x$ is integrable; it follows from 25.5 that $(f \triangledown g)_x$ is integrable, and

$$\int (f \triangledown g)_x \, d\mu = \int (\hat{f} \triangledown \hat{g})_x \, d\mu,$$

that is, $(f * g)(x)$ is defined and is equal to $h(x)$. ▮

The analogue of 88.3 holds also for Borel functions:

Theorem 2. *If f, g, f_1, g_1 are Borel functions such that $f = f_1$ a.e. and $g = g_1$ a.e., then the functions $f * g$ and $f_1 * g_1$ have identical domains of definition, and*

$$(f * g)(x) = (f_1 * g_1)(x)$$

for all x in their common domain; briefly, $f * g = f_1 * g_1$.

Proof. For each x in G, the functions $(f \triangledown g)_x$ and $(f_1 \triangledown g_1)_x$ are Borel functions (Lemma 2 to 83.1); since $(f \triangledown g)_x = (f_1 \triangledown g_1)_x$ a.e. by the lemma to Theorem 1, our assertion is clear from 25.5. ▮

The following corollaries indicate how the foregoing results allow one to deduce properties of convolution of Borel functions from the analogous properties already known for Baire functions:

Corollary 1. *If f and g are integrable Borel functions, then the domain of definition of $f * g$ is the complement of a Baire set of measure zero. If, moreover, one defines*

$$h(x) = \begin{cases} (f * g)(x) & \text{when } x \in D_{f*g}, \\ 0 & \text{when } x \in G - D_{f*g}, \end{cases}$$

then h is an integrable Baire function.

Proof. According to 68.1, we may choose (integrable) Baire functions f_1 and g_1 such that $f = f_1$ a.e. and $g = g_1$ a.e. Since $f * g = f_1 * g_1$ by Theorem 2, our assertion is immediate from 88.1 and 88.2. ▮

Corollary 2. *If f, g, h are integrable Borel functions such that $h = f * g$ a.e., then*

$$\int h \, d\mu = \left(\int f \, d\mu\right) \left(\int g \, d\mu\right).$$

Proof. Let f_1 and g_1 be (integrable) Baire functions such that $f = f_1$ a.e. and $g = g_1$ a.e. By 84.2, there exists an integrable Baire function h_1 such that $h_1 = f_1 * g_1$ a.e., and necessarily

$$\int h_1 \, d\mu = \left(\int f_1 \, d\mu\right) \left(\int g_1 \, d\mu\right).$$

Combining the relations $h = f * g$ a.e., $f * g = f_1 * g_1$ (Theorem 2), and $f_1 * g_1 = h_1$ a.e., we have $h = h_1$ a.e., and so

$$\int h\,d\mu = \int h_1\,d\mu = \left(\int f_1\,d\mu\right)\left(\int g_1\,d\mu\right) = \left(\int f\,d\mu\right)\left(\int g\,d\mu\right). \quad \blacksquare$$

Corollary 3 (Associative law). *Let f, g, h be integrable Borel functions, and choose (Theorem 1) integrable Borel functions k_1, k_2 so that $g * h = k_1$ a.e. and $f * g = k_2$ a.e. Then $f * k_1 = k_2 * h$ a.e.*

Proof. Let $\hat{f}$, $\hat{g}$, $\hat{h}$ be (integrable) Baire functions which are equal a.e. to f, g, h, respectively. By 84.2, we may choose integrable Baire functions $\hat{k}_1$ and $\hat{k}_2$ such that $\hat{g} * \hat{h} = \hat{k}_1$ a.e. and $\hat{f} * \hat{g} = \hat{k}_2$ a.e., and it follows from 85.2 that $\hat{f} * \hat{k}_1 = \hat{k}_2 * \hat{h}$ a.e. Combining the relations $k_1 = g * h$ a.e., $g * h = \hat{g} * \hat{h}$ (Theorem 2), and $\hat{g} * \hat{h} = \hat{k}_1$ a.e., we conclude that $k_1 = \hat{k}_1$ a.e. Similarly $k_2 = \hat{k}_2$ a.e., and so $f * k_1 = \hat{f} * \hat{k}_1$ and $k_2 * h = \hat{k}_2 * \hat{h}$; our assertion now follows from the relation $\hat{f} * \hat{k}_1 = \hat{k}_2 * \hat{h}$ a.e. $\blacksquare$

Corollary 4. *If f, g, h are integrable Borel functions such that $f * g = h$ a.e., then $\tilde{g} * \tilde{f} = \tilde{h}$ a.e.*

Proof. Let f_1 and g_1 be Baire functions such that $f = f_1$ a.e. and $g = g_1$ a.e., and let h_1 be an integrable Baire function such that $h_1 = f_1 * g_1$ a.e. Since $\tilde{h}_1 = \tilde{g}_1 * \tilde{f}_1$ a.e. by 84.3, and since $\tilde{g}_1 * \tilde{f}_1 = \tilde{g} * \tilde{f}$ (by the corollary of 82.2, and Theorem 2), we conclude that $\tilde{h}_1 = \tilde{g} * \tilde{f}$ a.e. Combining the relations $h = f * g$ a.e., $f * g = f_1 * g_1$, and $f_1 * g_1 = h_1$ a.e., we have $h = h_1$ a.e., and so $\tilde{h} = \tilde{h}_1$ a.e. Thus $\tilde{h} = \tilde{h}_1$ a.e. and $\tilde{h}_1 = \tilde{g} * \tilde{f}$ a.e., and therefore $\tilde{h} = \tilde{g} * \tilde{f}$ a.e. $\blacksquare$

It is also possible to extend the "group algebra" considerations of Sect. 86 to the Borel case. We refrain from writing out again the easy details, for it is clear from Theorem 2 that the resulting algebraic system will be essentially the same as (i.e., isomorphic to) the group algebra constructed in Sect. 86. From the point of view of normed algebras, it is unnecessary to consider Borel functions at all, Baire functions being entirely adequate to describe the group algebra.

EXERCISES

1. If f and g are integrable simple Borel functions, then $f * g$ is an (everywhere defined) integrable Baire function.

2. If G is Abelian, then $f * g = g * f$ for all integrable Borel functions f and g.

3. If G is unimodular, f and g are integrable Borel functions, and g is central in the sense that $g(yx) = g(xy)$ for all x, y in G, then $f * g = g * f$. It suffices to assume that for each fixed x, one has $g(yx) = g(xy)$ a.e.

*90. Complements on Haar Measure

Hovering over the relationship between Baire sets and Borel sets in the locally compact group G, there is the following remarkable fact: Haar measure μ is *completion regular* (in other words, restricted Haar measure ν is completion regular). [This is the last theorem in P. R. Halmos' *Measure Theory*, New York, 1950, to which the reader is referred for the proof.] It follows from 70.2 that ν is a *monogenic* Baire measure, thus every Borel measure on G which extends ν necessarily coincides with μ. Moreover, if E is any Borel set such that $\mu(E) = 0$, then there exists a Baire set F such that $E \subset F$ and $\nu(F) = 0$ (70.3). From this we see that the concepts "a.e. $[\mu]$" and "a.e. $[\nu]$" are equivalent.

Let us observe that the product measure $\pi = \nu \times \nu$ is restricted Haar measure for the product topological group $G \times G$. In any case, it is a Baire measure (78.5). If E and F are Baire sets in G, and $(s, t) \in G \times G$, then

$$\pi((s, t)(E \times F)) = \pi(sE \times tF) = \nu(sE)\nu(tF) = \nu(E)\nu(F) = \pi(E \times F);$$

it follows easily from 34.1 and the (UET) that $\pi((s, t)M) = \pi(M)$ for every Baire set M in $G \times G$. Since π is not identically zero, we conclude from 78.4 that π is restricted Haar measure for $G \times G$.

Since ν is completion regular, we know from 70.4 that $\mu \times \mu$ can be extended to a regular Borel measure on $G \times G$. However, since $\nu \times \nu$ is restricted Haar measure for $G \times G$, and is therefore monogenic, we can say more: if ρ is *any* Borel measure on $G \times G$ which extends $\nu \times \nu$, then ρ must be the unique regular Borel measure extending $\nu \times \nu$, and it follows that ρ is Haar measure for $G \times G$. Thus, not only can $\mu \times \mu$ be extended to a Borel measure on $G \times G$, but indeed *every* Borel measure extending $\nu \times \nu$ necessarily coincides with Haar measure for $G \times G$ and is an extension of $\mu \times \mu$.

EXERCISES

1. If ρ is a Borel measure on G such that $\rho(sD) = \rho(D)$ for all compact G_δ's D and $s \in G$, then ρ is proportional to Haar measure μ, and in particular ρ is regular.

2. If ρ is a Borel measure on $G \times G$ such that $\rho(C \times D) = \mu(C)\mu(D)$ for all compact G_δ's C and D, then ρ is Haar measure for $G \times G$, and $\rho(E \times F) = \mu(E)\mu(F)$ for all Borel sets E and F.

***3.** *WANTED:* An elementary proof that Haar measure is completion regular.

4. If μ_i is Haar measure for the locally compact group G_i $(i = 1, 2)$, it follows also from Exercise 62.13 that Haar measure for $G_1 \times G_2$ is an extension of $\mu_1 \times \mu_2$. (This argument avoids any reference to completion regularity.)

5. Let μ_i be Haar measure on G_i $(i = 1, 2)$, and let ρ be the Haar measure on $G_1 \times G_2$ which extends $\mu_1 \times \mu_2$. Let h be a Borel function on $G_1 \times G_2$.

(*i*) If h is ρ-integrable, then the iterated integrals

$$\int\int h \, d\mu_2 \, d\mu_1 \qquad \text{and} \qquad \int\int h \, d\mu_1 \, d\mu_2$$

exist and are equal to $\int h \, d\rho$.

(*ii*) Conversely, if $h \geq 0$ and the iterated integral

$$\int\int h \, d\mu_2 \, d\mu_1$$

exists, then h is ρ-integrable (cf. Exercise 70.15.)

6. If f is an integrable Borel function on G, then given any $\varepsilon > 0$ there exists a continuous Baire function g on G such that $\mu(N(f - g)) \leq \varepsilon$ (cf. Exercise 68.4).

References and Notes

(Numbers in brackets refer to the bibliography that follows)

§0. [21], [22].

§1. [21, §§4, 5].

§2. [51, p. 85], [21, 6.B].

§3. [20]. Incidentally, a wealth of bibliographical information may be found in [20] and [51].

§5. The name of Carathéodory is associated with the theory of outer measures, and in particular the key definition of "ν-measurable set" is due to him [8, p. 246], [9, p. 161].

§6. [20, Theorem 6.5.5], [41, Chapter X, §3], [21, 13.A].

§7. [21, §8]. For an informal discussion of a nonmeasurable set, see [23, p. 31]. The reader is referred to the book of Hildebrandt [26] for a masterly presentation of the theory of integration and differentiation of functions on the real line.

§8. [21, §12], [36, §12].

§9. [21, 13.B].

§10. [3]. The treatment of $0 \cdot \infty$ is a fascinating detail in itself. In the theory of Bourbaki, products of extended real numbers are defined with an eye on topological continuity, and so $0 \cdot \infty$ is not defined. The convention $0 \cdot \infty = 0$ pays its respects to the order structure of the nonnegative extended reals; the best argument in its favor is Lemma 2, which has a key application in the theory of product measures (cf. Lemma 5 of 39.2).

§11. Our definition of a measurable space $(X, \mathscr{S})$ differs from that of [21, §17] only in that we do not require that the union of $\mathscr{S}$ be equal to all of X. The extra generality is illusory (cf. Exercise 1). It may be of interest to record that the only essential use of the assumption $\bigcup \mathscr{S} = X$ in [21] occurs in the proof of 59.D.

§12. The definition of measurable function given here is that of Halmos [21, §18], whose treatment of $N(f)$ is quite unique.

If one wants the "simple" functions to be the linear combinations of characteristic functions of measurable sets, and if one wants every "measurable" function f to be the pointwise limit of a sequence of "simple" functions, then there is no way of getting around it: $N(f)$ *must* be a measurable set. Also, the Halmos definition is precisely what is needed for making the following assertion: the characteristic function of a set A is "measurable" if and only if

A is a measurable set. The inclusion of $N(f)$ in the definition of the measurability of f is an essential consequence of the decision to work with σ-rings (rather than the less general σ-algebras). As further justifications for this definition, we cite for example the proofs of 40.2 (Fubini's theorem), 41.2, and Corollaries 4 and 5 of 42.1 (cf. [54, p. 85, Theorem 2).

As remarked in Sect. 11, the value of the class of locally measurable sets is principally didactic. In discussing measurable functions relative to a measurable space $(X, \mathscr{S})$, one could get along with the smaller class $\mathscr{A}$ consisting of the algebra generated by $\mathscr{S}$ (cf. Exercise 1.3). For instance, it is easy to see that a real-valued function f on X is measurable with respect to $\mathscr{S}$ if and only if (*i*) $N(f) \in \mathscr{S}$, and (*ii*) $f^{-1}(M) \in \mathscr{A}$ for all Borel sets M. Similarly, the locally measurable sets which actually arise in Chapters 6 and 7 are in fact in $\mathscr{A}$ (cf. the proofs of Corollaries 3, 4, and 5 of 42.1, and of 47.4, 49.1, and Lemma 1 of 88.1).

For the theory of measurability of vector valued functions, see [27, Chapter III] or Bourbaki.

§13. [21, §19].

§14. [21, §20].

§15. The concept of locally measurable set is inspired by [21], where it is more or less explicit (see especially [21, p. 77 and 19.A]), and by [54], [55]. Naturally, the issue never comes up if one works only with σ-algebras.

§17. The definition of measure space given here is that in [21], modulo the slightly more general definition of measurable space given in §11. The extra generality is illusory (cf. Exercise 9).

The construction of μ_λ (Exercise 1) is given by Segal in [54, p. 8, Theorem 1], [55, Theorem 2.1].

Incidentally, the reader is referred to [54], [55] for a special approach to integration theory that has many advantages. Segal defines a "measure space" to be a triple $(X, \mathscr{R}, r)$, where $\mathscr{R}$ is a ring of subsets of X, r is a *finite* measure on $\mathscr{R}$, and the following condition is satisfied: if E_n is an increasing sequence of sets in $\mathscr{R}$ such that $r(E_n)$ is bounded, then $\bigcup_1^\infty E_n$ is also in $\mathscr{R}$. It follows easily that $\mathscr{R}_\lambda$ is a σ-algebra, and one can do business with $(X, \mathscr{R}_\lambda, r_\lambda)$, which is a measure space in the Halmos sense.

Conversely, suppose $(X, \mathscr{S}, \mu)$ is a measure space in the sense of Halmos, satisfying the following conditions: (a) $\mathscr{S}$ is a σ-algebra; indeed, (b) if $\mathscr{R}$ is the class of sets in $\mathscr{S}$ of finite measure, then $\mathscr{S} = \mathscr{R}_\lambda$ (this means that a set is measurable if and only if its intersection with every measurable set of finite measure is measurable); and (c) μ is *semifinite*, that is,

$$\mu(A) = \text{LUB}\,\{\mu(E) : E \subset A, E \in \mathscr{R}\}$$

for each A in $\mathscr{S}$. If r is the restriction of μ to $\mathscr{R}$, then the triple $(X, \mathscr{R}, r)$ is a "measure space" in the sense of Segal, and $(X, \mathscr{R}_\lambda, r_\lambda) = (X, \mathscr{S}, \mu)$.

See [6] for further comparisons of the Halmos and Segal theories.

§18. In [47], the Lebesgue theory is developed starting with the concept of "null set." Such an exposition was given by F. Riesz in [46].

§20. [44].

§**21.** [44], [60], [21, §22], [47, p. 100].

§**22.** [54, Chapter II, Section 1].

§**24.** [32, 12C], [62, p. 115], [12, p. 287], [45]. Cf. the definition of integrability in [49].

§**25.** [45], [21, §25].

§**26.** [21, §23].

§**27.** [47, §18], [51, p. 28].

§§**29–33.** The reader is referred to [36] for an especially lucid presentation of the convergence theorems (Chapter VI) and for an unforgettable diagram (p. 237).

§**29.** [20, p. 210], [36, p. 237].

§**30.** [51, p. 29].

§**31.** [47, p. 74].

§**32.** [51, p. 29].

§**33.** [47, p. 58]. For the classical form of the Riesz-Fischer theorem, see [36, p. 258]. The spaces $\mathscr{L}^p$: [21, §42].

§**34.** [19, p. 14, Theorem 3.3.22].

§**35.** Product measurable space: [21, §33].

§**36.** Sections: [21, §34].

§**39.** Product measure: [21, §35], [3].

§**40.** The concept of iterated integral is defined for possibly extended real valued functions simply to dramatize the fact that it is only the real values of such a function which count in the theory of integration. However, further restrictions on the function are often necessary (cf. 41.2, and Corollary 4 of 42.1).

§**41.** Fubini's theorem: [21, §36], [51, p. 77]. The name of Tonelli is associated with the converse (41.2) of Fubini's theorem (see [33, p. 145]); the power of this result is well illustrated in the theory of convolution (cf. 84.1).

§**42.** The Halmos formulation of product measure makes Fubini's theorem especially transparent. The reason for this is that the class of measurable sets in the product space is a highly restricted class of sets; it is, so to speak, "monotonely accessible" from the ring generated by the measurable rectangles.

§**46.** Theorem 2: [21, §29].

§**47.** Lebesgue decomposition: [51, p. 35].

§**48.** [21, 31.B].

§**49.** Jordan-Hahn decomposition: [21, §29], [51, p. 32].

§**52.** Radon-Nikodym theorem: [51, p. 36], [21, §31]. This theorem is often stated for extended real valued signed measures on a *totally* σ-finite measure space; our version is more restrictive on the signed measure, less restrictive on the underlying measure space. The key to our exposition is 45.2.

§**53.** [36, Section 35].

§**54.** Theorem 3: [28, p. 146].

§**55.** [21, §55].

§56. [21, §51], [54, p. 138]. Our definitions of "Baire set" and "Borel set" are taken from Halmos [21, §51].

The fact that every compact Baire set is a G_δ is proved in [21, 51.D], but this fact is not needed in the sequel since the compact Baire sets that occur are either visibly G_δ's, or may be assumed to be G_δ's (see, for example, the proofs of [21, 51.E, 55.A, 63.E, 64.E, 64.F, 64.G]).

Lemma 2 of Theorem 3 may be generalized for infinite products [39, Theorem 2.1]; incidentally, the product of a family of locally compact spaces is locally compact if and only if all but finitely many of the factor spaces are compact [28, Theorem 5.19].

§57. Theorem 2 is proved in [4]; a useful application occurs in Exercise 62.15.

§58. [21, 51.F].

§§59–62. [21, §52].

§§63–65. [21, §§53, 54], [53, Lemma 5.2].

§67. [21, §55].

§68. [54, p. 136].

§69. [21, §56], [53], [54, p. 126]. For a theory of measure and integration based on linear functionals, see [56], [5], [1], [37, Chapter I, §6], [25, Chapter III], [32, Chapter III].

§70. [21, p. 230].

§71. Topological groups: [42], [28, p. 105]. Theorem 1: [32, p. 109].

§§72–76. [59, p. 33]. A simultaneous proof of the existence and uniqueness of the Haar integral, avoiding Tychonoff's theorem (and other variants of the Axiom of Choice), is sketched in [10], and is written out in [25, §15].

In a *metrizable* locally compact group, the presence of a fundamental sequence of neighborhoods of the neutral element permits some simplifications [36, Section 17]. The idea, due to S. Banach [51, Appendix], is to use a "generalized limiting operation," whose existence is based on the Hahn-Banach theorem, and so, ultimately, on the Axiom of Choice.

Our exposition is a dilation of that of Weil [59].

§77. [32, §30]. Unimodular groups: [15].

§84. [43], [54, p. 149], [32, §31], [21, p. 269], [25, Chapter V].

§85. [43].

§86. [43], [52], [32, §31].

§90. [21, 64.I].

Bibliography

[1] H. BAUER, "Sur l'équivalence des théories de l'intégration selon N. Bourbaki et selon M. H. Stone," *Bull. Soc. Math. France*, **85** (1957) 51–75.

[2] S. K. BERBERIAN, *A First Course in Measure and Integration*, State University of Iowa Printing Service, Iowa City, 1962.

[3] S. K. BERBERIAN, "The product of two measures," *Amer. Math. Monthly*, **69** (1962) 961–968.

[4] S. K. BERBERIAN and J. F. JAKOBSEN, "A note on Borel sets," *Amer. Math. Monthly*, **70** (1963) 55.

[5] N. BOURBAKI, *Intégration*, ASI 1175 (Chs. 1–4), 1244 (Ch. 5), 1281 (Ch. 6), 1306 (Chs. 7, 8), Hermann & Cie., Paris, 1952–.

[6] A. BROWN, "On the Lebesgue convergence theorem," *Math. Nachrichten*, **23** (1961) 141–148.

[7] J. C. BURKILL, *The Lebesgue Integral*, Cambridge, Eng., 1951.

[8] C. CARATHÉODORY, *Vorlesungen über reelle Funktionen*, 3rd ed., Chelsea, New York, 1968.

[9] C. CARATHÉODORY, *Algebraic Theory of Measure and Integration*, Chelsea, New York, 1963.

[10] H. CARTAN, "Sur la mesure de Haar," *C. R. Acad. Sci. Paris*, **211** (1940) 759–762.

[11] C. CHEVALLEY, *Fundamental Concepts of Algebra*, Academic Press, New York, 1956.

[12] P. J. DANIELL, "A general form of integral," *Annals of Math.*, **19** (1917–1918) 279–294.

[13] N. DUNFORD and J. T. SCHWARTZ, *Linear Operators. Part I: General Theory*, Interscience, New York, 1958.

[14] R. GODEMENT, "Sur la théorie des représentations unitaires," *Annals of Math.*, **53** (1951) 68–124.

[15] R. GODEMENT, "Théorie des caractères II. Définition et propriétés générales des caractères," *Annals of Math.*, **59** (1954) 63–85.

[16] C. GOFFMAN, *Real Functions*, Rinehart, New York, 1953.

[17] L. M. GRAVES, *The Theory of Functions of Real Variables*, 2nd ed., McGraw-Hill, New York, 1956.

[18] A. HAAR, "Der Massbegriff in der Theorie der kontinuierlichen Gruppen," *Annals of Math.*, **34** (1933) 147–169.

[19] H. HAHN, *Reelle Funktionen, I. Teil: Punktfunktionen*, Akademische Verlagsgesellschaft, Leipzig, 1932. (Reprinted by Chelsea, New York, 1948.)

[20] H. HAHN and A. ROSENTHAL, *Set Functions*, University of New Mexico, Albuquerque, 1948.

[21] P. R. HALMOS, *Measure Theory*, Van Nostrand, New York, 1950.

[22] P. R. HALMOS, *Naive Set Theory*, Van Nostrand, New York, 1960.

[23] S. HARTMAN and J. MIKUSIŃSKI, *The Theory of Lebesgue Measure and Integration*, Pergamon Press, New York, 1961.

[24] E. HEWITT, *Theory of Functions of a Real Variable*, Holt, Rinehart & Winston, New York, 1960.

[25] E. HEWITT and K. A. ROSS, *Abstract Harmonic Analysis*, Vol. I, Springer-Verlag (Berlin) and Academic Press (New York), 1963.

[26] T. H. HILDEBRANDT, *Introduction to the Theory of Integration*, Academic Press, New York, 1963.

[27] E. HILLE and R. S. PHILLIPS, *Functional Analysis and Semi-groups*, Amer. Math. Soc. Colloquium Publ., Vol. 31, Providence, 1957.

[28] J. L. KELLEY, *General Topology*, Van Nostrand, New York, 1955.

[29] H. KESTELMAN, *Modern Theories of Integration*, 2nd rev. ed., Dover, New York, 1960.

[30] A. N. KOLMOGOROV and S. V. FOMIN, *Measure, Lebesgue Integrals, and Hilbert Space*, Academic Press, New York, 1961.

[31] H. LEBESGUE, *Leçons sur l'intégration et la recherche des fonctions primitives*, 2nd ed., Gauthier-Villars, Paris, 1928 (1950).

[32] L. H. LOOMIS, *An Introduction to Abstract Harmonic Analysis*, Van Nostrand, New York, 1953.

[33] E. J. McSHANE, *Integration*, Princeton University Press, Princeton, 1944.

[34] E. J. McSHANE and T. BOTTS, *Real Analysis*, Van Nostrand, New York, 1959.

[35] D. MONTGOMERY and L. ZIPPIN, *Topological Transformation Groups*, Interscience, New York, 1955.

[36] M. E. MUNROE, *Introduction to Measure and Integration*, Addison-Wesley, Cambridge, Mass., 1953.

[37] M. A. NAIMARK, *Normed Rings*, Noordhoff, Groningen, 1959.

[38] I. P. NATANSON, *Theory of Functions of a Real Variable*, 2 vols., Ungar, New York, 1955–1960.

[39] E. NELSON, "Regular probability measures on function space," *Annals of Math.*, **69** (1959) 630–643.

[40] J. V. NEUMANN, "The uniqueness of Haar's measure," *Mat. Sbornik*, **1** (1936) 721–734.

[41] J. V. NEUMANN, *Functional Operators I*, Annals of Math. Studies 21, Princeton University Press, Princeton, 1950.

[42] L. PONTRJAGIN, *Topological Groups*, Princeton University, Princeton, 1939.

[43] D. A. RAIKOV, "Harmonic analysis on commutative groups with the Haar measure and the theory of characters," *Trav. Inst. Math. Stekloff*, **14** (1945) 5–86. (Translated into German in *Sowjetische Arbeiten zur Funktionalanalysis*, Verlag Kultur und Fortschritt, Berlin, 1954.)

[44] F. RIESZ, "Sur les suites de fonctions mesurables," *Comptes Rendus* (Paris), **148** (1909) 1303–1305.

[45] F. RIESZ, "Démonstration nouvelle d'un théorème concernant les opérations fonctionelles linéaires," *Annales Sci. de l'Ecole Norm. Sup.*, **31** (1914) 9–14.

[46] F. RIESZ, "Sur l'intégrale de Lebesgue," *Acta Math.*, **42** (1920) 191–205.

[47] F. RIESZ and B. SZ.-NAGY, *Leçons d'analyse fonctionelle*, Académie des Sciences de Hongrie, Budapest, 1952.

[48] H. L. ROYDEN, *Real Analysis*, Macmillan, New York, 1963.

[49] W. RUDIN, *Principles of Mathematical Analysis*, McGraw-Hill, New York, 1953.

[50] W. RUDIN, *Fourier Analysis on Groups*, Interscience, New York, 1962.

[51] S. SAKS, *Theory of the Integral*, Warszawa-Lwów, 1937. (Reprinted by Hafner, New York.)

[52] I. E. SEGAL, "The group algebra of a locally compact group," *Trans. Amer. Math. Soc.*, **61** (1947) 69–105.

[53] I. E. SEGAL, "Invariant measures on locally compact spaces," *J. Indian Math. Soc.*, **13** (1949) 105–130.

[54] I. E. SEGAL, *Introduction to Modern Integration Theory*, University of Chicago mimeographed lecture notes, 1950.

[55] I. E. SEGAL, "Equivalences of measure spaces," *Amer. J. Math.*, **73** (1951) 275–313.

[56] M. H. STONE, "Notes on integration, I–IV," *Proc. Nat. Acad. Science U.S.A.:* Note I, **34** (1948) 336–342; Note II, **34** (1948) 447–455; Note III, **34** (1948) 483–490; Note IV, **35** (1949) 50–58.

[57] A. E. TAYLOR, *Introduction to Functional Analysis*, Wiley, New York, 1958.

[58] C. DE LA VALLÉE POUSSIN, *Cours d'analyse infinitésimale*, Vol. I, 3rd ed., Gauthier-Villars, Paris, 1914.

[59] A. WEIL, *L'intégration dans les groupes topologiques et ses applications*, ASI 1145, Hermann & Cie., Paris, 1940.

[60] H. WEYL, "Über die Konvergenz von Reihen, die nach Orthogonalfunktionen fortschreiten," *Math. Annalen*, **67** (1909) 225–245.

[61] J. H. WILLIAMSON, *Lebesgue Integration*, Holt, Rinehart & Winston, New York, 1962.

[62] W. H. YOUNG, "On integration with respect to a function of bounded variation," *Proc. London Math. Soc.*, **13** (1913–1914) 109–150.

[63] A. C. ZAANEN, *An Introduction to the Theory of Integration*, North-Holland, Amsterdam, 1958.

Index